Luboš Novák (Ed.)
**Electromembrane Processes**

## Also of Interest

*Ultrafiltration Membrane Cleaning Processes.*
*Optimization in Seawater Desalination Plants*
Guillem Gilabert-Oriol, 2021
ISBN 978-3-11-071507-1, e-ISBN (PDF) 978-3-11-071514-9,
e-ISBN (EPUB) 978-3-11-071516-3

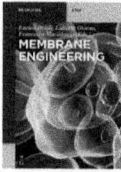

*Membrane Engineering*
Enrico Drioli, Lidietta Giorno, Francesca Macedonio (Eds.), 2019
ISBN 978-3-11-028140-8, e-ISBN (PDF) 978-3-11-028139-2,
e-ISBN (EPUB) 978-3-11-038154-2

*Membrane Systems in the Food Production.*
*Dairy, Wine and Oil Processing*
Alfredo Cassano, Enrico Drioli (Eds.), 2021
ISBN 978-3-11-074288-6, e-ISBN (PDF) 978-3-11-074299-2,
e-ISBN (EPUB) 978-3-11-074307-4

*Membrane Systems in the Food Production.*
*Ingredients and Juice Processing*
Alfredo Cassano, Enrico Drioli (Eds.), 2021
ISBN 978-3-11-071270-4, e-ISBN (PDF) 978-3-11-071271-1,
e-ISBN (EPUB) 978-3-11-071277-3

*Membrane Systems.*
*For Bioartificial Organs and Regenerative Medicine*
Loredana De Bartolo, Efrem Curcio, Enrico Drioli, 2017
ISBN 978-3-11-026798-3, e-ISBN (PDF) 978-3-11-026801-0,
e-ISBN (EPUB) 978-3-11-039088-9

# Electromembrane Processes

Theory and Applications

Edited by
Luboš Novák

DE GRUYTER

**Editor**
Ing. Luboš Novák, CSc.
MEGA a. s.
Pod Vinicí 87
CZ-471 27 Stráž pod Ralskem
Czech Republic
LNovak@mega.cz

ISBN 978-3-11-073945-9
e-ISBN (PDF) 978-3-11-073946-6
e-ISBN (EPUB) 978-3-11-073606-9

**Library of Congress Control Number: 2021949327**

**Bibliographic information published by the Deutsche Nationalbibliothek**
The Deutsche Nationalbibliothek lists this publication in the Deutsche Nationalbibliografie;
detailed bibliographic data are available on the Internet at http://dnb.dnb.de.

© 2022 Walter de Gruyter GmbH, Berlin/Boston
Cover image: Marco_de_Benedictis/iStock/Getty Images Plus
Typesetting: Integra Software Services Pvt. Ltd.
Printing and binding: CPI books GmbH, Leck

www.degruyter.com

# Preface

In 2013–2014, the Czech Membrane Platform, o. s. (CZEMP, o. s.) published the following compilations within the ARoMem project of the Operational Programme **Education for Competitiveness**:

1. **Pressure Membrane Processes** (in Czech, *Tlakové membránové procesy*). Petr Mikulášek et al. CZEMP and University of Chemistry and Technology Prague, 2013. ISBN 978-80-7080-862-7.
2. **Membrane Separation of Gases and Vapours** (in Czech, *Membránové dělení plynů a par*). Milan Šípek et al. CZEMP and University of Chemistry and Technology Prague, 2014. ISBN 978-80-7080-864-1.
3. **Electromembrane Processes** (in Czech, *Elektromembránové procesy*). Luboš Novák et al. CZEMP and University of Chemistry and Technology Praha, 2014. ISBN 978-80-7080-865-8.

At present, electromembrane processes are gaining importance in many industrial and research areas. Highly effective ion-selective membranes for various electromembrane processes including turnkey deliveries of comprehensive technologies are produced also in the Czech Republic, for example, RALEX® membranes and devices for electrodialysis, electrophoresis and electrodeionization in Stráž pod Ralskem.

For all these reasons, it was decided to prepare an extended and supplemented English book devoted to electromembrane processes, based on the mentioned Czech book.

The text is divided into three parts. After a brief summary of principles and historical survey of electromembrane processes, the first part of the book, **Theory and Materials**, contains chapters devoted to the basic thermodynamic principles, basic concepts and laws of electrochemistry, ion-selective materials for membranes, theory of transport through membranes and mathematic modelling of electromembrane processes.

The second part of the book, **Electromembrane Separation and Synthesis Processes**, discusses the most common processes: electrodialysis, electrodeionization, capacitive deionization, diffusion dialysis (diffusion dialysis does not belong to electromembrane processes but electrodialysis, an important electromembrane process, is included in this book and so the editor decided to include the description of diffusion dialysis due to logical reasons) and electrophoresis. The last chapter of this part describes the main industrial applications of electromembrane separation and synthesis processes.

The third part of the book, **Electromembrane Conversion Processes**, contains chapters devoted to membrane electrolysis and electromembrane processes for conversion of energy, which discusses hydrogen economics, redox flow batteries, reverse electrodialysis and supercapacitors.

The editor of this book, Luboš Novák, has devoted his professional life to electromembrane processes. His career began in 1976 in the Research Institute of the Uranium

https://doi.org/10.1515/9783110739466-202

Industry in former Czechoslovakia, where he sought after more efficient treatment of contaminated water from uranium mines.

He manufactured the first RALEX® membrane several years later, followed by complete electrodialysis units, small at first, industrial size later. In 1993, he became the MEGA company owner and led it to be one of the world's renowned suppliers of industrial electromembrane separation technologies, with over 200 running industrial installations around the world.

One of the key factors of his achievements and success is his vision and belief that the potential of electromembrane processes has not been fully understood and used yet. He has always been looking for novel applications and innovations across diverse industries.

To pursue his vision and objectives, Luboš Novák understands the need for broad and open infrastructure. He has been a member of European Membrane Society Council and he founded and built a Membrane Innovation Centre, focused mainly on electromembrane processes and operated by MemBrain, a daughter research branch of MEGA. As a member of a government Research, Development and Innovation Council, he is responsible for applied research. Among other activities, he founded the Czech Membrane Platform in 2008 to gather great minds of membrane academia and industry. Information must be shared and exchanged for the research to thrive – this publication is a pinnacle of state-of-the-art knowledge, available to everyone to learn and build upon.

This book is a comprehensive survey of the topic that could be written only in cooperation of many scientists and professionals: the theory of transport and its modelling (Miroslav Bleha, Roman Kodym, Milan Šípek, Dalimil Šnita and David Tvrzník), basic thermodynamic principles (Milan Šípek), basic concepts and laws of electrochemistry (Tomáš Bystroň and Dalimil Šnita), outline of ion-selective materials for membranes (Miroslav Bleha, Luboš Novák and Robert Válek), detailed description and survey of applications (Luboš Novák, David Tvrzník and Aleš Černín), information on diffusion dialysis (Zdeněk Palatý), information on method for energy conversion – a currently very significant problem (Karel Bouzek, Aleš Černín, Martin Paidar and Petr Mazúr). The editor would like to thank all of them for their cooperation. Special thanks to Eva Juláková and Milan Šípek. This publication would not be complete without their immense effort.

**Note:** Foreign literature on electromembrane processes often name the principal part of the processes as ion-exchange membranes. The authors of this book believe that it is more appropriate to use the term ion-selective membranes, since functional groups with the positive or negative charge, built in the membrane, allow selective transport of cations or anions through the membrane even without ion exchange. Some electromembrane processes do include ion exchange, for example, electrodeionization membrane separation processes but these require regeneration of ions, cation exchangers or anion exchangers.

# Contents

Part III: **Electromembrane conversion processes**

# List of contributing authors

**Ing. Miroslav Bleha, CSc.**
Institute of Macromolecular Chemistry AS CR
Heyrovského nám. 2
CZ-162 06 Prague
miroslav.bleha@czemp.cz

**prof. Dr. Ing. Karel Bouzek**
University of Chemistry and Technology
Technická 1905/5
CZ-160 00 Prague
Karel.Bouzek@vscht.cz

**doc. Ing. Tomáš Bystroň, Ph.D.**
University of Chemistry and Technology
Technická 1905/5
CZ-160 00 Prague
Tomas.Bystron@vscht.cz

**Ing. Aleš Černín, Ph.D.**
New Water Group s.r.o
Nádražní 312
CZ-407 56 Jiřetín pod Jedlovou
cerninales@gmail.com

**Ing. Roman Kodým, Ph.D.**
University of Chemistry and Technology
Technická 1905/5
CZ-160 00 Prague
Roman.Kodym@vscht.cz

**Ing. Petr Mazúr, Ph.D**
University of Chemistry and Technology
Technická 1905/5
CZ-160 00 Prague
Petr.Mazur@vscht.cz

**Ing. Luboš Novák, CSc.**
MEGA a. s.
Pod Vinicí 87
CZ-471 27 Stráž pod Ralskem
LNovak@mega.cz

**doc. Ing. Martin Paidar, Ph.D.**
University of Chemistry and Technology
Technická 1905/5
CZ-160 00 Prague
Martin.Paidar@vscht.cz

**doc. Ing. Zdeněk Palatý, CSc.**
University of Pardubice
Studentská 95
CZ-530 09 Pardubice
zdenek.palaty@upce.cz

**doc. Ing. Milan Šípek, CSc.**
University of Chemistry and Technology
Technická 1905/5
CZ-160 00 Prague
Milan.Sipek@vscht.cz

**prof. Ing. Dalimil Šnita, CSc.**
University of Chemistry and Technology
Technická 1905/5
CZ-160 00 Prague
Dalimil.Snita@vscht.cz

**Ing. David Tvrzník, Ph.D.**
MemBrain s. r. o.
Pod Vinicí 87
CZ-471 27 Stráž pod Ralskem
David.Tvrznik@membrain.cz

**Ing. Robert Válek, Ph.D.**
MemBrain s. r. o.
Pod Vinicí 87
CZ-471 27 Stráž pod Ralskem
Robert.Valek@membrain.cz

https://doi.org/10.1515/9783110739466-204

# List of symbols and abbreviations

| | | |
|---|---|---|
| $A$ | anion | – |
| $A$ | area | $m^2$ |
| $A$ | Helmholtz energy | J |
| $a$ | activity | dimensionless |
| $a_i$ | activity of component $i$ | dimensionless |
| $a_\pm$ | mean activity | dimensionless |
| $a_P$ | solid-phase specific surface area | $m^{-1}$ |
| $a_s$ | specific surface area of solid phase | $m^{-1}$ |
| $a_\pm$ | mean activity | dimensionless |
| $b$ | Tafel slope | dimensionless |
| $\boldsymbol{b}$ | position vector[1] | m |
| $C$ | cation | – |
| $C$ | capacitance | F |
| $C_{diff}$ | differential capacitance of the double layer | F |
| $c$ | molar concentration | $mol\ m^{-3}$ |
| $c_{anal}$ | analytic molar concentration | $mol\ m^{-3}$ |
| $c_i$ | concentration of the $i$-th ion | $mol\ m^{-3}$ |
| $D$ | diffusion coefficient | $m^2\ s^{-1}$ |
| $D_i$ | diffusion coefficient of the $i$-th ion | $m^2\ s^{-1}$ |
| $D_{i,M}$ | diffusion coefficient of the $i$-th component in membrane | $m^2\ s^{-1}$ |
| $D_{i,s}$ | diffusion coefficient of the $i$-th component in solid phase | $m^2\ s^{-1}$ |
| $D_0$ | pre-exponential factor | $m^2\ s^{-1}$ |
| $\boldsymbol{D(r)}$ | vector of electric displacement/induction | $C\ m^{-2}$ |
| $d$ | diameter, distance | m |
| $d_p$ | diameter of particle | m |
| $E$ | energy | J |
| $E$ | voltage (measured between the electrodes) of the galvanic cell | V |
| $E_e$ | electrode potential | V |
| $E^{o\prime}$ | formal electrode potential | V |
| $E^o{}_{Ox/Red}$ | standard electrode potential | V |
| $\boldsymbol{E(r)}$ | vector of electric field intensity | $V\ m^{-1}$ |
| $e^-$ | electron | – |
| $\boldsymbol{e}$ | elementary charge (charge of the electron), $\boldsymbol{e} \doteq 1.602 \times 10^{-19}$ C | C |
| $\boldsymbol{F}$ | Faraday's constant, $\boldsymbol{F} = 96{,}485.341$ C $mol^{-1}$ | $C\ mol^{-1}$ |
| $F/\boldsymbol{F}$ | force (scalar/vector) | N |
| $\boldsymbol{F_e(r)}$ | vector of electrostatic force | N |
| $f_{ji}$ | frictional force between molecules $j$ and $i$ | N |
| $f_r$ | friction coefficient | $m^{-2}\ V\ s$ |
| $\boldsymbol{f(r)}$ | vector of density of the electrostatic force | $N\ m^{-3}$ |
| $G$ | Gibbs energy | J |
| $G$ | conductance | $S \equiv \Omega^{-1}$ |
| $G^\#$ | Gibbs energy of the activated complex/transition state | $J\ mol^{-1}$ |
| $\Delta G^\#$ | molar activation Gibbs energy | $J\ mol^{-1}$ |
| $\Delta G_r$ | molar reaction Gibbs energy | $J\ mol^{-1}$ |

---

1 According to current usage, the vector quantities are marked with symbols in bold italics letters of regular fonts. Therefore, distinguish please between symbols of vectors and symbols of basic physical constants (only $\boldsymbol{R}$, $\boldsymbol{F}$, $\boldsymbol{e}$ and $\boldsymbol{\varepsilon}$) for which we used bold italics as well.

https://doi.org/10.1515/9783110739466-205

| | | |
|---|---|---|
| $G$ | tensor of specific conductivity of the environment | |
| $g$ | gravitational acceleration (standard gravitational acceleration $g_n$ = 9.823 m s$^{-1}$) | m s$^{-1}$ |
| $H$ | enthalpy | J |
| $\Delta H_r$ | molar reaction enthalpy | J mol$^{-1}$ |
| $h$ | height | m |
| $h_P$ | thickness of a single cell pair | m |
| $I$ | electric current | A |
| $I_E$ | current passing through active area | A |
| $I_G$ | ground fault current | A |
| $I_L$ | leakage electric current (shunt current) | A |
| $i$ | general symbol for $i$th component | – |
| $J_m/\boldsymbol{J_m}$ | mass flux (scalar/vector)[2] | kg m$^{-2}$ s$^{-1}$ |
| $J_n/\boldsymbol{J_n}$ | molar flux (scalar/vector)[2] | mol m$^{-2}$ s$^{-1}$ |
| $J_v/\boldsymbol{J_v}$ | volume flux (scalar/vector)[2] | m$^3$ m$^{-2}$ s$^{-1}$ |
| $J$ | local mass flux of the diluate[2] | kg m$^{-2}$ s$^{-1}$ |
| $J_i$ | local molar flux of the $i$th component[2] | mol m$^{-2}$ s$^{-1}$ |
| $J_{i,conv}$ | convective flux of the component $i$[2] | mol m$^{-2}$ s$^{-1}$ |
| $J_{i,dif}$ | diffusion flux of the component $i$[2] | mol m$^{-2}$ s$^{-1}$ |
| $J_{i,migr}$ | migration flux of the component $i$[2] | mol m$^{-2}$ s$^{-1}$ |
| $J_{i,M}$ | molar flux of the component $i$ in membrane[2] | mol m$^{-2}$ s$^{-1}$ |
| $j/\boldsymbol{j}$ | current density (scalar/vector) | A m$^{-2}$ |
| $j_0$ | exchange current density | A m$^{-2}$ |
| $K$ | equilibrium constant | dimensionless |
| $K$ | dialysis coefficient | m s$^{-1}$ |
| $K_{dis}$ | dissociation constant | dimensionless |
| $K'_{dis}$ | apparent (concentration) dissociation constant | dimensionless |
| $K_{SP}$ | solubility product | dimensionless |
| $K_w$ | autoprotolytic constant of water (ionic product of water) | dimensionless |
| $k$ | rate constant | s$^{-1}$ |
| $k$ | coefficient of the mass transfer | m s$^{-1}$ |
| $k_M$ | electrokinetic permeability of membrane | m$^2$ |
| $\boldsymbol{k}$ | Boltzmann constant, $\boldsymbol{k_B}$ = 1.38065 × 10$^{-23}$ J K$^{-1}$ | J K$^{-1}$ |
| $L$ | length, distance | m |
| $L_{ii}$ | diagonal phenomenological coefficient | dimensionless |
| $L_{ij}$ | interference phenomenological coefficient | dimensionless |
| $L_{MS}$ | thickness of the membrane stack | m |
| $l$ | thickness of the membrane | m |
| $M$ | molar mass | g mol$^{-1}$ |
| $M_c$ | molar mass of the polymer chain | g mol$^{-1}$ |
| $m$ | mass | kg |
| $\dot{m}$ | local mass flow | kg s$^{-1}$ |
| $N$ | number | dimensionless |
| $\boldsymbol{N_A}$ | avogadro constant, $\boldsymbol{N_A} \doteq 6.022142$ × 10$^{23}$ mol$^{-1}$ | mol$^{-1}$ |
| $N_i/\boldsymbol{N_i}$ | total flux of component $i$ (scalar/vector) | mol s$^{-1}$ |
| $\dot{N}$ | local molar flux | mol s$^{-1}$ |
| $\dot{N}_S$ | total molar flux of the electrolyte | mol s$^{-1}$ |

---

**2** All quantities of flux are also often named as intensity (density) of the flux, especially if they are related to square meters.

| | | |
|---|---|---|
| $n$ | amount of substance | mol |
| Ox | oxidized species | – |
| $P$ | concentration polarization level | dimensionless |
| $P$ | permeability (coefficient) | $m\,s^{-1}$ |
| $\boldsymbol{P}$ | tensor of hydraulic permeability | $m^2\,s^{-1}\,Pa^{-1}$ |
| $P_{el}$ | electric power dissipation (Joule's heat production) | C |
| $\boldsymbol{P(x)}$ | vector of electric polarization | $C\,m^{-2}$ |
| $p$ | pressure | Pa |
| $p_{STP}$ | standard pressure,[3] $p_{STP}$ = 101,325 Pa | Pa |
| $p_{el}(\boldsymbol{x})$ | electric power dissipation density | $W\,m^{-3}$ |
| Pe | Péclet number[4] | dimensionless |
| $Q$ | electric charge | C |
| $q$ | heat | J |
| $q$ | (volume) density of electric charge | $C\,m^{-3}$ |
| $\boldsymbol{q}$ | density of the heat flux | $W\,m^{-2}$ |
| $q(\boldsymbol{r})$ | density of the electric charge at the point $\boldsymbol{r}$ | $C\,m^{-3}$ |
| $R$ | resistance | $\Omega$ |
| $R$ | retention, retention coefficient | dimensionless |
| $R_{A,AM}$ | area resistance of anion-selective membrane | $\Omega\,cm^2$ |
| $R_{A,CM}$ | area resistance of cation-selective membrane | $\Omega\,cm^2$ |
| $R_w$ | water recovery | dimensionless |
| $\boldsymbol{R}$ | molar gas constant, $\boldsymbol{R}$ = 8,314 472 $J\,K^{-1}\,mol^{-1}$ | $J\,K^{-1}\,mol^{-1}$ |
| Re | Reynolds number[4] | dimensionless |
| Red | reduced species | – |
| $r$ | reaction rate | $mol\,m^{-3}\,s^{-1}$ |
| $r_i$ | diameter of the particle $i$ | m |
| $\boldsymbol{r}$ | position vector | m |
| $S$ | coefficient of sorption | dimensionless |
| $S$ | entropy | $J\,K^{-1}$ |
| $S$ | solubility coefficient | dimensionless |
| $S$ | supersaturation of the solution | dimensionless |
| Sc | Schmidt number[4] | dimensionless |
| Sh | Sherwood number[4] | dimensionless |
| $\Delta S_r$ | molar reaction entropy | $J\,K^{-1}\,mol^{-1}$ |
| $s$ | trajectory | m |
| $T$ | thermodynamic (absolute) temperature | K |
| $T_{STP}$ | standard temperature (= 273.15 K) | K |
| $t_i$ | transport (transference) number of the $i$th component | dimensionless |
| $\hat{t}_i$ | global (integral) transport number | dimensionless |
| $t'_i$ | local transport number of $i$th component | dimensionless |
| $t_{i,M}$ | transport number of $i$th component in membrane | dimensionless |
| $U$ | internal energy | J |
| $U$ | voltage | V |

---

**3** Since 1982, the standard pressure was defined as 100,000 Pa, according to the IUPAC recommendation.
**4** Reynolds number, as well as other similar quantities, is also called "criterion" (i.e. Péclet criterion, Reynolds criterion, Schmidt criterion, Sherwood criterion, etc.) in the modern literature.

| | | |
|---|---|---|
| $U_i$ | electrolytic (electrophoretic) mobility | $m^2\,V^{-1}\,s^{-1}$ |
| $u$ | specific internal energy | $J\,kg^{-1}$ |
| $u_i$ | mobility of component $i$ | $m\,mol\,N^{-1}\,s^{-1}$ |
| $u_{i.M}$ | mobility of ions $i$ in the membrane | $m\,mol\,N^{-1}\,s^{-1}$ |
| $V$ | volume | $m^3$ |
| $V_0$ | volume at 0 K | $m^3$ |
| $v$ | velocity | $m\,s^{-1}$ |
| $v/\mathbf{v}$ | (volume) flow rate (scalar/vector) | $m^3\,s^{-1}$ |
| $v_x$ | electrolyte flow velocity in the direction $x$ | $m\,s^{-1}$ |
| $W$ | width of flow channel, dilute chamber width | $m$ |
| $w$ | pressure–volume work | $J$ |
| $w_{el}$ | electrical work | $J$ |
| $w_{out}$ | performed work | $J$ |
| $w_{in}$ | consumed work | $J$ |
| $w^\star$ | work (other than pressure–volume) | $J$ |
| $w_i$ | mass fraction of the component $i$ | dimensionless |
| $X/\mathbf{X}$ | generalized force (scalar/vector) | N |
| $X_{H^+}$ | conversion of cation exchange resin | dimensionless |
| $x$ | length coordinate | $m$ |
| $x_i$ | molar fraction of the component $i$ | dimensionless |
| $Y$ | general extensive quantity | – |
| $\bar{Y}_i$ | partial molar quantity of the component $i$ | – |
| $\Delta Y$ | change of the general thermodynamic quantity | – |
| $y$ | length coordinate | $m$ |
| $z$ | number of electrons | dimensionless |
| $z$ | length coordinate | $m$ |
| $z_i$ | charge number of ion $i$ | dimensionless |
| $\alpha$ | phase $\alpha$ | – |
| $\alpha$ | degree of dissociation, separation factor | dimensionless |
| $\alpha$ | interphase area density | $m^2\,m^{-3}$ |
| $\alpha_{an}$ | charge transfer coefficient of the anodic reaction | dimensionless |
| $\alpha_{cat}$ | charge transfer coefficient of the cathodic reaction | dimensionless |
| $\beta$ | phase $\beta$ | – |
| $\beta$ | symmetry factor | dimensionless |
| $\gamma$ | surface tension | $N\,m^{-1}$ |
| $\gamma_i$ | activity coefficient of the component $i$ | dimensionless[5] |
| $\gamma_\pm$ | mean activity coefficient | dimensionless[3] |
| $\delta$ | thickness of the membrane | $m$ |
| $\delta$ | thickness of the separation layer (skin) | $m$ |
| $\delta$ | thickness of the diffusion layer | $m$ |
| $\varepsilon$ | void fraction | dimensionless |
| $\varepsilon$ | permittivity | $F\,m^{-1}$ |
| $\varepsilon_r$ | relative permittivity | dimensionless |
| $\boldsymbol{\varepsilon_0}$ | vacuum permittivity, $\boldsymbol{\varepsilon_0} = 8.854 \cdot 10^{-12}\,F\,m^{-1}$ | $F\,m^{-1}$ |
| $\eta$ | dynamic viscosity | Pa s |
| $\eta$ | current efficiency | dimensionless |

---

**5** See the note to eq. (2.54).

| | | |
|---|---|---|
| $\eta$ | overvoltage (overpotential) | V |
| $\eta_M$ | dynamic viscosity of the liquid phase in the membrane | Pa s |
| $\eta_T$ | thermodynamic efficiency of the electrochemical system | dimensionless |
| $\kappa$ | conductivity | S m$^{-1}$ |
| $\kappa_L$ | exchange resin bed conductivity | S m$^{-1}$ |
| $\Lambda$ | molar conductivity of the electrolyte | S m$^2$ mol$^{-1}$ |
| $\Lambda^0$ | limiting molar conductivity of the electrolyte | S m$^2$ mol$^{-1}$ |
| $\lambda$ | Donnan distribution coefficient | dimensionless |
| $\lambda_D$ | Debye length | m |
| $\lambda_Q$ | Heat conductivity | W m$^{-1}$ K$^{-1}$ |
| $\lambda_i$ | thermal conductivity | W m$^{-1}$ K$^{-1}$ |
| $\lambda_i$ | molar conductivity of the component $i$ | S m$^2$ mol$^{-1}$ |
| $\lambda_i^0$ | limiting molar conductivity of the component $i$ | S m$^2$ mol$^{-1}$ |
| $\mu_i$ | chemical potential of component $i$ | J mol$^{-1}$ |
| $\mu_i^0$ | standard chemical potential of component $i$ | J mol$^{-1}$ |
| $\tilde{\mu}_i$ | electrochemical potential of component $i$ | J mol$^{-1}$ |
| $v$ | number of cations or anions | dimensionless |
| $v_i$ | stoichiometric coefficient of the component $i$ | – |
| $v_{ij}$ | stoichiometric coefficient of the component $i$ in the $j$-th chemical reaction | – |
| $\pi$ | Ludolph's number | dimensionless |
| $\pi$ | osmotic pressure | Pa |
| $\rho$ | resistivity (specific resistance) | $\Omega$ m |
| $\rho$ | density | kg m$^{-3}$ |
| $\rho_Y$ | (volume) density of the quantity $Y$ | m$^3$ s$^{-1}$ |
| $\sigma(S)$ | entropy production | J K$^{-1}$ s$^{-1}$ |
| $\sigma_Y$ | (volume) density of the quantity $Y$ source | m$^3$ s$^{-1}$ |
| $\tau$ | time | s |
| $\tau_D$ | relaxation (Debye) time | s |
| $\tau_R$ | residence time | s |
| $\boldsymbol{\tau}$ | tensor of the tangential tension | |
| $\Phi_I$ | current (faradic) efficiency (charge yield) | dimensionless |
| $\Phi_Y$ | flux (surface density) of the quantity $Y$ | m$^3$ s$^{-1}$ |
| $\varphi\ (\varphi_e)$ | electric potential | V |
| $\varphi$ | volume fraction | dimensionless |
| $\varphi_\alpha$ | inner (Galvani) potential of the phase $\alpha$ | V |
| $\Delta\varphi_{Don}$ | Donnan potential | V |
| $\Delta\varphi_L$ | liquid-junction (diffusion) potential | V |
| $\Delta\varphi_M$ | membrane potential | V |
| $\nabla\varphi$ | gradient of potential | V m$^{-1}$ |
| $\chi$ | electric susceptibility | dimensionless |
| $\bar{\chi}_i$ | molar electric susceptibility of the component $i$ | mol$^{-1}$ |
| $\psi$ | permselectivity | dimensionless |
| $\psi_{A,B}$ | permselectivity for counter-ion A | dimensionless |

## Mathematic symbols

| | |
|---|---|
| Δ | difference |
| Σ | summation |
| Π | product |
| ∇ | linear operator nabla |
| $\nabla^2$ | Laplace operator |
| ∞ | infinity |

## Indexes

| | |
|---|---|
| A | anion |
| A | component A |
| AM | anion-selective membrane |
| ac | actual |
| an | anode, anodic |
| atm | atmospheric |
| av | average |
| B | component B |
| C | cation |
| c | critical |
| cat | cathode, cathodic |
| CM | cation-selective membrane |
| comb | combinatory member |
| conv | convection |
| CONC | concentrate |
| dif | diffusion |
| diff | differential |
| DIL | diluate |
| dis | dissociation |
| dissip | dissipative |
| Don | Donnan |
| eff | effective |
| EL | electrolyser |
| el | electrode; electrical |
| eq | equilibrium |
| F | feed |
| fix | fixed |
| g | gaseous phase |
| GC | galvanic cell |
| $i$ | component $i$ |
| if | interphase |
| in | internal, inside |
| $j$ | component $j$ |
| kin | kinetic |
| L | time lag, layer |
| l | liquid phase |
| m | molar |

| | |
|---|---|
| M | membrane, in membrane |
| $m$ | mass, related to the mass (specific) |
| max | maximal |
| Me | metal |
| migr | migration |
| $n$ | molar (related to the amount of substance) |
| out | outside |
| ox | oxidation |
| P | permeate, permeability, product |
| polym | polymer |
| r | reaction |
| red | reduction |
| ref | referent |
| rel | relative |
| res | residual |
| ret | retentate |
| S | at/near the surface, solute, solution, sorption |
| s | stationary |
| s | solid phase |
| sat | saturated |
| STP | standard temperature and pressure |
| surf | surface |
| theor | theoretical |
| tot | total |
| $V$ | volume (related to volume) |
| $\tau$ | time (related to time) |
| $^\circ$ | standard (quantity) |
| + | for positively charged ion (cation) |
| − | for negatively charged ion (anion) |
| ~ | (over the symbol) dimensionless quantity |
| $\infty$ | Infinity |

# Abbreviations

| | |
|---|---|
| AC | alternating current |
| AM | anion-selective membrane |
| BM | bipolar membrane |
| CCE | Calcium Carbonate Equivalent |
| CDI | capacitive deionization |
| CEDI | continuous electrodeionization |
| CIP | chemical cleaning |
| CM | cation-selective membrane |
| CZEMP | Czech Membrane Platform |
| DBL | diffusion-bound layer |
| DC | direct current |
| EC | electrocoating |
| ED | electrodialysis |

| | |
|---|---|
| EDBM (BMED) | electrodialysis with bipolar membranes |
| EDC | electrodialysis to concentrate electrolyte solutions |
| EDM | electrodialysis for demineralization |
| EDI | electrodeionization |
| EDR | electrodialysis reversal |
| EFC | electrophoresis |
| EL | electrolysis |
| EMP | electromembrane processes |
| FCE | feed conductivity equivalent |
| GDE | gas diffusion electrode |
| HWS | hot water sanitizable (module) |
| IEC | ion exchange capacity |
| IEX | ion exchange |
| LSI | Langelier saturation index |
| MBR | membrane bioreactor |
| MCDI | membrane capacitive deionization |
| ME | membrane electrolysis |
| MEA | membrane electrode assembly |
| NF | nanofiltration |
| P&ID | process and instrumentation diagram |
| PEM | proton exchange membrane, or polymer electrolyte membrane |
| PLC | programmable logic controller unit |
| RED | reverse electrodialysis |
| RFB | redox-flow battery |
| RO | reverse osmosis |
| SPE | solid polymer electrode |
| SWEDI | spiral wound EDI |
| T | thermodynamic |
| TDS | total dissolved solids |
| TEA | total exchangeable anions |
| TEC | total exchangeable cations |
| TN | thermoneutral |
| TOC | total organic carbon |
| UF | ultrafiltration |
| UV | ultraviolet |
| VRFB | vanadium redox flow battery |
| ZLD | zero liquid discharge |

Luboš Novák, Milan Šípek
# 1 Principles and historical survey of electromembrane processes

Compared to conventional separation methods (e.g. distillation, crystallization, extraction and sorption), membrane processes for separation, concentration or purification of substances are, in many cases, faster, more efficient and cheaper. A great advantage of membrane separation processes lies in the fact that the separation occurs at low temperatures, so there is no risk of change or decomposition of the separated substances sensitive to higher temperatures, and above all, the separation does not require the addition of other substances.

The most important element of every separation process is the **separation membrane** as a passive or active barrier separating two phases and allowing **selective transport** of individual components of the separated mixture due to different rates of their transport through the membrane. The transport rate depends on the intensity of driving forces, mobility and concentration of components in the membrane. The driving forces can be gradients of pressure, concentration, electric potential and temperature.

Separation membranes are **semipermeable (permselective)** and can be classified according to the **state of aggregation (phaseous state), state, origin and morphology**, as is shown in Fig. 1.1. Membranes can be produced using any material suitable for making a thin foil that permits penetration of certain substances and retains others. The most commonly used materials are **polymers**, but membranes of **inorganic origin** are increasingly being used, for example, ceramics, glass, metals or microporous carbon. Membrane materials will be discussed in detail in Chapter 4.

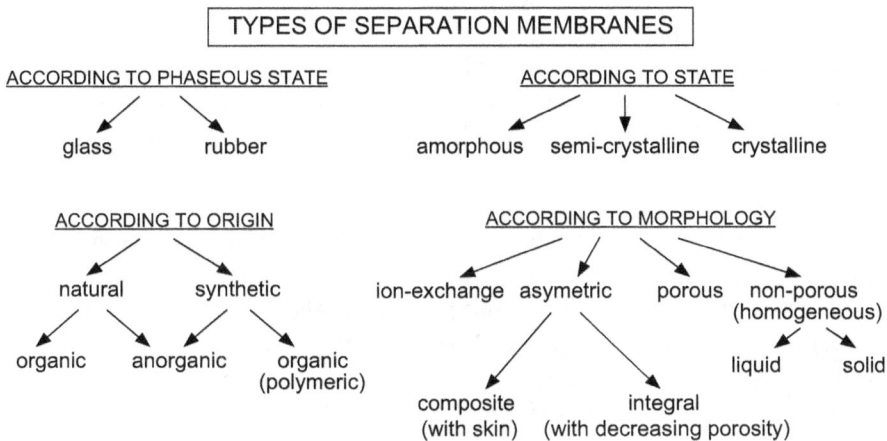

Fig. 1.1: Types of separation membranes.

https://doi.org/10.1515/9783110739466-001

Each membrane separation process is characterized by its physical principle of transport or retention.

The mixture of substances fed to the surface of separation membrane is called the **feed**. Substances that do not penetrate the membrane form the **retentate;** in the case of reverse osmosis, electrodialysis (ED), and electrodeionization (EDI), it is also called the **concentrate**. Substances passing through the membrane are called the **permeate** or the **pervaporate** in the case of pervaporation. This means that both the retentate and the permeate can be the product of a membrane process.

The separation membrane must have certain specific properties, that is, sufficient mechanical and chemical stability, but above all sufficient permeability (**performance**) and separation ability (**selectivity**). The selectivity is quantitatively expressed as a **separation factor**, which is determined by the ratio of the permeate content to the feed content of the given substance. The content can be expressed using concentrations or molar or mass fractions. The membrane selectivity can be also expressed by the **retention coefficient** $R$, given by the relation

$$R = 1 - (c_P/c_F)$$

where $c_P$ is the concentration of the separated component in the permeate and $c_F$ is the concentration of this component in the feed. The value of $R$ is a dimensionless variable and it varies from 0 to 1. If the value is 1, the retention is complete and the membrane is ideally semipermeable.

The oldest membrane materials, for example, for traditional filtration, were natural ones: animal intestines or bladders, textile fibres or, later, cellulose. Some progress was achieved with the introduction of chemical processes into the preparation of membrane materials, namely the treatment of cellulose to acetate or nitrate with improved separation effects and mechanical properties.

However, the turning point comes in the beginning of the twentieth century, when developing polymer chemistry was applied for the preparation of membrane materials. Without exaggeration, it can be stated that without polymer membrane materials, membrane separation processes could not have the present extent and the impact on industry evident today in many fields. Also, the development of inorganic chemistry contributed to the development of membrane separation processes but to a much lesser extent.

The main problem lies in the fact that, in general, **membranes with good separation properties demonstrate low permeability**. Polymer chemistry in the second half of the last century resolved this serious problem by allowing polymer chemists to produce separation membranes with good permeability and good separation properties, both homogeneous and composite, with the required structure or shape. Membranes with good separation properties must have a sufficient surface area and very thin separating layer (skin) and, at the same time, they must be mechanically and chemically stable. Therefore, for example, to separate mixtures or vapours and gases, polymer membranes are used in the form of hollow fibres placed in membrane modules or spirally wound membranes, meeting these strict requirements. We must not neglect the fact that in the second half of the twentieth century, experimental techniques have been developed significantly, which allow testing membranes by determining their transport parameters: **coefficients of permeability $P$, diffusion $D$ and sorption $S$.**

The transport of substances through the membrane depends on the properties of the passed substance and primarily on the nature of the membrane itself. Therefore, the transport of substances is different in cases of porous, non-porous or ion-selective membranes. Each of them follows a different mechanism. The mass transport through the membrane can be generally expressed in three ways, depending on the method of expressing the amount of the substance passed (**mass flux, flux of the amount of substance** and **volume flux**); see Chapter 5.

# 1.1 Description of electromembrane processes

**Electromembrane processes** are membrane processes in which the driving force of the transport of ions is the **electric potential gradient** applied in a system with ion-selective[1] or bipolar membranes (BM). They are very important, frequently used and widely applied in a number of industrial technologies. Their advantages are high efficiency of separation without phase changes, lower energy consumption compared to conventional methods, easy automation of the process and less space demands. Their disadvantage is the fact that the solutions require pre-treatment and the process itself leads to clogging the membrane with insoluble substances (the so-called **fouling**). Membranes applied in electromembrane processes must have low electrical resistance, high selectivity and good mechanical properties including chemical and thermal resistance.

Ion-selective membranes are divided into **anion-selective** and **cation-selective**. This division is based on the type of functional group in the membrane matrix. **Cation-selective membranes** (CM) contain negatively charged functional groups (e.g. $SO_3^-$), which allow the transport of positively charged ions (cations) through the membrane and prevent the passage of negatively charged ions (anions). **Anion-selective membranes** (AM) contain positively charged functional groups (primarily quaternary ammonium groups, e.g. $NH_3^+$ and $NRH_2^+$), allowing the transport of negatively charged ions (anions) and preventing the passage of positively charged ions (cations).

Besides these types of ion-selective membranes, there are also membranes containing both cation and anion functional groups. By combination and modification of the two basic types of ion-selective membranes, we obtain a BM, composed of two layers of polymer material, one of them contains only positively charged functional

---

1 For the terms of ion-selective membranes and ion-exchange membranes, see the note in Introduction, page VI.

groups and the other one only negatively charged functional groups. Thus, the BM consists of a cation-selective layer and an anion-selective layer.

**Electromembrane processes** can be divided into three groups:
a) electromembrane separation processes,
b) electromembrane synthesis processes and
c) electromembrane processes for conversion of energy and substances.

**Electromembrane separation processes** utilize ion-selective membrane to separate the electrolyte from one liquid phase and transfer it to the other liquid phase. They are based on the selective permeability of the membrane for cations and anions. These processes mainly include **electrodialysis**, ED (Chapter 7), which is used to demineralize or concentrate electrolyte solutions. The operating device for ED is called a **electrodialysis stack** (or an **electrodialyser**). **Electrodeionization**, EDI (Chapter 8) combines ED with classical ion exchangers. Further possibilities are **capacitive deionization**, CDI (Chapter 9) and continual electrodeionization, CEDI (Chapter 8) using also BMs.

**Electromembrane synthesis processes** include, besides other important methods, **membrane electrolysis**, using ion transport in electric field driven by selective separation ability of the ion-selective membrane in combination with electrochemical reaction for production of chemicals, for example, chlorine, sodium hydroxide and hydrogen by electrolysis of NaCl solution, or membrane production of hydrogen peroxide. This group of processes also include electrophoretic paint coating used for car bodies (Chapters 11 and 12).

## 1.2 History

The process of dialysis was first described in the mid-nineteenth century by Graham, who tried to separate large molecules of colloid substances from low-molecular substances using a semipermeable membrane [1, 2]. The permeability of membranes for electrolyte solutions was studied in the 1880s and 1890s by Ostwald [3], who also defined the **membrane potential** as the potential difference, created at the interface between the membrane and aqueous solution of ions due to different ion concentrations on both sides of the membrane. The existence of membrane potential was later confirmed also by Donnan [4], who studied and described mathematically the concentration equilibrium between an ion-selective membrane and electrolyte solution.

ED was also studied by Morse and Pierce [5], who concluded, based on experiments, that the transport of electrically charged particles through a non-charged, neutral dialysis membrane can be substantially accelerated by electric field. However, the practical use of electrolysis occurred only 30 years later [6].

An important contribution to the ED process was the use of **ion-selective membranes** [7] because they prevent the transport of co-ions, that is, ions with the same charge as the fixed charge of the membrane. As was already stated, the polymer chains of cation-selective membranes carry groups with negative electric charge, mostly sulpho-, phospho- and carboxyl groups. Conversely, the polymer chains of anion-selective membranes carry groups with positive electric charge, usually quaternary ammonium groups [8–10]. The use of these ion-selective membranes allows ED in a multi-chamber arrangement with regular alternation of anion-active and cation-active membrane in the so-called **stacks** [11]. In an electrodialyser arranged in this manner, only one pair of electrodes is necessary for many membrane stacks, which minimizes the irreversible energy losses due to secondary electrode reactions.

In the 1940s, interest in the wider industrial use of electrodialysis led to the development and production of sufficiently efficient synthetic ion-selective membranes. In the 1950s, research was focused on the preparation of stable and highly efficient ion-selective membranes with low electrical resistance [12, 13].

Since then, ED has become an important technological process in the desalination of water and concentration of electrolyte solutions [14]. Throughout the world, membranes with high selectivity and low osmotic or electroosmotic water permeability in contact with a diluted solution are more and more used to prepare drinking water by desalination of brackish water. These requirements are met by membranes with a heterogeneous structure, prepared by dispersing a finely ground ion exchanger in the polymeric matrix [15]. After treatment, stable ion-selective membranes can be obtained, having better mechanical, chemical and thermal stability, mainly for cation-selective membranes.

Since the 1960s, new ED units with more efficient ion-selective membranes have been operated in the USA and Europe, to obtain drinking water by desalination of seawater. Nowadays, the process of desalination of seawater, brackish water and industrial wastewater becomes increasingly important, since the lack of drinking water is imminent not only in Africa or Middle East but in many countries worldwide [16, 17]. In this context, it is appalling to realize that while people from rich countries flush their toilets with drinking water, people from poor African countries have to walk several kilometres for a sip, drop of drinking water.

In 1979, the company Ionics Inc. came with the possibility of so-called the **reversal electrodialysis process**, in which the electrodes in the electrodialyser can reverse polarity alternatively. This arrangement limits the formation of salt precipitates on the electrodes and membranes, increases the electrodialyser usability and reduces its cleaning time [18].

In Japan, ED is used to concentrate sodium chloride and recover table salt from seawater [19]. In the 1970s, ED also found many applications in the food-processing and pharmaceutical industries, using new ion-selective membranes with sufficient chemical and thermal stability. The American company DuPont launched Nafion® membrane and the Japanese company Asahi Glass Co. introduced Flemion® membrane. These are highly stabile ion-selective membranes on the basis of perfluorinated polymers, applied in brine electrolysis, where the resulting products are sodium hydroxide and chlorine [20–23]. The electrolysis of the NaCl solution to produce NaOH can be generally performed by three methods: amalgam, diaphragm and membrane. The first two methods are due to environmental and health risks practically not used in developed industrial countries. Japan was one of the first to adopt membrane technology and to stop producing NaOH using amalgam method – as serious health problems were met when people consumed fish contaminated with mercury.

In the former Czechoslovakia, in 1985, the first heterogeneous ion-exchange membrane was produced in the Research Institute of Uranium Industry. This led to the construction of RALEX electrodialysers for separation of sludge lake waters from the uranium mining, operating even today. Since 1989, research and business activities continue under the MEGA brand, owned and directed by the researcher who made the first RALEX® membrane, Luboš Novák.

Today, ion-selective membranes on the basis of perfluorinated polymers are widely applied in many other electrochemical processes and also serve as a solid electrolyte in **low-temperature fuel cells**. At present, research and development of these fuel cells draws much attention at universities and research centres [24]. Very interesting is also the use of membranes of perfluorinated polymers for **water electrodialysis** to obtain hydrogen. In connection with fuel cells, so-called **hydrogen economy** has been recently discussed as a solution of the problems of electricity storage and supply [25]. This issue is also closely related to the development and production of electric cars powered by hydrogen or methanol fuel cells.

An important benefit for electromembrane processes has brought the use of **bipolar membranes**, which consist of cation-selective and anion-selective membranes with an active catalytic interlayer. These membranes were developed in the 1970s and allow further applications, for example, in the chemical industry to produce acids and bases from their salts [26, 27].

It is worth mentioning that prof. Zdeněk Matějka from the UCT Prague was among the first to propose the combination of ED with conventional deionization on ion exchangers [28]. This process, now referred to as **continuous electrodeionization** (CEDI), is commercially used in the production of ultrapure water [29].

The development in the field of ion-selective membranes is far from being complete in terms of material, as well as their production and application. At present, research of the material and application of ion-selective membrane is focused on the reduction of energy demands and expanding application possibilities, particularly in energy conversion and storage. Current types of ion-selective membranes and their use in various electromembrane processes can be found in the relevant chapters of this book.

# References

[1]    T. Graham, Philos. Trans. 144: 177–182, 1854.
[2]    T. Graham, Philos. Trans. 151: 183–187, 1861.
[3]    W. Ostwald, Z. Phys. Chem. 6: 71–82, 1890.
[4]    F. G. Donnan, Z. Electrochem. 17: 572–581, 1911.
[5]    H. W. Morse and G. W. Pierce, Z. Phys. Chem. 45: 589–607, 1903.
[6]    J. R. Wilson, Design and Operation of Electrodialysis Plants. In Demineralization by
       Electrodialysis. (J. R. Wilson, Ed.). Butterworths, London, 1960.
[7]    L. Michaelis and A. Fujita, Biochem. Z. 158:28–37, 1925.
[8]    T. Sata, Ion Exchange Membranes. Preparation, Characterization, Modification and
       Application. RCS Advancing the Chemical Sciences, 2004.
[9]    H. Strathmann, Ion-Exchange Membranes. Separation Processes. Elsevier, Amsterdam, 2004.
       ISBN 0-44450236-X.
[10]   T. W. Xu, J. Membrane Sci. 263(1–2): 1–29, 2005.
[11]   K. H. Meyer and W. Strauss, Helv. Chim. Acta. 23: 795–798, 1940.
[12]   W. Juda and W. A. McRae, J. Am. Chem. Soc. 72: 1044, 1950.
[13]   A. G. Winger, G. W. Bodamer, and R. Kunin, J. Electrochem. Soc. 100: 178–184, 1953.
[14]   K. S. Spiegler, Ion-Exchange Technology. (F. C. Nachod and J. Schubert, Eds.). Academic
       Press, New York, 1956.
[15]   F. Bergsma and C. A. Kruissink, Fortschr. Hochpolym. Forsh. 21: 307–362, 1961.
[16]   H. G. Heitmann, Saline Water Processing – Desalination and Treatment Seawater, Brackish
       Water and Industrial Waste Water. VCH Verlag, Weinheim, 1990. ISBN 3-527-27826-5.
[17]   M. C. Porter, Handbook of Industrial Membrane Technology. In Electrodialysis. (T. A. Davis,
       Ed). Noyes publications, New Jersey, USA, 1990.
[18]   W. E. Katz, Desalination 28: 31–40, 1979.
[19]   T. Nishiwaki, Industrial Processing with Membranes. (R. E. Lacey and S. Loeb, Eds.).
       Wiley & Sons, New York, 1972.
[20]   W. G. Grot, Chem. Ing. Tech. 47: 617, 1975.
[21]   Y. Oda, M. Suhara, and E. Endo, Process for Producing Alkali Metal Hydroxide. US Patent
       4 065 366, 66, December 27.
[22]   K. Sato and H. Miyake, High performance ion exchange membrane for industrial use.
       Polym. J. 23(5): 531–540, 1991.
[23]   F. Barbir, PEM Fuel Cells – Theory and Practice. John Wiley &Sons, Chichester, 2000.
[24]   R. Soury, et. al. Functional fluoropolymers for fuel cell membranes. Solid State Ion. 176:
       2839–2848, 2005.
[25]   M. Conte, et al. Hydrogen economy for a sustainable development: State of the art and
       technological perspectives. J. Power Sources 100: 171–187, 2001.
[26]   H. Strathmann, Bipolar Membrane and Membrane Processes. In Membrane Separation.
       Academic Press, Twente, 2000.
[27]   X. Tongwen and Y. Weihua, Citric acid production by electrodialysis with bipolar membranes.
       Chem. Eng. Process. 41: 519–524, 2002.
[28]   Z. Matejka, J. Appl. Chem. Biotechnol. 21: 117–120, 1971.
[29]   A. Dey and J. Tate, Ultrapure Water J. 22: 20–29, 2005.

Part I: **Theory and materials**

After summarizing the basic principles of membrane processes, electromembrane processes and the historical survey of their application in opening Chapter 1, Part I introduces the chapters summarizing the basic necessary information from thermodynamics (Chapter 2) and electrochemistry (Chapter 3). Chapter 4 deals with materials for ion-selective membranes and their production and properties, and Chapter 5 describes the theory of transport through ion-selective membranes. The theoretical description and mathematical modelling of electromembrane processes, including balance equations and mathematical appendix, are given in Chapter 6.

https://doi.org/10.1515/9783110739466-002

Milan Šípek

# 2 Basic thermodynamic principles

In this chapter, only basic thermodynamic functions of classical thermodynamics are described and basic postulates of linear irreversible thermodynamics are given.[1] The explanation is limited to the extent necessary for understanding of the text in following chapters. The reader will find thermodynamic parts also in other parts of the book, for example, in Sections 3.4 and 3.5.

## 2.1 Basic concepts and definitions

A part of space enclosed in the continuous volume and containing, at a given time, a defined amount of matter is called a **system**. A part of space beyond this selected space is known as the **surroundings of the system**. The **boundary** separating the system from its surroundings might be quite abstract.

An important feature of the system is the requirement that the system contains a large number of mutually interacting **subsystems** whose dimensions are small in comparison with the volume of the entire system. Macroscopic systems in which subsystems interact mostly mechanically or electromagnetically are called **thermodynamic systems**.

Based on the relation of systems to their surroundings, we distinguish thermodynamic systems as follows:

- **isolated systems**, which exchange neither mass nor energy with their surroundings;
- **closed systems**, which exchange only energy (in the form of heat or energy) with their surroundings;
- **open systems**, which exchange both mass and energy with their surroundings;
- **adiabatic systems**, which do not exchange heat with their surroundings but may exchange energy.

A part of the system whose properties are either static or continuously changing in space is called a **phase**. According to their internal states, we distinguish **homogeneous** systems (consisting of a single phase) from **heterogeneous** systems (consisting of multiple phases).

---

1 General information on thermodynamics can be found in many physical chemistry textbooks, for example, P. Atkins and J. de Paula, *Physical Chemistry*. 9th edition. Oxford University Press, 2010. ISBN 978-0-19-954337-3; for *transport phenomena* see, for example, R. B. Bird, W. E. Stewart, and E. N. Lightfoot, *Transport Phenomena*. 2nd edition. Wiley, New York, 2002. ISBN 0-471-07392-X.

https://doi.org/10.1515/9783110739466-003

The quantities that characterize the properties of a thermodynamic system and relation of these quantities to the surroundings are called **thermodynamic quantities** (or **parameters**) of the system. Relations between thermodynamic quantities, characterizing the immediate state of the system, are expressed by **equations of state**, also called material relations.

Parameters of system can be **external** or **internal**, depending on one another. The **external parameters** describe the influence of surroundings on the system (volume of the system, number of its subsystems or intensity of electric or magnetic field, having the source outside the system). **Internal parameters** of the system describe the properties of its subsystems and their changes involved by processes going inside the system (concentration of components, temperature and pressure).

Thermodynamic quantities fall into the following groups:
-  **Extensive quantities (global parameters)** describe the system as a whole, change over time and are **additive**. Additivity means that the value of a given parameter for the entire system is equal to the sum of all individual subsystem values. Extensive quantities (global parameters) include volume, mass and amount of substance.
-  **Intensive quantities (local parameters)** depend on the time and spatial coordinates but do not depend on the size of the system and **are not additive**. Intensive quantities (local parameters) include temperature, pressure and chemical potential.

We can convert any extensive quantity $Y$ to an intensive quantity by relating the extensive quantity to a unit of mass $m$, to a unit of amount of substance $n$ or a unit of volume $V$. Thus, we get **specific quantities** $Y_m$, **molar quantities** $Y_n$ or **density of the quantity** $Y_V = \rho_Y$, and the relation between them is defined as

$$Y = mY_m = nY_n \tag{2.1}$$

where $m$ is the mass, $n$ is the amount of substance and $V$ is the volume.

If properties of the system change with time, at least one thermodynamic quantity is changing, there is some **process** occurring in the system. In thermodynamics, we mainly distinguish the following types of processes:
-  **isothermal process**, occurring at a constant temperature $[T]$;
-  **isobaric process**, occurring at a constant pressure $[p]$;
-  **isochoric process**, occurring at a constant volume $[V]$;
-  **adiabatic process**, occurring without any heat exchange with the surroundings $[dQ = 0]$.

The **state of the thermodynamic system** characterizes its arrangement at a given moment.

The **equilibrium state (state of thermodynamic equilibrium)** is a state in which there are no macroscopic changes in the isolated system and all quantities are constant in time. Such a state, corresponding to the most probable arrangement of the system, is reached if we isolate it from its surroundings and let it develop for a sufficient period of time. The time in which the system reaches the equilibrium state is called the **relaxation period**.

The state of a thermodynamic system outside the equilibrium state is a **non-equilibrium state**. The non-equilibrium state is usual for **closed and open systems** owing to their continuous interaction with surroundings.

## 2.2 Key thermodynamic concepts and quantities

The following quantities are important for the definition of thermodynamic laws and description of thermodynamic phenomena:
- temperature (thermodynamic temperature) $T$,
- heat $q$,
- work (e.g. pressure–volume, electrical and surface work) $w$,
- energy $E$,
- internal energy $U$,
- enthalpy $H$,
- entropy (thermodynamic) $S$,
- Helmholtz energy $A$,
- Gibbs energy $G$,
- chemical potential $\mu$,
- electrochemical potential $\tilde{\mu}$.

Descriptions of particular thermodynamic systems may also include other quantities.

## 2.3 Basic laws and postulates of thermodynamics

Thermodynamics is based on two postulates and four laws of thermodynamics.

The **first postulate of thermodynamics** states that any isolated system reaches the equilibrium state after a certain period of time. The equilibrium state is fully described by all independent external parameters influencing the system with at least one internal parameter (e.g. temperature). All other internal parameters of this system can be then determined by equations of state if the external parameters and selected internal parameter are known. The given assumption is called the **second postulate of thermodynamics**.

The **zeroth law of thermodynamics** states that the temperature, as an intensive state quantity, is the same in all systems with heat-permeable walls that are in equilibrium. **The first law of thermodynamics** expresses **the law of conservation of energy**. **The second law of thermodynamics** says that heat cannot spontaneously flow from a colder body to a hotter one. It introduces the concept of **entropy** as the characterization of energy degradation and the characterization of reversibility and irreversibility of a process. **The third law of thermodynamics** concerns the behaviour of substances near absolute thermodynamic zero. According to this law, the entropy of every ideally crystalline and perfectly pure substance is zero at 0 K. It also postulates that it is impossible to achieve the absolute zero temperature (0 K).

The first and second laws of thermodynamics are of special importance and will be described further in detail.

## 2.3.1 First law of thermodynamics

The mathematical formulation of the first law of thermodynamics in the differential form is the equation

$$\mathrm{d}U = \mathrm{d}q + \mathrm{d}w \qquad (2.2)$$

where $U$ is the internal energy, $q$ is the heat and $w$ is the pressure–volume work, defined by the equation

$$\mathrm{d}w = -p\,\mathrm{d}V \qquad (2.3)$$

where $p$ is the pressure and $V$ is the volume.

If the system does any work, for example, during expansion, then $\mathrm{d}V > 0$ and the work done is negative, $\mathrm{d}w < 0$. If, conversely, work is supplied to the system, for example, during compression, then $\mathrm{d}V < 0$ and the supplied work is positive, $\mathrm{d}w > 0$.

Note: Any work is a product of an intensive quantity and an extensive quantity. For instance, in the case of the pressure–volume work, the extensive quantity $V$ changes until the intensive quantity $p$ becomes equalized; for the electric work $\mathrm{d}w = E\,\mathrm{d}Q$, the electric charge $Q$ (extensive quantity) is transported until voltage $E$ (intensive quantity) becomes equalized. In the system doing work, there must exist a gradient of an intensive quantity as the driving force of the process.

By substituting relation (2.3) into relation (2.2), we get

$$\mathrm{d}U = \mathrm{d}q - p\,\mathrm{d}V \qquad (2.4)$$

According to this relation, the supplied heat at a constant volume leads to an increase of the internal energy of the system, as well as the removed heat to its decrease.

At a constant pressure $p$, the substitution $\mathrm{d}(pV) = p\,\mathrm{d}V$ and modification of relation (2.4) yields

$$\mathrm{d}q = \mathrm{d}U + p\,\mathrm{d}V = \mathrm{d}(U + pV) = \mathrm{d}H \qquad [p] \qquad (2.5)$$

The thermodynamic quantity $H$ (**enthalpy**) in this relation is defined by

$$H = U + pV \qquad (2.6)$$

which represents the heat that the system exchanges with the surroundings at a constant pressure.

## 2.3.2 Second law of thermodynamics

The second law of thermodynamics limits the validity of the first law of thermodynamics and introduces the quantity of **entropy** $S$, which is defined by the following relations:

**For reversible processes**

$$dS = dq/T \tag{2.7}$$

$$dq = T\,dS \tag{2.8}$$

**For irreversible processes**

$$dS > dq/T \tag{2.9}$$

$$dq < T\,ds \tag{2.10}$$

where $T$ is the thermodynamic (absolute) temperature (unit: kelvin, K).

For reversible processes, by substituting from (2.7) into relation (2.4), we get

$$dU = T\,dS - p\,dV \tag{2.11}$$

Since it applies that

$$dH = dU + p\,dV + V\,dp \tag{2.12}$$

by combining relations (2.11) and (2.12), we get

$$dH = T\,dS + V\,dp \tag{2.13}$$

If the system, besides the pressure–volume work, does also other work $w^*$ (e.g. chemical and electrical), then according to relation (2.4), we write

$$dU = dq - p\,dV + dw^* \tag{2.14}$$

$$dw^* = dU + p\,dV - dq = dU + p\,dV - T\,dS \tag{2.15}$$

$$dw^* = d(U - TS) = dA \quad [T, V] \tag{2.16}$$

From relation (2.16), it results that the maximum reversible work, other than the pressure–volume work which the system receives from the surroundings (or delivers to the surroundings) at a constant temperature and volume, is equal to the increase (or decrease) in its Helmholtz energy. The **Helmholtz energy** $A$ is then defined by the relation

$$A = U - TS \tag{2.17}$$

From relations (2.3) and (2.11), it follows that

$$dU = T\,dS + dw \tag{2.18}$$

and at a constant temperature

$$dU - d(TS) = d(U - TS) = dw \qquad [T] \tag{2.19}$$

With respect to relations (2.19) and (2.17), we can write

$$dw = dA \qquad [T] \tag{2.20}$$

This equation expresses the second meaning of the Helmholtz energy, as it shows that the maximum reversible work which the system receives from the surroundings (or delivers to the surroundings) at a constant temperature is equal to the increase (or decrease) in its Helmholtz energy.

If the system, besides the pressure–volume work, does also other work $w^*$ (e.g. chemical and electrical), then at a constant temperature and pressure, we get

$$dw^* = d(U + pV - TS) = dG \qquad [T, p] \tag{2.21}$$

With respect to the validity of relation (2.6) it holds

$$dw^* = d(H - TS) = dG \tag{2.22}$$

Relation (2.19) results in the fact that the maximum reversible work other than the pressure–volume work which the system receives from the surroundings (or delivers to the surroundings) at a constant temperature and pressure is equal to the increase (or decrease) of the Gibbs energy. The **Gibbs energy** $G$ is then defined by the relation

$$G = H - TS \tag{2.23}$$

Based on previous relations, we can write

$$dA = dU - T\,dS - S\,dT = T\,dS - p\,dV - T\,dS - S\,dT = -S\,dT - p\,dV \tag{2.24}$$

$$dG = dH - T\,dS - S\,dT = T\,dS + V\,dp - T\,dS - S\,dT = -S\,dT + V\,dp \tag{2.25}$$

The so-called **Gibbs equations** (joined formulations of the first and second laws of thermodynamics) for reversible processes, which are very important in thermodynamics and help us deduce other thermodynamic relations, are summarized as follows:

$$dU = T\,dS - p\,dV \tag{2.26}$$

$$dH = T\,dS + V\,dp \tag{2.27}$$

$$\mathrm{d}A = -S\,\mathrm{d}T - p\,\mathrm{d}V \qquad (2.28)$$

$$\mathrm{d}G = -S\,\mathrm{d}T + V\,\mathrm{d}p \qquad (2.29)$$

## 2.4 Conditions of thermodynamic equilibrium

For an **isolated system**, which exchanges neither mass nor energy with its surroundings, it holds with respect to relations (2.4), (2.7) and (2.9):

$$dS \geq 0 \qquad (2.30)$$

It states that **the entropy change is zero for a reversible process. For an irreversible process, the entropy increases and in equilibrium it reaches its maximum.**

For a closed system, which exchanges with its surroundings energy only, we get from relations (2.24), (2.7) and (2.9):

$$\mathrm{d}A = \mathrm{d}q - T\,\mathrm{d}S \qquad [T, V] \qquad (2.31)$$

$$\mathrm{d}A \leq 0 \qquad [T, V] \qquad (2.32)$$

**In a spontaneous, that is, irreversible process in a closed system at a constant temperature and volume, the value of its Helmholtz energy decreases and in equilibrium reaches its minimum.**

Similarly, at a constant temperature and pressure, according to relations (2.25), (2.7) and (2.9), it applies that

$$\mathrm{d}G = \mathrm{d}q - T\,\mathrm{d}S \qquad [T, p] \qquad (2.33)$$

$$\mathrm{d}G \leq 0 \qquad [T, p] \qquad (2.34)$$

**The Gibbs energy of a closed system decreases for irreversible processes at a constant temperature and pressure. In the state of thermodynamic equilibrium, it reaches its minimum.**

A special case of the system is the state of **local equilibrium**. The system can be in the state of local equilibrium, if we can divide it into a large number of subsystems, each of which is a thermodynamic system so small that its state can be considered with sufficient accuracy as equilibrated. The given subsystems must be large enough to contain a large number of particles and, at the same time, small enough that the intensive parameters of the resulting subsystems change negligibly and can be considered as spatially independent. The behaviour of each subsystem then can be described using the equations of state valid for equilibrated systems, although the system as a whole is not in equilibrium. This method allows studying of systems outside equilibrium using relations valid for equilibrated systems. Systems in the state of local equilibrium are called **locally equilibrated systems. The principle of local equilibrium is the first and basic postulate of linear irreversible thermodynamics.**

An important example of a non-equilibrium state is a **stationary state**. It is a state in which no parameter of the system is changing with time. For closed and open systems, no equilibrium state can occur – only a stationary state is possible when energy or mass (or both energy and mass) is exchanged with the surroundings. **Each equilibrium state is stationary**, while a **stationary non-equilibrium state**, in which irreversible processes occur, is called a **steady state**. A steady state must be kept by maintaining a constant gradient of energy or mass (or both energy and mass) between the system and the surroundings as the driving force of the process.

For example, the driving force in the case of diffusion is the gradient of concentration, and in the case of thermodiffusion, it is the gradient of concentration and the gradient of temperature as well.

Classical equilibrium thermodynamics is not appropriate for the description of dynamic processes in open systems because it does not introduce time changes of relevant thermodynamic quantities. It only describes the changes of thermodynamic quantities as functions of other thermodynamic quantities of a given system. As it was already mentioned, **irreversible thermodynamics** or **thermodynamics of irreversible processes** serve to describe processes in open and closed systems.

In the first approximation, the states outside the equilibrium and irreversible processes can be described by linear transport relations, which form the formal apparatus of **linear irreversible thermodynamics** or **linear non-equilibrium thermodynamics.**

The **linear irreversible thermodynamics** is based on several postulates. The first and basic postulate is the already described principle of local equilibrium. Another postulate is, according to Onsager, the linear dependence of generalized flows on generalized forces, expressed by the relation

$$J_i = \sum_{j=1}^{k} L_{ij} X_j, \qquad i, j = 1, 2, \ldots, k \tag{2.35}$$

where $J_i$ is the flow of component $i$, $X_j$ are the generalized forces as gradients of intensive quantities characterizing the system, and $L_{ij}$ are the phenomenological coefficients that do not depend on $J_i$ and $X_i$, but can be functions of state quantities. These linear relations are called **phenomenological laws** and they are valid near thermodynamic equilibrium only. Coefficients of the type $L_{ii}$ are called **diagonal coefficients,** while those of the type $L_{ij}$ ($i \neq j$) are called **interference coefficients.**

It must be emphasized that phenomenological coefficients must meet the Sylvester conditions; for example, for two irreversible processes, it must hold

$$L_{11} \geq 0; \quad L_{11} \cdot L_{22} - L_{21} \cdot L_{12} \geq 0$$

The third postulate of linear irreversible thermodynamics is the **Onsager theorem of reciprocity**, according to which it applies for phenomenological coefficients:

$$L_{ij} = L_{ji} \quad (i \neq j) \tag{2.36}$$

It was demonstrated that if the flow $J_i$ is influenced by the force $X_j$ causing another flow $J_j$, the influence of the force $X_i$ on the flow $J_j$ is expressed by the same phenomenological coefficient.

A stationary non-equilibrium state is called a **steady state**. In stationary states, dissipative processes occur; therefore, a stationary non-equilibrium state is characterized by a non-zero, positive value of the entropy production $\sigma(S)$. This entropy production results from internal processes in the system and represents the amount of entropy generated in unit volume of the system per time unit. The entropy production in linear region is

$$\sigma(S) = \sum_{i=1}^{k} \sum_{j=1}^{k} L_{ij} X_i X_j \geq 0, \quad i,j = 1, 2, \ldots, k \tag{2.37}$$

According to the **Prigogine theorem**, the entropy production reaches the minimum value in a stationary state in agreement with the specified boundary conditions corresponding to the given problem. This theorem applies if the phenomenological equations are linear, the phenomenological coefficients $L_{ij}$ are constant and the Onsager theorem of reciprocity is valid.

According to the **development criterion**, any open thermodynamic system develops itself into a stationary state so that the entropy production decreases and reaches its minimum just in the stationary state:

$$\frac{\partial \sigma(S)}{\partial \tau} \leq 0 \tag{2.38}$$

In the stationary state, the entropy production does not change with time.

## 2.5 Molar quantities and partial molar quantities

Molar quantities are defined only at a given constant temperature and pressure. To obtain them, we must divide any extensive thermodynamic quantity $Y$ by the **amount**

**of substance** $n$ (unit mole). For instance, the **molar volume** $V_n$ (unit: $m^3\ mol^{-1}$) is defined by the relation

$$V_n = V/n \qquad [T,p] \tag{2.39}$$

and similarly the molar Gibbs energy $G_n$ (unit: $J\ mol^{-1}$)

$$G_n = G/n \qquad [T,p] \tag{2.40}$$

A partial molar quantity $\bar{Y}_i$ of the component $i$ is defined by the relation

$$\bar{Y}_i = \left(\frac{\partial Y}{\partial n_i}\right)_{T,p,n_{j\neq i}} \tag{2.41}$$

where $n_i$ is the amount of substance of the component $i$, $T$ is the temperature and $p$ is the pressure. For instance,

$$\bar{G}_i = \left(\frac{\partial G}{\partial n_i}\right)_{T,p,n_{j\neq i}} \tag{2.42}$$

It was demonstrated that a partial molar quantity of the $i$-th component equals to the change of the relevant quantity of the solution caused by adding 1 mol of this component at a constant temperature and pressure into such amount of the solution that it does not change its composition.

All thermodynamic relations valid for systematic thermodynamic quantities also apply for molar and partial molar quantities. Thus, it applies for the Gibbs equations

$$d\bar{U}_i = Td\bar{S}_i - p\,d\bar{V}_i \tag{2.43}$$

$$d\bar{H}_i = T\,d\bar{S}_i + \bar{V}_i\,dp \tag{2.44}$$

$$d\bar{A}_i = -\bar{S}_i\,dT - p\,d\bar{V}_i \tag{2.45}$$

$$d\bar{G}_i = -\bar{S}_i\,dT + \bar{V}_i\,dp \tag{2.46}$$

In thermodynamics, the **chemical potential of the $i$-th component,** $\mu_i$, is defined by the relations

$$\mu_i = \left(\frac{\partial U}{\partial n_i}\right)_{S,V,n_{j\neq i}} = \left(\frac{\partial H}{\partial n_i}\right)_{S,p,n_{j\neq i}} = \left(\frac{\partial A}{\partial n_i}\right)_{T,V,n_{j\neq i}} = \left(\frac{\partial G}{\partial n_i}\right)_{T,p,n_{j\neq i}} \tag{2.47}$$

From relations (2.47) and definition (2.41), it is evident that only the partial molar Gibbs energy is equivalent to the chemical potential

$$\mu_i = \bar{G}_i \tag{2.48}$$

The exchange of mass with the surroundings also takes place in open systems, so the Gibbs energy, as the most important thermodynamic quantity, is not only a function of the temperature and pressure but also of the composition of the system. Therefore, its total differential has the form

$$dG = \left(\frac{\partial G}{\partial T}\right)_{p,n_i} dT + \left(\frac{\partial G}{\partial p}\right)_{T,n_i} dp + \sum_{i=1}^{k} \left(\frac{\partial G}{\partial n_i}\right)_{T,p,n_{j\neq i}} dn_j; \quad i,j=1,2,\ldots,k; \ i\neq j \quad (2.49)$$

With regard to relation (2.48), we can write for the chemical potential

$$d\mu_i = \left(\frac{\partial \mu_i}{\partial T}\right)_{p,n_i} dT + \left(\frac{\partial \mu_i}{\partial p}\right)_{T,n_i} dp + \sum_{j=1}^{k} \left(\frac{\partial \mu_i}{\partial n_j}\right)_{T,p,n_{i\neq j}} dn_j; \quad i,j=1,2,\ldots,k; \ i\neq j$$

$$(2.50)$$

From the Gibbs equation, relation (2.46), it results that

$$d\mu_i = -\bar{S}_i\, dT + \bar{V}_i\, dp + \sum_{j=1}^{k} \left(\frac{\partial \mu_i}{\partial n_j}\right)_{T,p,n_{i\neq j}} dn_j; \quad i,j=1,2,\ldots,k; \ i\neq j$$

$$(2.51)$$

Relation (2.51) describes the dependence of chemical potential of the component $i$ on temperature, pressure and composition of the mixture.

If we consider a pure component $i$ at the standard temperature and pressure as a standard state, then according to relation (2.51), we obtain for the chemical potential of the component $i$ as follows:

$$\mu_i = \mu_i^\circ + RT \ln x_i \qquad [T,p] \qquad (2.52)$$

where $\mu_i^\circ$ is the standard chemical potential of the component $i$, $R$ is molar gas constant, $R = 8.314\,462\,618$ J K$^{-1}$ mol$^{-1}$ and $x_i$ is the mole fraction of the component $i$ in the mixture.

In real systems, it holds

$$\mu_i = \mu_i^\circ + RT \ln a_i \qquad (2.53)$$

where $a_i$ is the activity of the component $i$.

Relation between $a_i$ of component $i$ and its molar fraction $x_i$ is given by the following equation[2]:

---

2 As the activity is a dimensionless quantity, as well as molar fraction, the activity coefficient in this equation is also dimensionless. Nevertheless, if the content of a component is expressed in other concentration quantities (molar concentration, molality), it is necessary to express it as a relative, divided by a standard value, for $c$ in mol L$^{-1}$ it is $c^\circ = 1$ mol L$^{-1}$ or for molality $m$ in mol kg$^{-1}$ it is $m^\circ = 1$ mol kg$^{-1}$. Then, the corresponding relations are $a_i = \gamma_i\, c_i/c^\circ$ or $a_i = \gamma_i\, m_i/m^\circ$, and the activity coefficient is again dimensionless.

$$a_i = \gamma_i \, x_i \tag{2.54}$$

where $\gamma_i$ is the activity coefficient of the component $i$.

In systems containing electrically charged particles (ions), the partial molar Gibbs energy does not depend only on the chemical potential of the component but also on the electric potential. In these systems, we introduce the concept of the **electrochemical potential** $\tilde{\mu}_i$ defined by the relation

$$\tilde{\mu}_i = \mu_i + z_i \boldsymbol{F} \varphi = \mu_i^\circ + \boldsymbol{R} T \ln a_i + z_i \boldsymbol{F} \varphi \qquad [T] \tag{2.55}$$

where $\mu_i$ is the chemical potential of the component $i$, $z_i$ is its charge number, $\boldsymbol{F}$ is Faraday's constant, $\boldsymbol{F} = 96{,}485.332\,12$ C mol$^{-1}$, and $\varphi$ is the electric potential.

On the assumption that only the temperature is constant, we write, based on relations (2.51) and (2.55), for real systems that

$$\mathrm{d}\tilde{\mu}_i = \bar{V}_i \, \mathrm{d}p + \boldsymbol{R} T \, \mathrm{d} \ln a_i + z_i \boldsymbol{F} \, \mathrm{d}\varphi \qquad [T] \tag{2.56}$$

For the change of the electrochemical potential at constant temperature and pressure, we can write

$$\mathrm{d}\tilde{\mu}_i = \boldsymbol{R} T \, \mathrm{d} \ln a_i + z_i \boldsymbol{F} \, d\varphi \qquad [T, p] \tag{2.57}$$

Note: From relations (2.42) and (2.48), it results that $\mathrm{d}G = \mu_i \, \mathrm{d}n_i$ at a constant temperature and pressure. A change of the Gibbs energy represents **chemical work**, connected with the change of the amount of substance (extensive quantity) of components occurring until the chemical potential (intensive quantity) of these components becomes identical. The identity of chemical potentials of components is an equilibrium criterion in multi-component and multi-phase systems. The driving force for chemical work is the gradient of the chemical potential.

For example, in two-phase system with components A and B, the chemical potential of component A in phase (1) has to be in equilibrium equal to the chemical potential of component A in phase (2). The same holds for the component B.

The same applies for the electrochemical potential. The gradient of the electrochemical potential allows transport of electrically charged particles until the electrochemical potentials become equilibrated. The electrochemical potential is then an equilibrium criterion in multi-component and multi-phase systems with electrically charged particles.

Tomáš Bystroň, Dalimil Šnita

# 3 Basic concepts and laws of electrochemistry

## 3.1 Introduction

It is difficult to give a precise and comprehensive definition for the word "electro-chemistry". **Electrochemistry** can be considered as an interdisciplinary scientific discipline that deals with mutual relations between chemical and electrical processes involving charge transport in the ion-conducting phase (electrolyte) and processes connected with the charge transfer and transport through the phase interface between an electron and ion conductor or the interface of two ion conductors. This might, at first sight, seem simple, but the charge transfer itself depends on a number of other phenomena. As the title implies, this chapter summarizes the basic concepts and laws of electrochemistry that are required to understand the following chapters.

Since electrochemistry also involves transport and accumulation of the electric charge, it might be beneficial to start with introduction of the basic physical view of electrostatic phenomena in Section 3.2.

## 3.2 Basics of electrostatics

### 3.2.1 Important quantities

The system of physical units SI is used in this book. Most equations of electrostatics were historically formulated in the CGS-ESU (centimetre–gram–second electrostatic units) system. In SI, the unit of **electric current**, the **ampere** (A), was historically defined such that the magnetic force exerted by two infinitely long, thin, parallel wires one meter apart and carrying a current of one ampere is exactly $2 \times 10^{-7}$ N m$^{-1}$. According to the valid SI definition, ampere is a base unit, and it is obtained by fixing the elementary charge value, $e = 1.602\ 176\ 634 \times 10^{-19}$ C. **Electric charge** (unit: **coulomb**, C) is defined as a derived unit, as a product of current and time: C = A s.

The systems of unit used presently also resulted in different forms of basic relations [1, 2].

Let us now consider two point charges, $Q_1(x_1)$ and $Q_2(x_2)$, placed at a distance $x^1$ from each other. The charges $Q_1$ and $Q_2$ will act upon each other with the **electro-static force** $F_{e,12}$ (unit: newton, N), defined by **Coulomb's law**, as follows:

---

1 See the note on page XI.

https://doi.org/10.1515/9783110739466-004

$$F_{e,12} = \frac{1}{4\pi\varepsilon} \frac{Q_1(x_1)\,Q_2(x_2)}{(x_2 - x_1)^3} (x_2 - x_1) \tag{3.1}$$

that is, the point charge $Q_1$ placed at the position $x_1$ **attracts** by electrostatic force $F_{e,12}$ ($Q_1Q_2 < 0 \Rightarrow |F_{e,12}| > 0$, charges of the different sign) or **repulses** ($Q_1Q_2 > 0 \Rightarrow |F_{e,12}| < 0$, charges with the same sign) the point charge $Q_2$ placed at the position $x_2$. In this equation, $\varepsilon$ is **permittivity** (unit: farad per meter, F m$^{-1}$). Permittivity is often relative, $\varepsilon_r = \varepsilon/\varepsilon_0$, where $\varepsilon_0$ is the **permittivity of vacuum**, $\varepsilon_0 = 8.854\,187\,812\,2$ F m$^{-1}$.

The vector field of **electric field intensity**, $E(x)$ (unit: V m$^{-1}$) in each point of the space is defined by the electrostatic force $F_e(x)$ affecting the unitary point charge $Q(x)$ as follows:

$$F_e(x) = E(r)Q(x) \tag{3.2}$$

$$E_e(x) = \frac{1}{4\pi\varepsilon} \frac{Q_1(x_1)}{(r_2 - r_1)^3} (x_2 - x_1)$$

$$F_{e,ij}(x_{i \neq j}) = E_i(x_j)Q_j(x_j) \tag{3.3}$$

From a mathematical point of view, the electric field intensity is represented as the **vector field**. The field created by point charge is **spherically symmetric** (radial field). Such a field is **rotation-free** (the position of the charge is the centre of spherical symmetry).

$$\nabla \times E_i = 0 \tag{3.4}$$

The overall field created by multiple point charges is simply the sum of the electric fields created by individual point charges present in the system (the principle of **superposition**):

$$E_e(x) = \sum_i E_{e,i}(x) = \frac{1}{4\pi\varepsilon} \sum_i \frac{Q_i(x_i)}{(x - x_i)^3} (x - x_i); \qquad F_e(x) = E(r)Q(x) \tag{3.5}$$

The overall field is also rotation-free, since sum of the rotation-free fields is the rotation-free field (one of the **Maxwell equations**), see eq. (3.4). Hence, there exists a scalar field of the **electric potential** $\varphi$ (unit: volt, V), such that its negative gradient is equal to the electric field strength, as follows:

$$E(r) = -\nabla\varphi \tag{3.6}$$

From another point of view, an electric potential is the amount of work needed to move a unit charge from a reference point (usually infinity) to a specific point inside the field, as follows:

$$dw_e = F_e(r) \cdot ds = -QE(r) \cdot ds = \nabla\varphi \cdot ds \tag{3.7}$$

$$-w_e = \int_\infty^r \nabla\varphi \cdot ds = \varphi(r) - \varphi(\infty) = \varphi(r) \tag{3.8}$$

as $\varphi(\infty)$ is defined as zero. In the **continuum theory**, the distribution of the charge in space is defined by the **scalar field, charge density** $q(\mathbf{r})$ (unit: $C\,m^{-3}$) as

$$q(\mathbf{r}) = \lim_{\Delta V \to \infty} \frac{\Delta Q}{\Delta V} \qquad \text{where } \mathbf{r} \in \Delta V \tag{3.9}$$

Here, $\Delta Q$ is the charge inside of the volume element, $\Delta V$. The superposition of the fields created by all elements of charge, $dq = q\,dV$ placed in infinite space $R^3$ is then written in the **integral form** as

$$\boldsymbol{E}_e(\boldsymbol{x}) = \frac{1}{4\pi\varepsilon} \int\limits_{R^3} \frac{q(\boldsymbol{r})}{(\boldsymbol{x} - \boldsymbol{r}_i)^3} (\boldsymbol{x} - \boldsymbol{r}_i)\,dV \tag{3.10}$$

The local vector field of **electrostatic force density** $f_e(\boldsymbol{x})$ (unit: $N\,m^{-3}$) is then given as:

$$\boldsymbol{f}_e(\boldsymbol{x}) = \boldsymbol{E}(\boldsymbol{x})\,q(\boldsymbol{x}) \tag{3.11}$$

**Gauss law of electrostatics** says that there exists a vector field of **electric displacement, $D(\boldsymbol{x})$** (unit: $C\,m^{-2}$), such that its flow through any closed surface ($S$) is equal to the electric charge placed in the volume ($V$) enclosed inside this surface:

$$\oiint\limits_{S} \boldsymbol{D} \cdot d\boldsymbol{s} = \iiint\limits_{V} q\,dV \tag{3.12}$$

The local form of **Gauss law of electrostatics** (one of the **Maxwell equations**) follows from Gauss law of divergence of vector field (purely mathematic law) as follows:

$$\nabla \cdot \boldsymbol{D} = q \tag{3.13}$$

Coupling between electric displacement and electric field intensity in a system of **point charges placed in vacuum** is given by the equation:

$$\boldsymbol{D} = \varepsilon_0 \boldsymbol{E} \tag{3.14}$$

In reality, the charge is distributed inside of atoms and molecules in a complex way that is governed by quantum theory. This distribution can be approximated by the vector field of **electric polarization, $P(\boldsymbol{x})$**, which represents the local **density of electric dipoles** as follows:

$$\boldsymbol{D} = \varepsilon_0 \boldsymbol{E} + \boldsymbol{P} \tag{3.15}$$

For the first approximation, we consider that the polarization is proportional to the electric field intensity and polarization of the mixture can be expressed as the sum of the polarizations of the components:

$$\boldsymbol{P} = \sum_i \boldsymbol{P}_i = \sum_i \varepsilon_0 c_i \bar{\chi}_i = \varepsilon_0 \sum_i (c_i \bar{\chi}_i)\boldsymbol{E} = \varepsilon_0 \chi \boldsymbol{E} = \varepsilon_0 (\varepsilon_r - 1)\boldsymbol{E} \tag{3.16}$$

$$D = \varepsilon E = \varepsilon_0 E + P = \varepsilon_0(1 + \chi)E = \varepsilon_0 \varepsilon_r E \qquad (3.17)$$

The **permittivity** $\varepsilon$ ($\varepsilon = \varepsilon_0 \varepsilon_r$), the **relative permittivity** $\varepsilon_r$, **the susceptibility** $\chi$ of the mixture and the **molar susceptibilities** $\bar{\chi}_i$ of components represent the phenomenological parameters. This approach is acceptable for relatively **weak fields** and relatively **slow processes**. The most common solvent, water, exhibits extremely high permittivity ($\varepsilon_r \approx 80$). For **constant permittivity**, the Gauss law of electrostatics can be written in the form of **Poisson equation** as follows:

$$\nabla \cdot D = \nabla \cdot \varepsilon E = \varepsilon \nabla \cdot E = \varepsilon \nabla \cdot (-\nabla \varphi) \Rightarrow \nabla^2 \varphi = -\frac{q}{\varepsilon} \qquad (3.18)$$

To describe processes in electrochemistry, concepts of ideal electrotechnical components such as resistor and capacitor are often used. A **resistor** is a passive two-terminal electrical component that implements **electrical resistance** $R$ as a circuit element. **Important parts of electromembrane systems behave like resistors. Ohm's law** says that

$$I = GU = U/R \qquad (3.19)$$

that is, the **electric current** $I$ is proportional to the **voltage** $U$ and to the **conductance** $G$ or inversely proportional to the **resistance** $R$. The local **Ohm's law** in 3D can be written in the following form:

$$j = -\kappa \, \nabla \varphi = \kappa E = -\frac{1}{\rho} \nabla \varphi = +\frac{1}{\rho} E \qquad (3.20)$$

The **current density** $j(x)$ is proportional to the **electric field intensity**, $E(x) = -\nabla \varphi$, and to the **conductivity** $\kappa(x)$ or inversely proportional to **resistivity** $\rho(x)$.

When electric current flows through system with internal resistivity, electric power is dissipated in the form of Joule's heat. The **electric power dissipation** $P_{el}$ (Joule's heat production) and the **electric power dissipation density** $p_{el}(x)$ are given by the equations:

$$P_{el} = IU; \qquad p_{el}(x) = j \cdot E \qquad (3.21)$$

A **capacitor** is a passive two-terminal electronic component that stores electric charge. The change of the accumulated charge $dQ$ is proportional to the **differential capacitance** $C_{diff}$ and to the **change of voltage** $dU$. The overall amount of accumulated charge is proportional to the **capacitance** $C$ and to the **voltage** $U$:

$$dQ = C_{diff} \, dU; \qquad Q = CU \qquad \text{for } C_{diff} = \text{const} \qquad (3.22)$$

**Important parts of electromembrane systems behave like capacitors:**

$$U = \int_0^{\Delta x} E \, dx = E \, \Delta x = Q \frac{\Delta x}{\varepsilon A} = \frac{Q}{C} \Rightarrow C = \frac{\varepsilon A}{\Delta x} \Rightarrow C = C_{\text{diff}} = \text{const for } \varepsilon = \text{const} \qquad (3.23)$$

where $\Delta x$ is the distance between the oppositely charged capacitor plates. The capacitor **energy** $E_{\text{cap}}$ and the capacitor **energy density** $e_{\text{cap}}(x)$ are given by the equations:

$$C \, dU = dQ = I \, dt; \quad CU \, dU = UI \, dt = P_{\text{el}} \, dt \qquad (3.24)$$

$$C = \int_0^U U \, dU = UI \, dt = \int_0^t P_{\text{el}} \, dt = W_{\text{el}} = E_{\text{cap}} = C \frac{U^2}{2} \qquad (3.25)$$

The **capacitor energy** $E_{\text{cap}}$ related to volume gives the capacitor **energy density** $e_{\text{cap}}(x)$ as shown below:

$$\frac{C \frac{U^2}{2}}{\Delta x A} = \frac{\left(\frac{\varepsilon A}{\Delta x}\right) \frac{U^2}{2}}{\Delta x A} = \frac{\varepsilon}{(\Delta x)^2} \frac{U^2}{2} = \varepsilon \frac{E^2}{2} = e_{\text{cap}} = \frac{DE}{2} \qquad (3.26)$$

The last equation can be generalized as the energy density of electrostatic field, shown below:

$$e_{\text{cap}}(x) = \frac{D(x) \cdot E(x)}{2} \qquad (3.27)$$

## 3.3 Basic concepts and laws of electrochemistry

### 3.3.1 Basic concepts of electrochemistry

For detailed information, see specialized books, for example [3, 4]. Many processes occurring in nature are connected with the transport of charge, for example, formation of lightning, corrosion of metal objects or transmission of stimuli in neurons. All these actions have something in common: they occur in the presence of electric fields and require electrically conductive environments to take place. Based on the type of the charge carrier facilitating its transport, we divide electrically conductive environments (electrical conductors) into two groups:

– **Electron conductors** (first-class conductors), in which the transport of charge is achieved by movement of electrons (or holes); these mainly include metals, semiconductors, graphite, glassy carbon or conductive polymers in oxidized/reduced form.
– **Ion conductors** (**electrolytes**, second-class conductors), in which charge is transported by movement of ions; liquid electrolytes are mainly solutions of dissociated substances and melts of ionic compounds; solid electrolytes can be

crystalline and ceramic materials with mobile ion; solvated polymer electrolytes possess groups capable of dissociation; similarly conductive polymers in an oxidized/reduced form contain mobile counter ions.

Now, let us, for example, consider an aqueous alkaline solution of sodium sulphite that **dissociates** in this solution according to the equation:

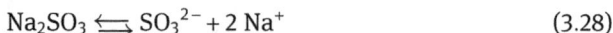

$$Na_2SO_3 \leftrightarrows SO_3^{2-} + 2\,Na^+ \tag{3.28}$$

We further immerse into the solution, two platinum electrodes (so that they do not touch each other) and connect them to a source of constant voltage. Due to polarization of electrodes, an electric field is created around them. If the voltage between the electrodes is sufficient, the electric current starts to flow continuously through the outer electric circuit. This also means that electrons must be transferred from one electrode to the other through the solution phase, where charge is transported by ions. In an electric field, the electric forces affect previously randomly moving charged particles, and the dissociated ions are forced to preferentially move towards respective electrodes, that is, in this case, $SO_3^{2-}$ towards the positively polarized electrode and $Na^+$ towards the negatively polarized electrode. On the surface of the positively polarized electrode, electrons are extracted from the component – an **oxidation** of $SO_3^{2-}$ takes place according to the equation:

$$SO_3^{2-} + 2\,OH^- \leftrightarrows SO_4^{2-} + H_2O + 2\,e^- \tag{3.29}$$

On the surface of the negatively polarized electrode, a **reduction** (adding electrons to the component) of $H_2O$ molecules occurs according to the equation:

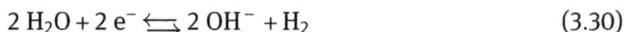

$$2\,H_2O + 2\,e^- \leftrightarrows 2\,OH^- + H_2 \tag{3.30}$$

By summing up the half-reactions (3.29) and (3.30), we get the overall cell reaction:

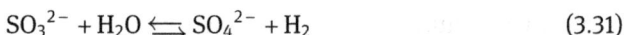

$$SO_3^{2-} + H_2O \leftrightarrows SO_4^{2-} + H_2 \tag{3.31}$$

The electrode on which oxidation is predominantly taking place is called an **anode**; the electrode on which reduction dominates is called a **cathode**.[2] It is important to note that the oxidation of each $SO_3^{2-}$ ion is connected with transporting two electrons from the solution phase to the anode and two electrons from the cathode to the solution phase (and formation of $H_2$). Hence, to describe the reaction rate on an electrode (assuming that no side reactions take place), it is sufficient to measure the **electric**

---

2 This holds for electrolytic or galvanic cell only. For an electrochemical cell in equilibrium state, it is more convenient to use the term negative electrode for the electrode attracting cations and positive electrode for the one attracting anions.

**current** $I$ (unit: A) flowing through the circuit. The electric current is a physical phenomenon describing an organized movement of the electric charge carriers. It is defined as the amount of charge $dQ$ passing through a selected surface in the time $d\tau$ as follows:

$$I = dQ/d\tau \tag{3.32}$$

The direction of the electric current is defined as the direction of movement of the positively charged electric charge carriers. Thus, as anions in the solution move in the opposite direction to cations, the electric current flows in the direction of cation movement, and for the overall transported charge $dQ$ it applies that $dQ = dQ_C + |dQ_A|$, where $dQ_C$ and $dQ_A$ are charges transported as a result of the movement of cations and anions, respectively. Since the electric current is an extensive quantity, its size depends on the dimensions of the system. It is, therefore, convenient to introduce a corresponding intensive quantity $j$ called an **electric current density** (unit: A m$^{-2}$), independent on the area of the surface, $A$, through which the electric current flows:

$$I = j \cdot A \tag{3.33}$$

Electrochemical reactions are typical heterogeneous processes because they always occur on the phase interface. Therefore, we are usually interested in the current density at this interface. If we consider that the current flows perpendicularly to the interface and the interface is flat (vectors of current density and surface are then parallel), the current in the eq. (3.33) can be written simply as product of two scalars, that is, $I = jA$.

Let us return to the example of $Na_2SO_3$ solution discussed above and consider what will happen to the $SO_3^{2-}$ ion (or with general electroactive species) located in the solution bulk at the moment of connection of the electrodes to a voltage source:
- ion will be transported towards the anode surface,
- depending on the properties of the anode surface, ion can be adsorbed on the electrode surface,
- (adsorbed) ion will be anodically oxidized, transferring the electron(s) to the anode,
- if the reactant has been adsorbed, the oxidation product will be desorbed from the anode surface,
- products will be transported from the electrode surface back to the electrolyte bulk,
- any of these steps can be preceded or followed by chemical reaction step.

All the relevant processes must be considered when describing the electrode reaction. Analogous processes occur upon reduction of the substance, in our case, the electrically neutral $H_2O$ molecules on the cathode. As already mentioned, in the case of more complicated mechanisms, the electrode reaction can be preceded or followed by a subsequent chemical reaction. The total rate of the electrode reaction is, of course, influenced by rates of all these processes. In the following sections, the principles governing individual steps will be discussed in more detail.

## 3.3.2 Basic laws of electrochemistry

The first of the pioneers of modern electrochemistry was Michael Faraday, whose name is often cited in this discipline. He formulated two laws of electrochemistry that have been combined into **Faraday's law** (3.34), as we know it today:

$$\Delta n_i = v_i\, Q/z\mathbf{F} \tag{3.34}$$

where $\Delta n_i$ is the amount of component $i$ (unit: mol) transformed in the electrochemical reaction, $v_i$ is the stoichiometric coefficient of component $i$, $Q$ (unit: C) is the charge passed through the system, $z$ is the number of electrons exchanged by one reaction turnover and $\mathbf{F}$ is the Faraday constant ($\mathbf{F} = 96\,485$ C mol$^{-1}$) corresponding to the charge of 1 mole of protons. The importance of Faraday's law is that it links the laws of conservation of charge and matter (the amount of substance produced and consumed is directly proportional to the charge passed/transported through the system) [3, 4].

Another basic law of electrochemistry (and also of physics) is the **law of conservation of charge**. It states that electric charge cannot be destroyed (it is preserved even in nuclear reactions). Therefore, the total charge remains constant in an isolated system.

The third postulate is a **macroscopic law of electroneutrality**: the total charge in the phase bulk (with the exception of the surface) is zero. Mathematically, it can be formulated by the relation:

$$\mathbf{F} \int_V \sum_i z_i c_i \, \mathrm{d}V = 0 \tag{3.35}$$

where $c_i$ is the molar concentration[3] of charge carriers (e.g. ions) $i$ and $z_i$ is the **charge number** of the charge carrier $i$, that is, the ratio between the charge carrier charge and proton charge. This law applies on a macroscopic scale as a result of the thermodynamic work needed to increase distance between the opposite charges or to approach the same charges. On the other hand, it is obvious, that this law does not apply at the microscopic level. In particular, electroneutrality is often disrupted on surfaces and

---

**3** This quantity should be more accurately named the amount-of-substance concentration, also briefly the amount of concentration, but in practice, the term molar concentration is still preferred.

phase interfaces. An example is the phase interface of electrolyte and electrode (electron conductor), which will be discussed in more detail in Section 3.5.5.

Let us return to the concept of the electrode that has two possible interpretations. The electrode might be understood as an electron conductor connected to an external current lead (e.g. wire of platinum or, often, of copper). In a more appropriate concept, it is considered a half-cell: the **electrode** consists of two or more electrically conductive phases connected in series and can exchange charge carriers (ions or electrons); while one end phase is the electron conductor, the other end phase is the electrolyte. The basic description of electrolyte behaviour is given in Section 3.4 and that of interface of electron and ion conductor in Sections 3.5.5 and 3.5.6.

## 3.4 Thermodynamics of electrolytes

As already suggested above, the term **electrolyte** refers to substances that exhibit ionic conductivity. These can be compounds undergoing dissociation to ions during melting or dissolution in a solvent, as well as those that are ion-conductive in the solid state. In the presence of charge carriers, it is not sufficient to operate with a **chemical potential** $\mu_{i,\alpha}$ (unit: J mol$^{-1}$) of the component $i$ in the phase $\alpha$, which we use for description of equilibrium in systems containing uncharged particles. A more detailed discussion of the chemical potential and its definition is presented in Chapter 2. For details, see [3–6].

To describe equilibria in systems containing phases with components carrying electric charge, Guggenheim introduced a quantity called **electrochemical potential** $\tilde{\mu}_{i,\alpha}$ (unit: J mol$^{-1}$) of the component $i$ in the phase $\alpha$ containing (mobile) charge carriers. It is defined by the relation:

$$\tilde{\mu}_{i,\alpha} = \left(\frac{\partial G_\alpha}{\partial n_i}\right)_{T,p,n_{j \neq i}} = \mu_{i,\alpha} + z_i F \varphi_\alpha = \mu^\circ_{i,\alpha} + RT \ln a_{i,\alpha} + z_i F \varphi_\alpha \qquad (3.36)$$

where $G_\alpha$ (unit: J) is the Gibbs energy of the phase $\alpha$, $\varphi_\alpha$ is an **inner (Galvani) potential** of the phase $\alpha$, $\mu^\circ_{i,\alpha}$ is the standard chemical potential of the component $i$ in the phase $\alpha$ and $a_{i,\alpha}$ represents the activity of the component $i$ in the phase $\alpha$. In accordance with the definition of electric potential in Section 3.2, the inner potential equals the electrostatic work connected with transport of a unit (positive) charge from infinity into the bulk of given phase. It is worth mentioning that it is not possible to distinguish between electrostatic and "chemical" influences on the properties of matter. Therefore, separation of electrochemical potential into its chemical and electrostatic components is purely artificial. Still, the approach allows formal introduction of an effect of the electric potential on the charge carriers present in the phase.

## 3.4.1 Behaviour of electrolyte solutions

In context of the whole book, it is probably more interesting to look at the (aqueous) electrolyte solutions. For the purposes of the present chapter, the discussion below is limited to the case of single binary electrolyte solution phase. As already mentioned, the electrolyte in an aqueous solution **dissociates** to ions as follows:

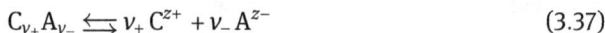

$$C_{v_+}A_{v_-} \leftrightharpoons v_+ C^{z+} + v_- A^{z-} \tag{3.37}$$

Ionic substances in solid phase consist of individual ions interacting strongly due to the presence of electrostatic forces. The result of attractive forces between oppositely charged ions and repulsive forces between coincidentally charged ions is the formation of a crystal lattice whose stability is characterized by **lattice energy**. During dissolution, individual ions are separated from the surface of the crystal and they are immediately surrounded by polar molecules of water (solvent). Electrostatic behaviour of water molecules will be approximated here, using a concept of electric dipoles.[4] Due to the mutual interaction of ions and water in the solution phase, the water molecules get oriented towards the ion by the end of the dipole possessing the opposite charge of the ion. Thus, they partially shield the charge of this ion leading to its stabilization in the solution. The described process is called **hydration of ions**. If we talk about a general solvent, we call this process **solvation**. It is necessary to note that if the salt is to be dissolved and dissociated in the solution, the solvation energy must be higher than the lattice energy. It is just the solvation energy that keeps the oppositely charged ions present in the solution separated from each other. It can be stated that any foreign particle (molecule, ion, colloidal particle) present in the solution is, to some extent, solvated, affecting the structure of the solvent in its surroundings.

Depending on the degree to which they dissociate in the given solvent, we divide electrolytes into strong and weak. In the case of **strong electrolytes**, all ions are present in the solution only in the form of dissociated solvated ions. **Weak electrolytes** dissociate only partially, and large part of ions in the solution is present in the form of **ion pairs** that do not transport charge. The degree of dissociation depends on the analytical concentration of the electrolyte. Equilibrium activities of individual components can be described by the following general dissociation equation:

---

4 At the microscopic level, we often encounter pairs of oppositely charged but equally sized charges placed at a certain distance, which we call an **electric dipole**. The concept of the dipole is often used to approximate the charge distribution in the molecule of polar substances.

$$
\begin{aligned}
K_{\text{dis}} &= \frac{a(\text{C}^{z+})^{v_+}\, a(\text{A}^{z-})^{v_-}}{a(\text{C}_{v_+}\text{A}_{v_-})} \\[2mm]
&= \frac{c(\text{C}^{z+})^{v_+}\, c(\text{A}^{z-})^{v_-}}{c(\text{C}_{v_+}\text{A}_{v_-})}\, \frac{\gamma(\text{C}^{z+})^{v_+}\, \gamma(\text{A}^{z-})^{v_-}}{\gamma(\text{C}_{v_+}\text{A}_{v_-})}\, \frac{1}{c^{\circ\,(-v_+ - v_- + 1)}} \\[2mm]
&= K'_{\text{dis}}\, \frac{\gamma(\text{C}^{z+})^{v_+}\, \gamma(\text{A}^{z-})^{v_-}}{\gamma(\text{C}_{v_+}\text{A}_{v_-})}
\end{aligned}
\tag{3.38}
$$

where $K_{\text{dis}}$ is the **dissociation constant** (dimensionless), $a$ is the **activity** (dimensionless), $\gamma$ is the **activity coefficient** (dimensionless) and $c$ is the concentration of the component, while $c^{\circ}$ is the unit concentration ($c^{\circ} = 1$ mol dm$^{-3}$); $K'_{\text{dis}}$ is the **apparent (concentration) dissociation constant** (dimensionless), which can be expressed using concentrations of components in the electrolyte solution, while neglecting their activity coefficients.

If we return to the eq. (3.37) that describes the electrolyte dissociation, it is evident that solutions never contain only one type of ions. Therefore, we usually need not know the activities or activity coefficients of individual ions, because equations like (3.38) only contain their products. It is, therefore, common to use the **mean activity** $a_{\pm}$, or the **mean activity coefficient** $\gamma_{\pm}$ of the electrolyte as shown:

$$
a_{\pm} = \sqrt[(v_+ + v_-)]{a_{\text{C}^{z+}}^{v_+}\, a_{\text{A}^{z-}}^{v_-}}, \qquad \gamma_{\pm} = \sqrt[(v_+ + v_-)]{\gamma_{\text{C}^{z+}}^{v_+}\, \gamma_{\text{A}^{z-}}^{v_-}}
\tag{3.39}
$$

Using relations (3.39), the relation (3.38) transforms to:

$$
K_{\text{dis}} = \frac{a_{\pm}^{(v_+ + v_-)}}{a_{\text{C}_{v_+}\text{A}_{v_-}}} = K'_{\text{dis}}\, \frac{\gamma_{\pm}^{(v_+ + v_-)}}{\gamma_{\text{C}_{v_+}\text{A}_{v_-}}}
\tag{3.40}
$$

It follows from the definition that the activity coefficients attain values in the interval between 0 and 1. The situation is simple in the case of very diluted electrolyte solutions, where the activity coefficient values are close to unity. However, when electrolyte solutions of practically relevant concentration are treated, the values of the activity coefficients are substantially lower. In such a case, the activity coefficients have to be considered in order to avoid unacceptable error. The activity coefficients of binary electrolytes can be relatively easily estimated based on theories such as the one of Debye and Hückel (and can be found in tables). More advanced and more complicated theories exist for ternary electrolytes. However, it might be easier and more precise to determine the activity coefficient values for the given conditions by means of appropriate experiment.

Polar amphiprotic solvents (e.g. water, methanol or ammonia) behave like very weak electrolytes. They are, to a certain degree, able to dissociate while transferring the proton from one molecule of solvent to another; see eq. (3.41). This process is called **autoprotolysis**; for water it is

$$2\,H_2O \rightleftarrows OH^- + H_3O^+ \tag{3.41}$$

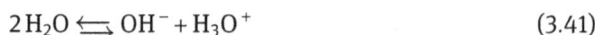

The equilibrium of autoprotolytic dissociation of water (3.41) can be described by the relation

$$K_{dis} = \frac{a_{OH^-}a_{H_3O^+}}{a_{H_2O}^2} \tag{3.42}$$

Since the activity of water as a pure solvent is constant and equal to 1, the **autoprotolytic constant of water** (sometimes called the **ionic product of water**) $K_w$ can be expressed as

$$K_w = K_{dis}a_{H_2O}^2 = a_{OH^-}a_{H_3O^+} \tag{3.43}$$

At the temperature of 25 °C, its value is $K_w$ (25 °C) $\doteq 1.00 \times 10^{-14}$ and it grows with rising temperature; for example, $K_w$(60 °C) $\doteq 9.614 \times 10^{-14}$.

To describe the extent of dissociation, the **degree of dissociation** $\alpha$ is also often used. It expresses the ratio of ions dissolved in the electrolyte solution that are dissociated as follows:

$$\alpha = \frac{c_{C^{z+}}}{\nu_+ c_{anal}} = \frac{c_{A^{z-}}}{\nu_- c_{anal}} \tag{3.44}$$

Here, $c_{anal}$ represents analytical concentration of the dissolved electrolyte, $C_{\nu_+}A_{\nu_-}$.

## 3.4.2 Mass transport in electrolyte solutions

Let us consider a closed thermally isolated system represented by the electrolyte $C_{\nu_+}A_{\nu_-}$ dissociated in the solution phase $\alpha$, according to eq. (3.37). The thermodynamic equilibrium in such a system is achieved when there are no gradients in electrochemical potential of all present components $i$, as well as in temperature and pressure (effect of hydrostatic pressure is not considered) [3–6, 8]. In such an equilibrated uniform phase, no net mass transport takes place. Any (external) action leading to local change in any of these internal quantities will lead to transport of components of the phase. For the isothermal phase, the **total molar flux of the component** $i$ in the phase $\alpha$, $J_i$ (unit: mol m$^{-2}$ s$^{-1}$) is described by

$$J_i = -c_i\left(\frac{D_i}{RT}\nabla\tilde{\mu}_{i,\alpha} - v\right) \tag{3.45}$$

where $D_i$ (unit: m$^2$ s$^{-1}$) is the **diffusion coefficient** of component $i$ in the phase $\alpha$, $\nabla\tilde{\mu}_{i,\alpha}$ is the gradient of electrochemical potential of component $i$ in the phase $\alpha$ and the vector $v$ (unit: m s$^{-1}$) represents the local **flow velocity** of the electrolyte solution.

After substituting for $\tilde{\mu}_{i,\alpha}$ from eq. (3.36), the relation (3.45) can be rewritten, leading to a general expression of the total flux of the component $J_i$ as follows:

$$J_i = -D_i \left[1 + \left(\partial \ln \gamma_i / \partial \ln c_i\right)\right] \nabla c_i - z_i\, c_i\, u_i\, F\, \nabla\varphi + c_i \mathbf{v} \tag{3.46}$$

The above formula is valid for any real solution. For ideal systems, where term $(\partial \ln \gamma_i / \partial \ln c_i) \approx 0$ (i.e. for diluted solutions, $c_i < 10^{-3}$ mol dm$^{-3}$), it can be simplified to a formula known as the **Nernst–Planck equation:**

$$J_i = -D_i \nabla c_i - z_i\, c_i\, u_i\, F\, \nabla\varphi + c_i \mathbf{v} = J_{i,\text{dif}} + J_{i,\text{migr}} + J_{i,\text{conv}} \tag{3.47}$$

At higher concentrations, this formula can be used as a more or less reasonable approximation.

The total flux of the component $J_i$ defined by eq. (3.46) or (3.47) for diluted electrolyte solutions can be expressed as the sum of individual contributions of mass transport (diffusion, convection and migration) corresponding to a different mechanism due to the respective driving forces.

**Diffusion** occurs due to the gradients in the chemical potentials (concentration gradients) of individual components. The diffusion flux of the component $i$, $J_{i,\text{dif}}$ (unit: mol m$^{-2}$ s$^{-1}$), is described by Fick's law as

$$J_{i,\text{dif}} = -D_i \left(\frac{\partial c_i}{\partial x}, \frac{\partial c_i}{\partial y}, \frac{\partial c_i}{\partial z}\right) = -D_i\, \nabla c_i \tag{3.48}$$

**Migration** of charged components (ions) occurs due to the presence of an electric field between polarized electrodes. The driving force of the migration movement is the electrostatic force (electric potential gradient) that affects components carrying electric charge. The migration flux of the component $i$ carrying the charge $z_i$ can be described as:

$$J_{i,\text{migr}} = -z_i\, c_i\, u_i\, F\, \nabla\varphi \tag{3.49}$$

where $u_i$ is the **mobility** of the component $i$ (unit: m mol s$^{-1}$ N$^{-1}$).

**Convection** occurs due to the flow of the electrolyte caused by the pressure gradient; this gradient can be exerted by external mechanical force (forced convection) or density gradient in the solution due to local differences in composition or temperature (natural convection). Therefore, each infinitesimal phase element moves as a whole. Therefore, the convective flux of the component $i$, $J_{i,\text{conv}}$, equals

$$J_{i,\text{conv}} = c_i \mathbf{v} \tag{3.50}$$

In the bulk of liquid electrolytes, where the macroscopic law of electroneutrality (3.35) applies, the charge is not transported by convection because the total charge contained in every solution bulk element is zero.

Since each mole of ions $i$ carries the charge $z_i\,F$, the current density $\boldsymbol{j}$ can be written as

$$
\begin{aligned}
\boldsymbol{j} &= \sum_i z_i F\left(\boldsymbol{J}_{i,\mathrm{dif}} + \boldsymbol{J}_{i,\mathrm{migr}} + \boldsymbol{J}_{i,\mathrm{conv}}\right) = \sum_i z_i \boldsymbol{F}\left(-D_i\,\nabla c_i - z_i c_i u_i \boldsymbol{F}\,\nabla\varphi + c_i \boldsymbol{v}\right) \\
&= -\boldsymbol{F}\sum_i z_i D_i\,\nabla c_i - \boldsymbol{F}^2\,\nabla\varphi \sum_i z_i^2 c_i u_i + \boldsymbol{v}\boldsymbol{F}\sum_i z_i c_i
\end{aligned}
\tag{3.51}
$$

The current density is a vector quantity. However, to avoid vector notation in the rest of Chapter 3, we will assume its direction to be perpendicular to the surface. This allows considering only the current density vector magnitude, $\|\boldsymbol{j}\| = j$.

The contribution of the component $i$ to the total transported charge $\Sigma Q_i$, or to the total current density $j$, under conditions where convection and diffusion transports are negligible, is expressed by the ion $i$ **transport (transference) number** $t_i$ (dimensionless), defined by the relation:

$$
t_i = \frac{Q_i}{\sum_i Q_i}, \qquad t_i = \frac{j_i}{\sum_i j_i} = \frac{j_i}{j}, \qquad \sum_i t_i = 1
\tag{3.52}
$$

where $Q_i$ is the charge carried by the component $i$ and $j_i$ is the (migration) current density due to the transport of the component $i$. It is good to note that experimentally determined transport numbers are always (to a certain degree) influenced by diffusion and convection.

Now, let us consider a well-mixed (uniform) electrolyte solution placed in a vessel of cylindrical shape, where the two opposite cylinder bases represent two electrodes. As there are no ion concentration gradients in the electrolyte solution, there is also no net ion diffusion flux in the system. Convection does not have any influence on the transport of charge in electrolyte solutions, either. As a result, the potential in the solution bulk between electrodes is evenly distributed allowing simplification of **Ohm's law** (3.20). Then, the resistance of the electrolyte $R$ between the electrodes in uniform solution with constant cross section can be described using the formula:

$$
R = \frac{L}{A}\rho = \frac{L}{A}\frac{1}{\kappa}
\tag{3.53}
$$

where $L$ is the distance between electrodes (length of conductor, unit m), $A$ is the cross-sectional area of the electrolyte (conductor, unit m$^2$), $\rho$ (unit: $\Omega$ m) is the **resistivity** (specific resistance) of the electrolyte (conductor) and $\kappa = 1/\rho$ (unit: $\Omega^{-1}\,\mathrm{m}^{-1} = \mathrm{S\,m^{-1}}$) is the **conductivity** of the electrolyte. The conductivity of the electrolyte is influenced by the mobility of ions, which is reflected in the migration term of the eq. (3.49). It is defined by the relation:

$$
\kappa = \sum_i z_i^2 \boldsymbol{F}^2 c_i u_i = \sum_i \lambda_i c_i
\tag{3.54}
$$

where $\lambda_i$ represents the **molar conductivity** of the component $i$. If we relate the conductivity $\kappa$ to the electrolyte concentration, we get the **molar conductivity** of the electrolyte, $\Lambda$ (unit: S m$^2$ mol$^{-1}$). Since it applies for the concentration of the component $i$ in a strong electrolyte solution (which is, by definition, fully dissociated to ions) with the concentration $c$, such that $c_i = v_i\, c$, the molar conductivity can be expressed as

$$\Lambda = \frac{\kappa}{c} = \frac{\sum_i \lambda_i v_i c}{c} = \sum_i \lambda_i v_i \qquad (3.55)$$

However, both $\lambda_i$ and $\Lambda$ still depend on the electrolyte concentration in real systems. Thus, eq. (3.55) has practical meaning after rewriting it in the form (3.56) referred to as the **Kohlrausch law of independent migration of ions**. It applies exactly only at very low electrolyte concentrations in the solution when ions do not interact with each other.

$$\Lambda^0 = \sum_i \lambda_i^0 v_i \qquad (3.56)$$

Here, $\Lambda^0$ represents the **limiting molar conductivity** of the infinitely diluted electrolyte and $\lambda_i^0$ the corresponding **limiting molar conductivity** of the component $i$ (limiting molar conductivities are tabulated for individual ions). For solutions of weak electrolytes with degree of dissociation $\alpha$, it applies analogically that $c_i = \alpha\, v_i\, c$, and Kohlrausch law then takes the form:

$$\Lambda^0 = \alpha \sum_i \lambda_i^0 v_i \qquad (3.57)$$

At higher concentrations of strong electrolyte, individual ions begin to interact and the molar conductivities of the electrolyte decreases. The very same trend can be seen in molar conductivities of weak electrolytes. There, however, the main reason for molar conductivity changes is the change in the degree of dissociation of the electrolyte.

To reduce resistance (increase conductivity) of the electrolyte solution and suppress the migration flux of the electroactive component, a strong electrolyte can be added to the solution where it dissociates. Such electrolyte that does not participate in any redox or other reactions in the system is referred to as a **supporting, basic** or **inert electrolyte**.

Discussion of all phenomena connected with a mass transport in electrolyte solutions was, up to this point, limited to constant temperature. Thus, it would be worth mentioning shortly also the effect of temperature on mass transport, in short. This effect can be quite severe, since temperature influences many physical and physical–chemical parameters of the system. In principle, there are two ways by which temperature affects the intensity of mass transport. Firstly, temperature influences physical properties of the solvent phase (electric permittivity, viscosity, etc.). For example, in the case of viscosity, increasing temperature decreases viscosity of the solution (and also viscous friction between components in the solution) leading to rise of diffusion coefficients of all compounds as well as mobilities of the ions. Secondly, temperature influences all the equilibria in the system causing changes in solubility or dissociation constants. The extent of temperature dependence on the equilibria can be, in the first approximation, estimated by the size of the entropic effects on the corresponding $\Delta G$ of the process, via formula $\Delta G = \Delta H - T\,\Delta S$.

## 3.5 Multiphase systems and interfaces

Up to this point, we have considered systems consisting of a single homogeneous phase (electrolyte solution). As already mentioned earlier, thermodynamic equilibrium within single phase $\alpha$ is established when there are no gradients in electrochemical potentials $\tilde{\mu}_{i,\alpha}$ of all present components $i$ as well as in temperature $T_\alpha$ and pressure $p_\alpha$. Real systems and practically useful devices are, however, structured and they consist of several phases in mutual contact. The region where two different phases are in contact is called an **interface**. Since properties of the two phases in contact are (often dramatically) different, the interfaces are regions that possess very interesting properties, where large change in the system behaviour takes place on a very small distance ($\sim 10^{-9}$ m). Therefore, the interfaces are of primary interest in research [3–5, 9].

Here, the general discussion will be limited to two interesting cases of two phases in contact that will be useful in following parts of the book. The first is when two different phases ($\alpha$ and $\beta$) are directly contacted and equilibrated. The equilibrium in the two-phase system is achieved when there is equilibrium within both phases as well as between them. In particular, there are no gradients of temperature and pressure, and electrochemical potentials of all joint components including joint charge carriers $i$ present in both phases are equal, that is, $\tilde{\mu}_{i,\alpha} = \tilde{\mu}_{i,\beta}$ (for electroneutral components $j$ this simplifies to $\mu_{j,\alpha} = \mu_{j,\beta}$). Then, the difference between inner potentials of these phases can be expressed by eq. (3.58). This means that difference between chemical potentials of joint charge carrier $i$ in phases $\alpha$ and $\beta$ is proportional to difference in inner potentials of these phases.

$$\varphi_\alpha - \varphi_\beta = \frac{\mu_{i,\beta} - \mu_{i,\alpha}}{z_i \mathbf{F}} \tag{3.58}$$

The second interesting case is when the two phases ($\alpha$ and $\alpha'$) of the same composition, temperature and pressure are in indirect contact and not in equilibrium. This, for example, occurs when two copper wires are connected by a resistor with high resistance preventing them from getting in equilibrium with each other. Then, it applies that $\tilde{\mu}_{i,\alpha} \neq \tilde{\mu}_{i,\alpha'}$; $\mu_{i,\alpha} = \mu_{i,\alpha'}$ and thus, it can be easily shown that inner potential difference between these phases is directly proportional to electrochemical potential difference between the phases, as follows:

$$\varphi_\alpha - \varphi_{\alpha'} = \frac{\tilde{\mu}_{i,\alpha} - \tilde{\mu}_{i,\alpha'}}{z\mathbf{F}} \tag{3.59}$$

In other words, the inner potential difference between the phases $\alpha$ and $\alpha'$ (measured by, e.g., voltmeter in the case of two copper wires) is proportional to the difference between the electrochemical potentials of the components $i$ in phases $\alpha$ and $\alpha'$ (and in turn, changes in difference of electrochemical potentials of components $i$ in phases in equilibrium with phases $\alpha$ and $\alpha'$).

## 3.5.1 Liquid-junction potential

Let us consider two solution phases ($\alpha$ and $\beta$) differing in concentrations of strong electrolyte $C_{\nu_+}A_{\nu_-}$ and separated by a **diaphragm**. The term diaphragm refers to a semipermeable mechanical barrier that minimizes free mixing of electrolytes (due to convection). An example of the diaphragm can be porous inert ceramic or polymeric material. In this section, diaphragm represents the interface between solutions $\alpha$ and $\beta$. At the moment of contact of these solutions, the ions start diffusing through the diaphragm in the direction of their decreasing chemical potential/activity (concentration). The diaphragm thickness defines, in the first approximation, thickness of the interface. If the products of diffusion coefficients (mobilities) and charge numbers, $z_iD_i$, of ions $C^{z+}$ and $A^{z-}$ are not identical, the electroneutrality at the interface is almost immediately disrupted. The opposite interface sides get charged with opposite charges and a **liquid-junction (diffusion) potential** $\Delta\varphi_L$ is generated on the interface. This charge separation at the interface generates electric field, causing the migration flux of ions to be directed against their diffusion flux. This influences the total ion fluxes through the interface so that there is no further disruption of the electroneutrality. Over time, the activities/concentrations of the corresponding ions in both solution phases become equal and the liquid-junction potential disappears. However, the levelling of the activities/concentrations of components in solutions is controlled by the rate of ion transport through the charged liquid interface. Therefore, this might be, depending on the geometry of the system (ratio of solution phase volumes to diaphragm cross section), slow process. In any case, it is important that the liquid-junction potential is not an equilibrium quantity though it might appear to be stable on the timescale of the experiment [5–8].

   If no electric current flows through the system ($j = 0$) and pressures and temperatures of solution phases $\alpha$ and $\beta$ are equal, we can omit the convection term in the eq. (3.51). After expressing the potential gradient $\nabla\varphi$ on the interface (diaphragm), we get

$$\nabla\varphi = -\frac{\sum_i z_i F D_i}{\sum_i z_i^2 F^2 u_i c_i}\nabla c_i \tag{3.60}$$

For the integration of this equation, it is necessary to know the activity/concentration profiles of ions $i$ in the liquid interface. For example, in the best-known Henderson model, it is assumed that the concentration of ions in the interface changes linearly from $c_{i,\alpha}$ (concentration of $i$ in solution $\alpha$ in the diaphragm vicinity) to $c_{i,\beta}$ (concentration of $i$ in solution $\beta$ in the diaphragm vicinity). After integrating the eq. (3.60) using this assumption (and considering constant thickness of the interface), we get the expression for liquid-junction potential on the diaphragm in the following form:

$$\Delta\varphi_L = \varphi_\beta - \varphi_\alpha = -\frac{RT}{F}\frac{\sum_i z_i u_i(c_{i,\beta}-c_{i,\alpha})}{\sum_i z_i^2 u_i(c_{i,\beta}-c_{i,\alpha})}\ln\frac{\sum_i z_i^2 u_i c_{i,\beta}}{\sum_i z_i^2 u_i c_{i,\alpha}} \tag{3.61}$$

If the transport of ions in solutions is more intense than in the diaphragm, for example, due to convection, the concentrations $c_{i,\alpha}$ and $c_{i,\beta}$ practically represent the volume concentrations of ions $i$ in solutions.

Generally, the liquid-junction potential is generated on every interface of two electrolyte solutions of different chemical compositions or different concentrations. It can be suppressed with a **salt bridge** that contains a concentrated solution of strong electrolyte whose both ions have similar mobilities.

## 3.5.2 Polymer electrolytes – ion-selective membranes

From the definition of the electrolyte given in Section 3.3.1 it follows that the electrolyte can be represented by an **ion-conductive polymer material**. While all ions present in the electrolyte solution are mobile, part of the charge carriers in the polymer electrolyte is immobilized in its structure. If the immobile ions are of the same charge sign, this electrolyte is difficult to permeate for ions of the same charge as that of ions fixed in the polymer electrolyte structure (**co-ions**). On the contrary, ions with the opposite charge (**counter-ions**) can be transported through the polymer electrolyte easily. This is the basis of ion-selectivity of the polymer electrolyte. A membrane made of the polymer electrolyte is called an **ion-selective (conductive) polymer membrane**. Based on the signs of immobile ion charges, we distinguish **cation-selective membranes** (containing immobile anions) and **anion-selective membranes** (containing immobile cations). The fact that transport characteristics of individual ion types in ion-selective membrane are fundamentally changed is significant for practical application of these membranes [4, 5].

It is necessary to realize that treatment of these systems is complicated by the fact that polymer membrane absorbs the solvent. In other words, it consists of polymer phase (which is often also not completely homogenous) and solution phase, with an interface between them being represented by a complex 3D structure. From a macroscopic point of view, the phases can often be assumed as sufficiently mixed in order to allow looking at/treating the "soaked" polymer electrolyte as one pseudo homogenous phase. Chemical composition and properties of ion-selective (polymer) membranes will be discussed in Chapter 4.

The transport of a matter in an ion-selective membrane follows the same principles as in a liquid electrolyte. Therefore, it can be also described using the Nernst–Planck eq. (3.47). However, the transport characteristics of the transported components in the "soaked" membrane (diffusion coefficients, mobilities) will attain different values than in electrolyte solutions. The mass transport in ion-selective membrane will be discussed in more detail in Chapter 5.

## 3.5.3 Equilibrium at the interface between ion-selective membrane and liquid electrolyte solution

Let us consider the same system as in Section 3.5.1, except that a cation-selective or anion-selective membrane (instead of diaphragm) is placed between solution phases $\alpha$ and $\beta$. Again, a driving force exists for the transport of individual components in the direction of their decreasing activity, but the membrane is nearly impermeable for co-ions. Therefore, the transport of counter-ions will greatly prevail, which will lead to charging of opposing sides of the membrane surfaces with opposite charges and establishment of a **membrane potential**, $\Delta\varphi_M$. Membrane potential can be experimentally measured. In the case of ideal ion-selective membrane (totally impermeable for counter-ions), it is given by two **Donnan potentials** $\Delta\varphi_{Don}$ resulting from establishing equilibria between the two solution phases $\alpha$ and $\beta$ and the adjacent sides of the membrane phase. In the case of real ion-selective membranes that are partly permeable also for co-ions, a contribution of liquid-junction (diffusion) potential has to be taken into account, as well. This is especially important in more concentrated solutions, where ion association takes place and charges of the ions immobilized in the membrane are largely screened by the counter-ions. Consequently, even co-ions are partially allowed to enter the membrane structure. The equilibria on both interfaces of the membrane with the solutions $\alpha$ and $\beta$ are established due to balancing of the chemical and electrochemical potentials of common electroneutral and charged components, respectively. This will be discussed in more detail in the following text.

## 3.5.4 Interfaces in a simple electrochemical system

It follows from the definition of electrode in Section 3.3.1 that the electrode contains at least one interface between electron and ionic conductor (electrolyte). However, practically interesting systems contain more phase interfaces. The schemes of common electrochemical systems are depicted in Fig. 3.1. Scheme in Fig. 3.1a consists of two electrodes ($E_1$ and $E_2$) immersed in the individual solutions, $S_1$ and $S_2$, respectively. The electrodes, $E_1$ and $E_2$ are contacted by copper current leads, $Cu_1$ and $Cu_2$, respectively, and the solutions $S_1$ and $S_2$ are in electrical contact via liquid junction.

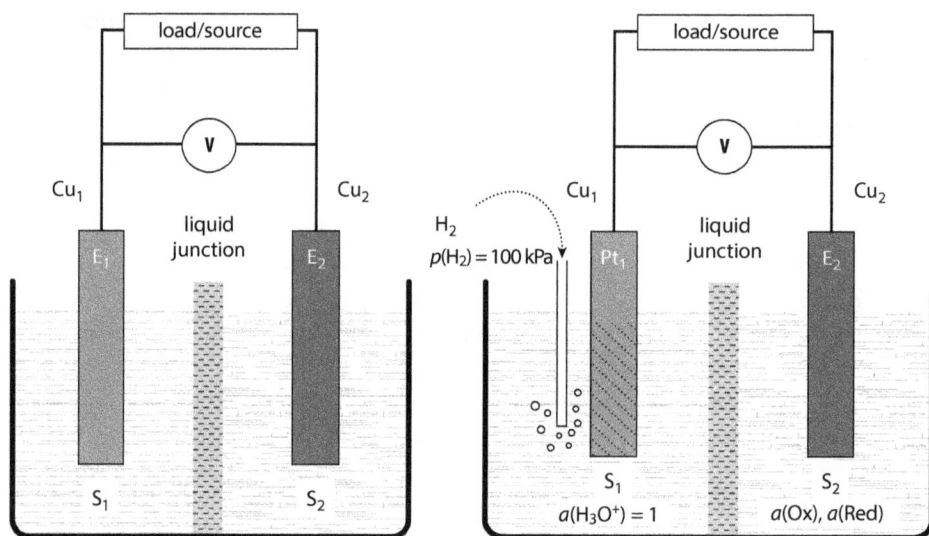

**Fig. 3.1:** Scheme of a general two-electrode electrochemical system with a liquid junction (a) and a two-electrode electrochemical system consisting of standard hydrogen electrode and redox electrode with a liquid junction (b).

In this system, there are numerous phase interfaces: $Cu_1 \mid E_1$, $E_1 \mid S_1$, $S_1 \vdots S_2$, $S_2 \mid E_2$ and $E_2 \mid Cu_2$. For description of electrochemical cells, a simplified notation is often used, where phase boundaries are represented by a vertical bar, |, and a liquid junction (interface between two miscible solution phases) is represented by a single dashed vertical bar, ¦. As discussed above, we distinguish two types of interfaces – those where equilibrium can establish (i.e. interfaces $Cu_1 \mid E_1$, $E_1 \mid S_1$, $S_2 \mid E_2$, and $E_2 \mid Cu_2$) and those where it cannot (on the timescale of the experiment), that is interface $S_1 \vdots S_2$. A potential difference at the $S_1 \vdots S_2$ interface, that is, liquid-junction potential ($\Delta\varphi_L$), can be eliminated or calculated. There is one additional interface to mention (i.e. $Cu_1 \mid Cu_2$), where equilibrium is not established, and it is used for measuring the voltage (electrode potential difference) in the electrochemical system by, for example, a voltmeter. Measuring potential differences allows determining changes in chemical potentials of reacting components in the system. This can be expressed as $\sum_j \nu_j \mu_j$, where $j$ represents all components involved in an overall reaction taking place in the cell. This expression turns out to correspond to Gibbs reaction energy $\Delta G_r$ of the overall reaction process. A slightly more specific electrochemical system with one half-cell represented by a standard hydrogen electrode and another half-cell by a redox electrode is shown in Fig. 3.1b. This will be discussed in Section 3.5.6. Another example of a typical interface in electrochemical system is the one of ion-selective membranes and electrolyte solutions as introduced briefly in Section 3.5.3 and discussed in more detail in the following text; see [3, 5, 6, 8, 10].

## 3.5.5 Interface of electron conductor and electrolyte solution

Before looking more closely at the description of electrochemical reactions at the phase interface electron conductor | electrolyte solution, it is worth looking at the structure of this interesting region that fundamentally predetermines the behaviour of the entire electrochemical system. First, as was already stated in Section 3.3.2, the macroscopic law of electroneutrality does not hold in this region in general, both from the side of the solution and from the electron conductor phase side. However, if we add up all charges on both sides of the interface, the entire region is electroneutral. In particular, the surplus of charge $Q$ near the interface in one phase is compensated by the opposite charge $-Q$ on the other side of the interface in the other phase. Due to the presence of these two oppositely charged layers, this region is called an **electrical (electrode) double layer**. The local charge separation (the resulting differences in inner potentials of the phases in contact, $\Delta\varphi_{el}$, can reach even more than 1 V) at a very short distance ($10^{-9}$ to $10^{-8}$ m) is the source of a very strong local electric field (up to about $10^9$ V m$^{-1}$). This is all that can be concluded based on thermodynamics. Development of any model describing a more detailed structure of the interface requires accepting some non-thermodynamic assumptions. In the first approximation, the interface region behaves as a simple capacitor. The concept of a parallel plate capacitor (see Section 3.2) was also included in the oldest double layer model proposed by Helmholtz. In this simple model, it is assumed that a value of the differential capacitance of the double layer $C_{ed(diff)}$ (unit: C V$^{-1}$) defined by eq. (3.22) is constant, that is, independent of the electrolyte solution composition and electrode potential.

Another complication is given by the fact that the interface composition changes dramatically if some component present in the electrolyte solution adsorbs at the electrode surface (in the interface). Moreover, the surplus charge at the interface in the solution is not, due to the thermal movement of the molecules, aligned in a plane parallel to the electron conductor phase. Instead, its concentration decreases in the direction from the electron conductor phase, giving rise to its name, **diffuse**

**electrical double layer**. More recent and much more complicated models that also take into account the presence of the diffuse electrical double layer and eventually adsorption, describe the structure and behaviour of the interface better than the original Helmholtz model.

A simplified structure of an electrical double layer is schematically depicted in Fig. 3.2. Close to the electrode, there is an **inner layer of the electrical double layer**, containing mainly the polarized solvent molecules and (specifically adsorbed) ions (partially) freed from their solvation shells. The centres of these specifically adsorbed ions define a virtual plane called an **inner Helmholtz plane** (IHP). Further from the electrode, we find already fully solvated (not specifically adsorbed) ions, their centres define the **outer Helmholtz plane** (OHP). The **diffuse part of the double layer** stretches from the OHP towards the electrolyte volume. Its thickness can, depending on the electrolyte solution composition and concentration, vary by several orders of magnitude. It contains the rest of the electrolyte surplus charge that is not present within the Helmholtz part of the double layer; see [3–6, 11].

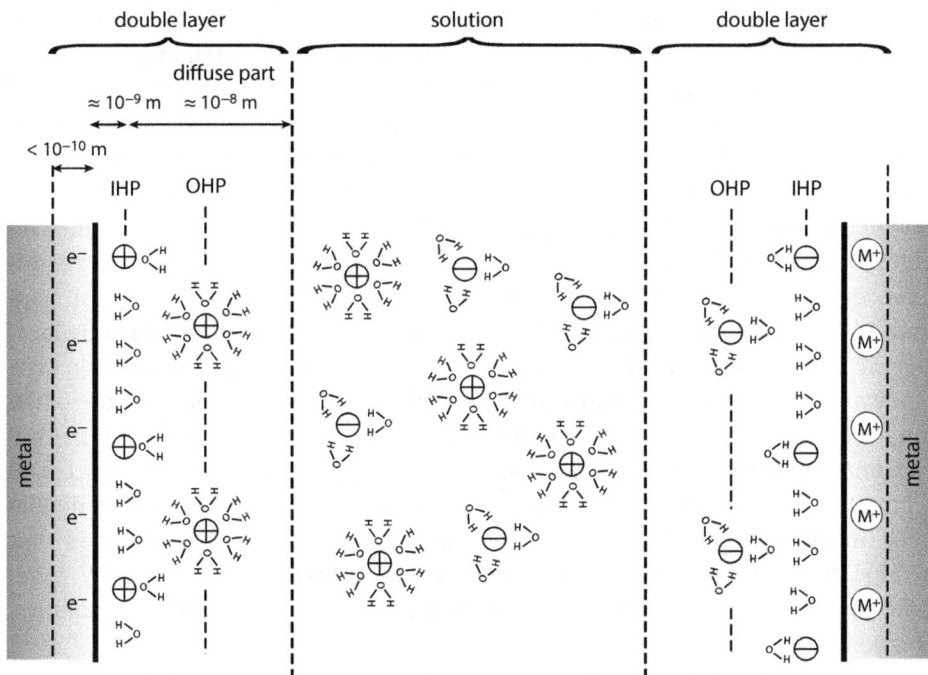

**Fig. 3.2:** Scheme of the electrical double layer. IHP, inner Helmholtz plane; OHP, outer Helmholtz plane.

## 3.5.6 Electrochemical equilibrium at the interface of electron and ion conducting phases

As discussed in the Section 3.5, establishing equilibrium between phases possessing mobile charge carriers involves transport of the joint charge carrier(s) (such as electron or ion) across the interface (from one to the other phase), in order to equalize electrochemical potentials of charge carrier(s) in both phases (at equal constant temperature and pressure everywhere in both phases). This interfacial charge carrier(s) transport is, in the case of electron conductor | ion conductor interface, connected with electrochemical reaction at the interface; see also [3–6, 9–13].

Let us return to the system presented in Fig. 3.1b in Section 3.5.4 and focus our attention at the interface $S_2 \mid E_2$. This represents interface of an inert electron conductor immersed in an electrolyte solution containing dissolved electrochemically active components $Ox^{z(Ox)}$ and $Red^{z(Red)}$. An electrochemical reaction taking place at this interface can be described by the equation:

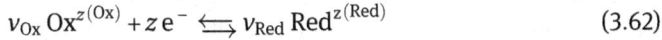

$$\nu_{Ox}\, Ox^{z\,(Ox)} + z\,e^- \rightleftharpoons \nu_{Red}\, Red^{z\,(Red)} \tag{3.62}$$

To simplify our notation, the components $Ox^{z(Ox)}$ and $Red^{z(Red)}$ will be referred to as Ox and Red, respectively. Since the electrochemical reaction takes place at the interface, a sum of the electrochemical potentials of all involved components in the state of equilibrium must be zero. Therefore, for this system, we can write (considering $\nu_{Ox} = -1$, $\nu_{Red} = 1$) as follows:

$$-\tilde{\mu}_{Ox,S_2} - z\tilde{\mu}_{e,E_2} + \tilde{\mu}_{Red,S_2} = 0 \tag{3.63}$$

Furthermore, by expanding the electrochemical potentials of individual components according to the eq. (3.36) and keeping in mind that $z_{Ox} - z_{Red} = z$ and that a charge number of the electron is $z_e = -1$, we get

$$zF\varphi_{S_2} - zF\varphi_{E_2} = \mu_{Red,S_2} - \mu_{Ox,S_2} - z\mu_{e,E_2} \tag{3.64}$$

and the difference of inner potentials, $\varphi_{E_2} - \varphi_{S_2}$, at the phase interface $S_2 \mid E_2$ is:

$$\varphi_{E_2} - \varphi_{S_2} = \frac{\mu_{Ox,S_2} + z\mu_{e,E_2} - \mu_{Red,S_2}}{zF} = \frac{\mu^\circ_{Ox,S_2} - \mu^\circ_{Red,S_2}}{zF} + \frac{\mu_{e,E_2}}{F} - \frac{RT}{zF}\ln\frac{a_{Red,S_2}}{a_{Ox,S_2}} \tag{3.65}$$

The chemical potential of an electron in any electron conductor, $\mu_{e,E_2}$, is unknown (and material specific) constant. Therefore, the difference of inner potentials at this single interface is, in principle, unknown. Moreover, any attempt to measure potentials at this single interface inevitably leads to formation of new interfaces, leading to a system with closed electric circuit. Let the overall current flowing in the system be infinitesimally low and measure cell voltage by the voltmeter. The measured voltage or, more precisely, the **equilibrium voltage** of the cell (formerly known as the **electromotive force**), $E_{eq}$, corresponds to the sum of all potential differences on all interfaces in the system. In our case, this corresponds to

$$-E_{eq} = \left( \varphi_{Cu_2} - \varphi_{Cu_1} \right) = -\left( \varphi_{E_1} - \varphi_{S_1} \right) - \left( \varphi_{S_1} - \varphi_{S_2} \right) - \left( \varphi_{S_2} - \varphi_{E_2} \right) \tag{3.66}$$

The potential difference at the interface $S_1 \mid S_2$ corresponding to liquid-junction potential $\Delta\varphi_L$ can be eliminated or calculated. The contact potentials at the interfaces between two external electron conductors ($Cu_1$ and $Cu_2$) and electrodes ($E_1$ and $E_2$), respectively, are constant, independent of the current flowing through the system and usually negligible. Therefore, they were omitted in the text. If it is moreover ensured that interface $E_1 \mid S_1$ is (at the given temperature) always the same and with constant inner potential difference, the measured changes in $E_{eq}$ can be attributed to the changes at the interface of interest, that is, $S_2 \mid E_2$. The last mentioned condition can be fulfilled if a general interface, $E_1 \mid S_1$ is always represented by a suitable **reference electrode**. A **standard hydrogen electrode** (SHE) was selected as a standard reference electrode (see left half-cell in Fig. 3.1b). This system allows us to define an **equilibrium electrode potential**, $E_{eq,Ox/Red}$ (of the given redox process) at the interface, $S_2 \mid E_2$. It is the equilibrium potential of the cell where one half cell is the studied electrode (e.g. $S_2 \mid E_2$ interface as in our case) and the second half cell is the already mentioned standard hydrogen electrode, where reaction takes place in the direction of $H_2$ oxidation.[5] In other words, the equilibrium voltage measured under these conditions equals to the equilibrium electrode potential that can be (in this case, both external electron conductors are copper wires) calculated by means of the **Nernst equation** obtained by combining the eq. (3.65) with (3.66)[4]:

$$E_{eq,Ox/Red} = -\left( \varphi_{Cu_1} - \varphi_{Cu_2} \right) =$$

$$= -\left( \varphi_{Cu_1} - \varphi_{S_1} \right) + \frac{\mu^\circ_{Ox,S_2} - \mu^\circ_{Red,S_2}}{zF} + \frac{\mu_{e,Cu_1}}{F} - \frac{RT}{zF} \ln \frac{a_{Red,S_2}}{a_{Ox,S_2}} \tag{3.67}$$

$$= E^\circ_{Ox/Red} - \frac{RT}{zF} \ln \frac{a_{Red,S_2}}{a_{Ox,S_2}}$$

where $E^\circ_{Ox/Red}$ is the **standard electrode potential** of the redox couple.

It can be seen that $E_{eq,Ox/Red}$ depends on the composition of the solution $S_2$, that is, on the activities of the electrochemically active components $i$ near the electrode surface $a_{i,S2}$ (on OHP level, see Section 3.5.5).

---

5 The liquid-junction potential must be subtracted from the measured equilibrium voltage of the cell.

The potential of the standard hydrogen electrode is defined to be zero at all temperatures, $E_{eq}(SHE) = E^\circ(H^+/H_2) = 0$ V. This is because equilibrium voltage of the electrochemical cell, where both electrodes are represented by the standard hydrogen electrodes, has to be zero, regardless of the temperature. For practical purposes the standard hydrogen electrode is rarely used, and it is often replaced by more suitable and easier to handle reference electrodes. A number of them are described in most electrochemical textbooks.

The above-derived Nernst equation (3.67) is valid only for the electrochemical reaction (3.62) (considering $v_{Ox} = -1$, $v_{Red} = 1$). Therefore, at this point, it is necessary to provide the generally valid form of the Nernst equation. Every electrode reaction can be written in the form:

$$0 = \sum_i v_i X_i - z\, e^-$$ (3.68)

where $v_i$ is the stoichiometric coefficient of the component $X_i$. According to IUPAC recommendation, electrochemical reactions should be written as reductions. Then, it follows that $v_i < 0$ for oxidized components and $v_i > 0$ for reduced components. For this general reaction, the general form of Nernst equation can be derived as

$$E_{eq,Ox/Red} = E^\circ_{Ox/Red} - \frac{RT}{zF} \ln \prod_i a_i^{v_i}$$ (3.69)

Activities of the electrochemically active components are not of practical use, and it is more convenient to define some type of an electrode potential, for which the knowledge of activity coefficients is not required. Therefore, a **formal electrode potential** of the redox couple $E^{\circ\prime}_{Ox/Red}$ is defined in eq. (3.70), so that it already contains activity coefficients of all components involved (for the used composition of the system). Such a value can then be easily measured:

$$E^{\circ\prime}_{Ox/Red} = E^\circ_{Ox/Red} - \frac{RT}{zF} \ln \prod_i \gamma_i^{v_i}$$ (3.70)

Finally, using previous equations and definitions it can be shown that Gibbs reaction energy of an electrode reaction at equilibrium is

$$\Delta G_r = -zFE_{eq,Ox/Red}$$ (3.71)

and the standard Gibbs reaction energy of the electrode reaction is

$$\Delta G^\circ_r = -z\,FE^\circ_{Ox/Red}$$ (3.72)

# 3.6 Classification of electrochemical systems

Electrochemical systems (processes) were traditionally divided into two groups: those that are a source of electricity and those that consume electricity. Let us have a closer look at this division. The spontaneity of the overall cell reaction at a constant temperature and pressure can be evaluated by the **molar reaction Gibbs energy of the overall cell reaction**, $\Delta G_{r,cell}$ (unit: J mol$^{-1}$) (see Section 2.4):

$$\Delta G_{r,cell} = \Delta H_{r,cell} - T\,\Delta S_{r,cell} \tag{3.73}$$

where $\Delta H_r$ is a **molar reaction enthalpy** (unit: J mol$^{-1}$) and $\Delta S_r$ is a **molar reaction entropy** (unit: J K$^{-1}$ mol$^{-1}$), for the overall cell reaction. The cell reaction takes place spontaneously if $\Delta G_{r,cell} < 0$. The electrochemical system is then a source of electrical work and we call it a **galvanic cell (system)**. Conversely, if $\Delta G_{r,cell} > 0$, the reaction requires external electrical work to proceed and the system is called an **electrolyser (electrolytic system)**.

Before advancing further, it is first useful to consider a general overall cell reaction as follows:

$$z_{i,red} \sum_{i,ox} v_{i,ox}\, X_{i,ox} = z_{i,ox} \sum_{i,red} v_{i,red}\, X_{i,red}$$

as a sum of both half reactions, one written as oxidation as follows:

$$\sum_{i,ox} v_{i,ox}\, X_{i,ox} - z_{i,ox}\, e^- = 0,$$

and the second as reduction, as follows:

$$0 = \sum_{i,red} v_{i,red}\, X_{i,red} - z_{i,red}\, e^-,$$

taking place at individual electrodes in a direction as written (though infinitesimally slowly). The choice of which electrode reaction will proceed in reduction and which in oxidation direction is purely arbitrary; therefore electrodes can be either assigned as positive/negative when no current is flowing or anode/cathode when current is flowing. The subscripts ox and red in these equations refer to the assumed oxidation and reduction processes, respectively.

The **electrical work** $w_{el}$ connected with carrying $z$ moles of electrons between the electrodes through the electrolyte equals the change in the Gibbs energy, $\Delta G$. Moreover, in order to operate with positive values of the both electrolyser and galvanic cell voltage, it follows that

$$w_{el} = \Delta G = zFU > 0 \tag{3.74a}$$

$$w_{el} = \Delta G = -zFE < 0 \tag{3.74b}$$

where $U$ is the **voltage of the electrolyser** (applied between electrodes from an external power source) and $E$ is the **voltage** (as measured between the electrodes) **of the galvanic cell**. The voltage of the electrolyser and the voltage of the galvanic cell are given by the electrode potentials of the cathode $E_{cat}$ and anode $E_{an}$ and by the ohmic voltage loss resulting from the passage of the current $I$ through the system with internal resistance $R_{in}$. Therefore, we can write

$$U = E_{an} - E_{cat} + R_{in}I, \qquad R_{in}I > 0 \qquad (3.75a)$$

$$E = E_{cat} - E_{an} - R_{in}|I|, \qquad R_{in}|I| > 0 \qquad (3.75b)$$

It follows from these equations that for the current $I$ to pass through the system, we must apply the voltage $U$ to the electrolyser. Alternatively, in the case of a galvanic cell, the voltage between the electrodes will be equal to $E$.

With no current ($I = 0$), the ohmic voltage loss term is zero and can be omitted and the concepts of electrolyser and galvanic cell or anode and cathode make no sense, anymore. Under such conditions, the potentials at the electrodes acquire their equilibrium values and we distinguish positive (+) and negative (–) electrode, with equilibrium potentials $E_{(+),eq}$ and $E_{(-),eq}$, respectively. Voltage measured in the system under these conditions is called the **equilibrium voltage of the cell**, $E_{eq}$. Please note that the term "equilibrium voltage" is somewhat misleading, creating a false impression that it relates to the equilibrium conditions in the whole system. This is far from the truth. Instead, it relates to a state, where equilibrium conditions are achieved at both electrodes (electron conductor | electrolyte interfaces) present. This was discussed in Section 3.5.6, where equilibrium electrode potential $E_{eq,Ox/Red}$ was defined. The equilibrium voltage of the electrochemical cell can be calculated by substituting the Nernst eq. (3.69) into the equations (3.75) and expressed in the form[6]

$$E_{eq} = E_{red, eq} - E_{ox, eq} =$$

$$= E^\circ_{red} - E^\circ_{ox} - \frac{RT}{zF} \ln \prod_{i,red} a_{i,red}^{v_{i,red}} + \frac{RT}{zF} \ln \prod_{i,ox} a_{i,ox}^{v_{i,ox}} \qquad (3.76)$$

If the calculated $E_{eq}$ is negative (corresponding to $\Delta G_{r,cell} < 0$ of the cell reaction as written), the cell reaction will proceed spontaneously. For $E_{eq} < 0$ ($\Delta G_{r,cell} > 0$ of the cell reaction as written), the cell reaction would spontaneously proceed in the opposite direction than originally assumed. In the latter case, the cell reaction (as written) can be forced to take place only if external voltage $U > E_{eq}$ is applied.

Under equilibrium at both electrodes, it also applies for the overall cell reaction that

---

6 If we consider, for instance, the electrode reactions described by eqs. (3.29) and (3.30) or the total reaction (3.31) taking place in the electrolyser, we get for the equilibrium cell voltage, $E_{eq} = E_{red,eq} - E_{ox,eq} = E^\circ(H_2O/H_2) - E^\circ(SO_4^{2-}/SO_3^{2-}) - \frac{RT}{2F} \ln \frac{a_{SO_4^{2-}} \, a_{H_2}}{a_{SO_3^{2-}} \, a_{H_2O}}$.

$$\Delta G_{r,\text{cell}} = -zFE_{\text{eq}} \tag{3.77}$$

A difference between the potential of the current-loaded electrode $E_{\text{an/cat}}$ and potential of the corresponding electrode at equilibrium, $E_{(+)/(-),\text{eq}}$, is called the **overpotential** $\eta$:

for electrolyser $\qquad \eta_{\text{an}} = E_{\text{an}} - E_{(+),\text{eq}}$ and $\eta_{\text{cat}} = E_{\text{cat}} - E_{(-),\text{eq}} \tag{3.78a}$

for galvanic cell $\qquad \eta_{\text{an}} = E_{\text{an}} - E_{(-),\text{eq}}$ and $\eta_{\text{cat}} = E_{\text{cat}} - E_{(+),\text{eq}} \tag{3.78b}$

For both the electrolyser and the galvanic cell, it applies that $\eta_{\text{an}} \geq 0$, $\eta_{\text{cat}} \leq 0$. By substituting for $E_{\text{cat}}$ and $E_{\text{an}}$ from eqs. (3.78) into (3.75), we get:

$$U = E_{(+),\text{eq}} - E_{(-),\text{eq}} + \eta_{\text{an}} - \eta_{\text{cat}} + R_{\text{in}}I = E_{(+),\text{eq}} - E_{(-),\text{eq}} + \eta_{\text{EL}} \tag{3.79a}$$

$$E = E_{(+),\text{eq}} - E_{(-),\text{eq}} - \eta_{\text{an}} + \eta_{\text{cat}} - R_{\text{in}}I = E_{(+),\text{eq}} - E_{(-),\text{eq}} + \eta_{\text{GC}} \tag{3.79b}$$

where $\eta_{\text{EL}}$ ($> 0$) represents the **overvoltage** on the electrolytic cell and $\eta_{\text{GC}}$ ($< 0$) is overvoltage on the galvanic cell.

It is evident that the overvoltage and its individual components (overpotential on electrodes and ohmic voltage loss inside the electrochemical systems) represent energy losses, decreasing the **thermodynamic efficiency of the electrochemical system** $\eta_{\text{T}}$ (dimensionless), defined by the ratio of the effectively used work $w_{\text{out}}$ and consumed work $w_{\text{in}}$ as

$$\eta_{\text{T}} = -w_{\text{out}}/w_{\text{in}} \tag{3.80}$$

The following treatment is very simplified and does not take into account the peculiarities introduced by a possible dominating effect of the entropic term, $T \Delta S_{r,\text{cell}}$. Chemicals are supplied to the galvanic cell, and energy is released in the form of Gibbs (electric) energy. For the reaction extent of $\xi = 1$ mol, the enthalpy released from chemical bonds equals the molar reaction enthalpy, $\Delta H_{r,\text{cell}}$. According to the first thermodynamic law, it is possible to use only that part of the energy that is not released in the form of Joule heat, $Q_{\text{J}}$. The thermodynamic efficiency of the galvanic cell $\eta_{\text{T,GC}}$ (dimensionless) is then

$$\eta_{\text{T, GC}} = -\frac{w_{\text{out}}}{-\Delta H_{r,\text{cell}}} = \frac{\Delta H_{r,\text{cell}} - Q_{\text{J}}}{\Delta H_{r,\text{cell}}} \tag{3.81}$$

If the cell is operated reversibly and isothermally at the temperature $T$ (then $Q_{\text{J}} = T \Delta S_{r,\text{cell}}$), the thermodynamic efficiency can be expressed by the ratio of the molar cell reaction Gibbs energy $\Delta G_{r,\text{cell}}$ and the molar cell reaction enthalpy $\Delta H_{r,\text{cell}}$ as:

$$\eta_{\text{T, GC}} = \frac{\Delta H_{r,\text{cell}} - T\Delta S_{r,\text{cell}}}{\Delta H_{r,\text{cell}}} = \frac{\Delta G_{r,\text{cell}}}{\Delta H_{r,\text{cell}}} \tag{3.82}$$

To the electrolyser, the work is supplied in the form of electric energy that according to the eq. (3.74) corresponds to the Gibbs energy $\Delta G$. Useful performed work is stored in the form of chemical bonds (chemicals) and its amount can be expressed in terms of enthalpy. For the reaction extent of $\xi = 1$ mol, the stored enthalpy equals $\Delta H_{r,cell}$. The thermodynamic efficiency of the electrolyser $\eta_{T,EL}$ can be expressed by the relation (3.83) using the ratio of $\Delta H_{r,cell}$ and $\Delta G$, or alternatively the ratio of **thermoneutral voltage** $U_{TN}$ and the actual voltage applied to the electrolyser $U$ as follows:

$$\eta_{T,EL} = \frac{\Delta H_{r,cell}}{\Delta G} = \frac{zFU_{TN}}{zFU} = \frac{U_{TN}}{U} \tag{3.83}$$

The thermoneutral voltage is the voltage at which the electrolytic process can be operated isothermally, without the need to withdraw/supply any other (than electric) energy from/to the system (isothermal adiabatic operation). It is defined by the relation

$$U_{TN} = \Delta H_{r,cell}/zF \tag{3.84}$$

Unfortunately, not all current flowing through the system is always used to perform the desired reactions. This lowers **current (faradic) efficiency** or **charge yield** of the process, $\Phi_I$ (dimensionless). In the case of the galvanic system, the current efficiency $\Phi_{I,GC,j}$ is determined by the ratio of the overall charge that passed the galvanic cell, $Q_{tot,GC}$ (or external circuit) and the theoretical charge $Q_{theor,GC,j}$ calculated using Faraday's law from the amount of reactant $\Delta n_{R,GC}$ consumed in the cell during its discharge via reaction $j$ (with respect to the number of electrons exchanged during the reaction $j$, turnover $z_j$, and stoichiometric coefficient of the reactant component $\nu_{R,j}$). In the case of electrolyser current efficiency, $\Phi_{I,EL}$, one has to compare the theoretical charge $Q_{theor,EL}$ required for synthesis of the desired amount of product, $n_{P,EL}$, via reaction $j$ and the total charge passed through the system, $Q_{tot,EL}$. After substituting from Faraday's law, we can write $\Phi_I$ of the process $j$ as follows:

$$\Phi_{I,GC,j} = Q_{tot,GC}/Q_{theor,GC,j} = Q_{tot,GC}\nu_{R,j}/n_{R,GC}z_j F \tag{3.85}$$

$$\Phi_{I,EL,j} = Q_{theor,EL,j}/Q_{tot,EL} = n_{P,EL}z_j F/\nu_{P,j} Q_{tot,EL} \tag{3.86}$$

If more electrode reactions take place in the system (and reactants and products take part in no undesired chemical reactions), it applies that $\sum\Phi_{I,GC,j} = 1$ and $\sum\Phi_{I,EL,j} = 1$.

The **total efficiency** of the process $j$ in electrochemical system $\eta_{tot}$ (dimensionless) is determined by the relation

$$\eta_{tot} = \eta_T \Phi_{I,j} \tag{3.87}$$

## 3.7 Kinetics of electron transfer reaction at electron conductor/electrolyte interface

In this section of the text, we will deal with the description of the rates of electrochemical reactions at the electron conductor / electrolyte interface. Let us again consider the reaction (3.62) written in the form (3.68) with $v_{Ox} = -1$, $v_{Red} = 1$, taking place at the interface $S_2 \mid E_2$ of the electrochemical system depicted in Fig. 3.1b. Additionally, we will assume that the reaction (3.62) is elementary, that is, it proceeds in one step (with a single transition state). Besides, let us remind that the solution $S_2$ also contains the supporting electrolyte at sufficient concentration, so we can neglect the influence of the migration and accumulation of electroactive components near the electrode surface in the electric field.

Due to a microscopic reversibility, all reaction processes are, in principle, reversible. This also means that the process described by the eq. (3.62) occurs in both the forward (reduction) and reverse (oxidation) directions, respectively. The total current flowing through the external circuit is given by the sum of **partial currents**, that is, the partial reduction (**cathodic**) $i_{cat}$ and partial oxidation (**anodic**) $i_{an}$ currents at the considered electrode interface[7] as follows:

$$i = i_{cat} + i_{an} \tag{3.88}$$

The rate of elementary electrode reactions can usually be described by the first-order kinetics. The values of **partial current densities** are then given by the equations:

$$j_{cat} = i_{cat}/A = - zFk_{cat}\, c_{Ox,S}, \quad j_{an} = i_{an}/A = zFk_{an}\, c_{Red,S} \tag{3.89}$$

where $c_{Ox,S}$ and $c_{Red,S}$ are the concentrations of the corresponding components near the electrode surface (on OHP level, see Section 3.5.5), see Fig. 3.2; $k_{cat}$ and $k_{an}$ are (heterogeneous electron transfer) **rate constants of reduction** and **oxidation reactions**, respectively. These rate constants follow the **Arrhenius equation** (3.90) that represents a transition between thermodynamics and kinetics of processes as shown below:

$$k = D_0 \exp\left(-\frac{\Delta G^{\#}}{RT}\right) \tag{3.90}$$

Here, $\Delta G^{\#}$, a **molar activation Gibbs energy**, is given by the difference in the molar Gibbs energies of reactants in the initial state $G^o$ and the Gibbs energy of a transition state $G^{\#}$; $D_0$ is the pre-exponential factor.

---

7 The terms cathodic and anodic are often misused in connection with negative or positive electrode potential. The adjectives cathodic and anodic should be strictly used only for processes involving reduction and oxidation, respectively.

The relationship between the change of the Gibbs energy of the activation of the partial reduction, $\Delta(\Delta G^{\#}_{cat})$, and oxidation reaction, $\Delta(\Delta G^{\#}_{an})$, and the change of the molar reaction Gibbs energy $\Delta(\Delta G_r)$ is according to Polányi's rule given by the equations

$$\Delta\left(\Delta G^{\#}_{cat}\right) = -\beta\,\Delta(\Delta G_r); \quad \Delta\left(\Delta G^{\#}_{an}\right) = (1-\beta)\Delta(\Delta G_r) \tag{3.91}$$

where $\beta$ is the **symmetry factor** (dimensionless), whose meaning will be discussed later in the text. It has already been shown that there is a direct relationship between cell voltage (electrode potential) and Gibbs energy, eqs. (3.74). Also, one has to be able to eliminate the influence of $R_{in}I$ term and control the potential at the $S_2$ | $M_2$ interface. Then, for the particular case of the reaction (3.62), it applies that changing the electrode potential from $E_1$ to $E_2$ (i.e. by $\Delta E = E_2 - E_1$) causes a change in the molar Gibbs energy of reactants $\Delta G_{Ox+ze}$ as well as the change in the molar reaction Gibbs energy $\Delta(\Delta G_r)$:

$$\Delta(\Delta G_r) = \Delta(\Delta G_{Ox+ze}) = -z\mathbf{F}\,\Delta E \tag{3.92}$$

Substituting from eq. (3.92) onto (3.91) allows expressing a corresponding dependence of the **activation energies** of the partial reactions on the electrode potential:

$$\Delta(\Delta G^{\#}_{cat}) = \beta z\mathbf{F}\,\Delta E = -\beta\,\Delta(\Delta G_r); \tag{3.93}$$

$$\Delta(\Delta G^{\#}_{an}) = -(1-\beta)z\mathbf{F}\,\Delta E = (1-\beta)\Delta(\Delta G_r)$$

Combining eqs. (3.88), (3.89), (3.90) with (3.93) and Nernst eq. (3.69) allows expressing a **general equation of the polarization curve** that describes the dependence of the current density on the electrode potential as

$$j = j_0\left\{\frac{c_{Red,S}}{c_{Red,bulk}}\exp\left[\frac{(1-\beta)z\mathbf{F}(E_2 - E_{eq,\,Ox/Red})}{RT}\right] - \frac{c_{Ox,S}}{c_{Ox,bulk}}\exp\left[\frac{-\beta z\mathbf{F}(E_2 - E_{eq,Ox/Red})}{RT}\right]\right\}$$

$$\tag{3.94}$$

where $c_{Red,bulk}$ and $c_{Ox,bulk}$ correspond to the bulk concentrations of Red and Ox, respectively and $E_{eq,Ox/Red}$ is the equilibrium potential at the electrode, given by the redox couple activities as defined by eq. (3.69) and, finally, $j_0$ is an **exchange current density**. It is defined as a partial current density at equilibrium potential. Under such conditions, no current flows through the external circuit, that is, $j = 0$ and $-j_C = j_A = j_0$. Combining the relation (3.94) with the Nernst equation allows deriving a formula for $j_0$ as follows:

$$j_0 = z\mathbf{F}k^{\circ}c_{Ox,\,bulk}\exp\left[-\beta\ln\frac{c_{Ox,bulk}}{c_{Red,bulk}}\right] = z\mathbf{F}k^{\circ}c_{Ox,\,bulk}\exp\left[\ln\left(\frac{c_{Red,bulk}}{c_{Ox,bulk}}\right)^{\beta}\right] \tag{3.95}$$

$$= z\mathbf{F}k^{\circ}\left(c_{Ox,\,bulk}\right)^{(1-\beta)}\left(c_{Red,\,bulk}\right)^{\beta} = j^{\circ}{}_0\left(c_{Ox,\,bulk}\right)^{(1-\beta)}\left(c_{Red,\,bulk}\right)^{\beta}$$

Here, $k°$ represents a **standard rate constant**. It is obvious that $j_0$ depends on the system composition.

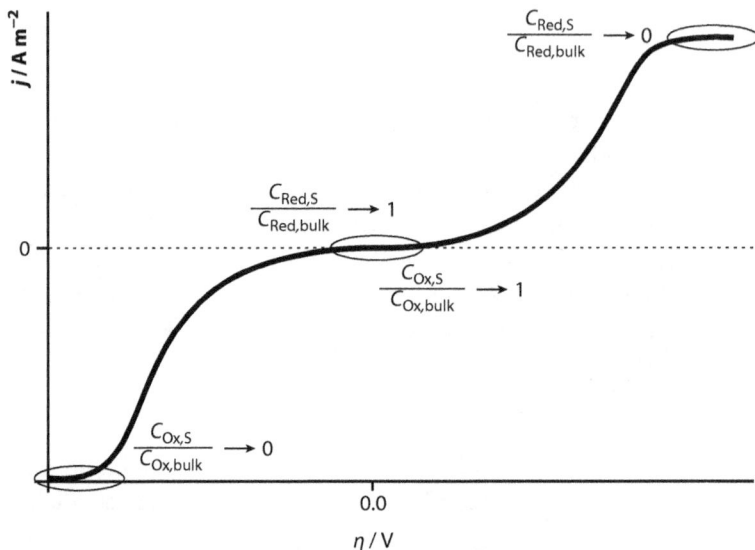

**Fig. 3.3:** A general shape of a polarization curve for an electrode reaction. Calculated according to eq. (3.94) for reaction $Ox^{z(Ox)} + e^- \rightleftharpoons DRed^{z(Red)}$ taking place without adsorption of the reactant or product and considering a constant hydrodynamic boundary layer thickness, $c_{Ox,bulk} = c_{Red,bulk}$, $D_{Ox} = D_{Red}$ and $\beta = 0.5$.

Shape of the polarization curve described by eq. (3.94) is shown in Fig. 3.3. It is apparent that close to $E_{eq,Ox/Red}$ it applies for the concentration of the both components, Ox and Red, that $c_S/c_{bulk} = 1$. Under such conditions, any change in electrode potential influences current density. On the other hand, at sufficiently negative/positive overpotential, $\eta$, the concentrations of the reactants near the electrode surface significantly drop, reaching the zero value eventually, that is, $c_S/c_{bulk} \approx 0$. This suggests that the rate of electron transfer is so high that every specie Ox (at $\eta \ll 0$) or specie Red (at $\eta \gg 0$) reaching the electrode surface is immediately reduced or oxidized, respectively. The current density in these potential regions is largely potential independent, since it depends mainly on the intensity of mass transport towards/from the electrode surface (mass transport control).

Several simplifications of the eq. (3.94) are widely used. Firstly, if the rate of the electron transfer reaction is negligible compared to the rate of electroactive species Ox and Red transport towards/from the electrode, it is reasonable to assume that $c_{Ox,S}/c_{Ox,bulk} \approx c_{Red,S}/c_{Red,bulk}, \approx 1$. Moreover, the difference between the actual and equilibrium electrode potential under such conditions, that is, $E_2 - E_{eq,Ox/Red}$, represents the **activation overvoltage (overpotential)** $\eta$. Then, the eq. (3.94) can be simplified to the form known historically as a **Butler–Volmer equation** that describes kinetics of an electron transfer reaction assuming infinitely fast mass transport as

$$j = j_0 \left\{ \exp\left[\frac{(1-\beta)zF\eta}{RT}\right] - \exp\left[\frac{-\beta zF\eta}{RT}\right] \right\} \tag{3.96}$$

With sufficiently high values of overvoltage $\eta$ (still under the condition that the mass transport is much faster than the electron transfer), when the rate of one of the partial reactions at the electrode is negligible compared to the reverse process, we can omit one of the exponential terms.[8] Then, it applies for the oxidation and reduction as follows:

$$j_{an} = j_0 \exp\left[\frac{(1-\beta)zF\eta}{RT}\right]; \quad j_{cat} = j_0 \exp\left[\frac{-\beta zF\eta}{RT}\right] \tag{3.97}$$

After expressing the logarithm and rearrangement, we get a **Tafel equation** for the oxidation reaction at high positive overvoltage as shown below:

$$\eta = -\frac{2,3RT}{(1-\beta)zF} \log j_0 + \frac{2,3RT}{(1-\beta)zF} \log j_{an} = a_{an} + b_{an} \log j_{an} \tag{3.98}$$

and for the reduction reaction at high negative overvoltage, as follows:

$$\eta = \frac{2,3RT}{\beta zF} \log j_0 - \frac{2,3RT}{\beta zF} \log j_{cat} = a_{cat} + b_{cat} \log |j_{cat}| \tag{3.99}$$

The terms $a_{an}$ and $a_{cat}$ (not to be confused with activities) are determined particularly by a nature of the process and the electrode surface properties/composition (to what extent is the process catalysed at the electrode surface) and they are used to determine a value of the exchange current density. The terms $b_{an}$ and $b_{cat}$ are called **Tafel slopes** of the oxidation and reduction reactions, respectively. They determine how the current density changes with the value of the electrode potential (overpotential). The value of the Tafel slope is also sometimes useful when determining the mechanism of the electrode reaction. The polarization curves expressed by the Butler–Volmer equation and Tafel equation (in the form $\log j = f(\eta)$) are shown in Fig. 3.4.

It might be also useful to look at the current densities at the vicinity of $E_{eq,Ox/Red}$ when $|\eta| \approx 0$. In this region, it is possible to use linearization of the Butler–Volmer equation and the current density can be approximated by the formula

$$j = j_0 \frac{zF\eta}{RT} \tag{3.100}$$

---

8 Error in the current density introduced by the application of Tafel instead of Butler–Volmer equation is below 1 %, if $|\eta| > RT/(zF) \ln (100)$.

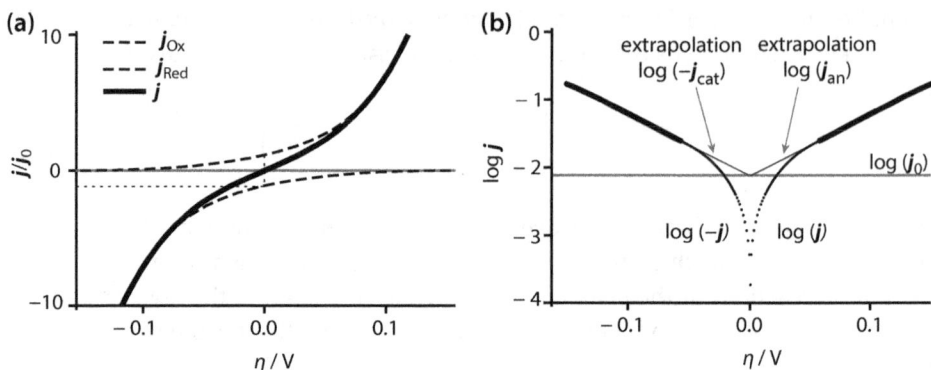

**Fig. 3.4:** Polarization curve expressed using the Butler–Volmer (A) and the Tafel (B) equations. $j_0$ – exchange current density, $j_{an}$ – oxidation (anodic) partial current density, $j_{cat}$ – reduction (cathodic) partial current density, $j$ – total current density at the electrode, calculated for $\beta = 0.5$, $z = 1$, $j_0 = 8.3 \times 10^{-3}$ A m$^{-2}$.

Let us return to the **symmetry factor $\beta$** discussed above. In the first approximation, it is a proportionality constant that takes values from 0 to 1 and, very often, it applies that $\beta \approx 0.5$. Its value determines what part of the supplied electrical energy, $\Delta G = -zFE = \Delta(\Delta G_r)$ (i.e. $\Delta(\Delta G_r)\beta$) is used to decrease (increase) the activation energy of the reduction reaction. The remaining part of the supplied energy, $\Delta(\Delta G_r)(1 - \beta)$, increases (decreases) the activation energy of the oxidation reaction; see eq. (3.92). The rigorous definition of the symmetry factor is given by the relation

$$\beta = \frac{\partial \Delta G^{\#}_{cat}}{\partial \Delta G_r} \tag{3.101}$$

Speaking about the symmetry factor makes sense only in the case of an elementary electrode reaction. For elementary reaction, it almost certainly applies that $z = 1$. Simultaneous transfer of more than one electron is energetically very unlikely. If the electrode reaction has a complex mechanism involving a sequence of several electrochemical reactions and/or chemical steps (or when we are not sure that the reaction is elementary), the term $z\beta$ in kinetic equations is substituted with the **cathodic charge transfer coefficient** $\alpha_{cat}$ and the term, $z(1 - \beta)$ with the **anodic charge transfer coefficient** $\alpha_{an}$. These are defined by the following relations:

$$\alpha_{an} = \frac{RT}{F}\frac{\partial \ln j_{an}}{\partial E}, \quad \alpha_{cat} = \frac{RT}{F}\frac{\partial \ln|j_{cat}|}{\partial E} \tag{3.102}$$

For example, the Butler–Volmer equation, then, has the following form:

$$j = j_0\left\{\exp\left[\frac{\alpha_{an}F(E_2 - E_{eq,\,Ox/Red})}{RT}\right] - \exp\left[\frac{-\alpha_{cat}F(E_2 - E_{eq,\,Ox/Red})}{RT}\right]\right\} \tag{3.103}$$

It can be seen from the above definitions of the charge transfer coefficients and symmetry factor that $\alpha_{cat} \neq z\beta$ and $\alpha_{an} \neq z(1 - \beta)$. Instead, it usually applies that $\alpha_{cat} + \alpha_{an} = z$. Moreover, the charge transfer coefficient can theoretically acquire any non-negative value (in most cases, $\alpha < 2$).

# References

[1]   http://web.mit.edu/sahughes/www/8.022/index.html; accessed 18. 3. 2021.
[2]   The Feynman Lectures on Physics, Vol. II. Mainly Electromagnetism and Matter. The New Millennium Edition. Addison–Wesley, 1964. ISBN-13: 978-0465024940. ISBN-10: 0465024947.
[3]   E. Gileadi, Physical Electrochemistry, Fundamentals, Techniques and Applications. Wiley-VCH, Weinheim, 2011. ISBN: 978-3-527-31970-1.
[4]   J. A. Bard, G. Inzelt, and F. Scholz Eds., Electrochemical Dictionary. Springer-Verlag, Berlin Heidelberg, 2008. ISBN 978-3-540-74597-6.
[5]   H. H. Girault, Analytical and Physical Electrochemistry. 1st Edition. EPFL Press, Lausanne, 2004. ISBN 9780824753573. https://www.crcpress.com/Analytical-and-Physical-Electrochemistry/Girault/p/book/9780824753573.
[6]   A. Bard and L. R. Faulkner, Electrochemical Methods, Fundamentals and Applications. 2nd Edition. John Wiley & Sons, Inc., New York, 2001.
[7]   J. Koryta, J. Dvořák, and L. Kavan, Principles of Electrochemistry. 2nd Edition. Wiley, Chichester, 1993. ISBN-13 978-0471938385, ISBN-10 0471938386. https://www.amazon.com/Principles-Electrochemistry-Jiri-Koryta/dp/0471938386.
[8]   J. O. M. Bockris and A. K. N. Reddy, Modern Electrochemistry. Volume 1. 2nd Edition. Plenum Press, New York, 1998. ISBN 0-3006-45554-4.
[9]   J. O. M. Bockris, A. K. N. Reddy, and M. E. Gamboa-Aldeco, Modern Electrochemistry. Volume 2A. 2nd Edition. Kluwer Academic/Plenum Publishers, New York, 2000. ISBN 0-306-46166-8.
[10]  G. Inzelt, A. Lewenstam, and F. Scholz Eds., Handbook of Reference Electrodes. Springer-Verlag, Berlin Heidelberg, 2013. ISBN 978-3-642-36187-6. https://link.springer.com/book/10.1007/978-3-642-36188-3.
[11]  J. O. M. Bockris and A. K. N. Reddy, Modern Electrochemistry. Volume 2B. 2nd Edition. Kluwer Academic/Plenum Publishers, New York, 2000. ISBN 0-306-46324-5.
[12]  A. Rao, J. Maclay, and S. Samuelsen, Efficiency of electrochemical systems. J. Power Sources 134: 181, 2004.
[13]  Ø. Ulleberg, Modelling of advanced alkaline electrolyzers: A system simulation approach. Int. J. Hydrogen Energy 28: 21, 2003.

Luboš Novák, Miroslav Bleha, Aleš Černín, Robert Válek

# 4 Ion-selective materials and membranes

## 4.1 Ion-selective materials

The basic and essential characteristic of all ion-selective materials is their ionic nature, determined by a functional group of acid or basic type, which is a part of the structural skeleton of the material either independently or jointly [1]. The polar nature of the modified material is determined by the type and amount of bound functional groups and can cover a wide range of properties from strongly acid to strongly basic, and the available ones are even combined materials. The carriers of functional groups are of organic and inorganic origin. In the case of organic ones, these are synthetic polymers of various types as well as cellulose or dextran derivatives, in which the functional group is covalently bound to the polymer chain. Cross-linking methylene or divinylbenzene bonds forms a three-dimensional (3D) skeleton limiting the mobility of polymer chains in an insoluble structure which will swell in a suitable solvent. Inorganic materials of natural and synthetic type have functional groups embedded in the crystalline structure of the compound. Natural materials include aluminosilicates of various types; an example of synthetic materials is zirconyl phosphate.

Based on their external form, ion exchangers are granular and non-granular. In the first case, particles are spherical, beaded or irregular. Among non-granular, we rank films, membranes, fibres, desks or frits. Granular ion-selective materials are used in ion-exchange procedures under equilibrium or dynamic conditions, implemented mainly in column systems. Ion-selective (ion-exchange[1] – see also the note in **Introduction**) membranes are the most widespread type of non-granular materials.

Nowadays, the main application representatives of ion-selective materials are polymer ion-exchange particles and polymer cation-selective and anion-selective membranes. Ion exchangers are part of preparative columns for removal of unwanted substances from mostly aqueous solutions and also of analytical column systems used in standard and special chromatographic separation procedures. Many of these conventional ion-exchange methods are and will be replaced by more effective membrane operations with ion-selective membranes.

---

**1** The literature on electromembrane processes often uses the term ion-exchange membranes. The authors of this book believe that it is more appropriate to use the term ion-selective membranes, since functional groups with the positive or negative charge, built in the membrane, allow selective transport of cations or anions through the membrane without ion exchange. Ion exchange really occurs only in polymer matrix membranes with the cation exchangers or anion exchangers which require regeneration of ions after the separation process (e.g. after the electrodeionization).

https://doi.org/10.1515/9783110739466-005

# 4.2 Ion-selective membranes

Ion-selective membranes allow selective transport of ions or molecules carrying certain charge by creating a spatial electric charge in the membrane. The membrane is usually a foil made of ion exchanger, which is a macromolecular substance containing tightly bound functional groups of a certain charge capable of dissociation in aqueous environment. The membrane is thus a polyelectrolyte with ionizable groups in the cross-linked matrix structure, in which the charge of the bound group is compensated by an oppositely charged mobile ion. For mobile ions, referred to as counter-ions, the membrane is selectively permeable. If the functional groups of the polyelectrolyte are not sufficiently distributed in its structure, a chance is created for the penetration of ions with the same charge (co-ions). Ion-selective membranes are divided into groups based on several criteria (see also Fig. 1.1). The basic division is into homogeneous and heterogeneous. Traditionally, membranes are categorized according to the chemical nature of the functional group and may by distinguished according to the basic material type, its structure or membrane format [2].

## 4.2.1 Functionality of ion-selective membranes

Membranes are divided into two basic groups, anion-selective and cation-selective, based on the type of functional group bound to the membrane matrix. The cation-selective membranes of the general pattern $R_pX^-$ ($R_p$ is the basic polymer skeleton of the membrane) contain negatively charged functional groups, where $X^-$ can be $SO_3^-$, $COO^-$, $PO_3^{2-}$, $PO_3H^-$, $AsO_3^{2-}$, $C_6H_4O^-$ or other suitable group allowing passing of positively charged particles through the membrane and preventing transport of anions. The anion-selective membranes of the general pattern $R_pX^+$ ($R_p$ is the basic polymer skeleton of the membrane) contain positively charge groups, usually on the basis of quaternary amine and other groups, where $X^+$ can be $NH_3^+$, $NRH_2^+$, $NR_2H^+$, $NR_3^+$, $PR_3^+$, $SR_2^+$ (where R is alkyl group or hydrogen), allowing passing of negatively charged particles and preventing transport of cations. These functional groups substantially influence the essential utility properties of the membranes, their selectivity and electrical resistance. The sulphone group $-SO_3^-$ is in the cation-selective membrane fully dissociated in almost the entire pH range, unlike the carboxyl group $-COO^-$, not dissociating in the pH range below 3. The quaternary amine group $-NR_3^+$ dissociates in a wide pH range and the secondary amine group $-NH_2R^+$ is slightly dissociable. The ion-selective membranes used therefore have mostly sulphonic, carboxyl and quaternary amine groups in their structure [3].

Ion-selective membranes retain the ability of ion exchangers, that is, the ability to receive from the solution positively or negatively charged counter-ions and in return release to the solution an equivalent amount of other counter-ions from the membrane phase, while maintaining the total charge balance. This property is directly used only exceptionally, for example, in determining their ion-exchange capacity, which denotes the number of functional (exchange) groups in the mass or volume unit of the membrane, or industrially at using ion-exchange technologies. The membranes used in electromembrane processes must meet the following operating requirements: low electrical resistance, high selectivity, good mechanical properties and chemical and thermal resistance. It is practically impossible to meet all these preferred properties at the same time because they act in opposition. Therefore, it is necessary to seek an acceptable optimum in the characteristic properties of these membranes. Formerly, all experts and producers tried to prepare a universal membrane, suitable for all categories of electromembrane processes; nowadays, the approach is directly opposite – properties are adapted to the type of process and its optimization [4].

In addition to the basic type of ion-selective membranes with one type of functionality, there are membranes containing both cation- and anion-functional groups. These are bipolar, amphoteric and mosaic membranes. By combination and certain modification of both basic types of ion-selective membranes, we obtain a bipolar membrane composed of two layers of membrane material, which contains only one type of fixed functional groups. Using suitable techniques (pressing, bonding with conductive polymer, co-extrusion etc.), these layers are joined together and the "transition zone" between the layers is very narrow (2–5 nm). Bipolar membranes do not have the separating nature but serve as the source of $H^+$ and $OH^-$ ions (Fig. 4.1), produced in the transition zone by dissociation of water, which is utilized in their application in electrodialysis to prepare acids and bases from their corresponding salts. To increase the efficiency of this reaction, various types of catalysts can be added to the interlayer to reduce the electric potential needed to splitting of water (they reduce the electrical resistance of the bipolar membrane). Then, it is possible to reduce the energy demands while maintaining the current density or, on the other hand, to increase the current density while maintaining the driving force level, that is, the electric potential in electromembrane process. Since the activation energy of water dissociation is very high, the catalyst allows reducing this energy due to the formation of very reactive activation complexes.

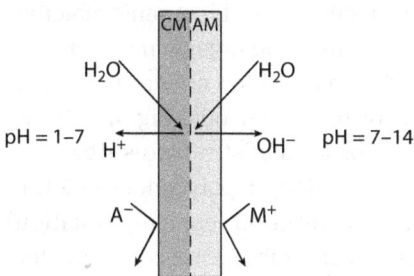

**Fig. 4.1:** Schematic representation of water dissociation on bipolar membrane.

Amphoteric ion-selective membranes contain weakly acidic and weakly basic functional groups that are statistically distributed throughout the membrane matrix. Such materials with carboxyl and amine groups determine the interactions between bound polyampholyte groups and counter-ions in the surrounding solution, thereby influencing and controlling its pH. Amphoteric membranes do not belong yet among the materials used and it is expected that they will be used in future due to their ability to control the charge level in the system by changing the pH of the surrounding solution in the preparation of anti-fouling materials, preventing sorption of organic substances and biological macromolecules on their surface. Application areas are membranes for haemodialysis, separation of ionogenic drugs and proteins in general or membranes for piezodialysis, allowing effective desalination of solutions using high-quality membranes.

Mosaic ion-selective membranes differ from amphoteric by their geometrically defined arrangement of cation- and anion-functional groups in the membrane matrix structure. The regular parallel alternation of groups forms a continuous transport path of their counter-ions through the membrane, and if the electrolyte solutions create a concentration gradient on the membrane, parallel transport of anions and cations can occur, leading to a current circuit between ion-exchange elements. The influence of the flowing current results in a significantly higher permeability of the mosaic membrane for salts that for non-electrolytes. The preparation of thin mosaic layer with the optimum size of delimited functional domains is a way to solve complex separation problems of separating water-soluble organic substances from salts.

## 4.2.2 Material and structural characteristics of ion-selective membranes

The second key parameter of the preparation of ion-selective membranes is the basic material, which mainly determines the physical properties of membrane, mechanical and thermal stability as well as chemical resistance. Based on this criterion, membranes are divided into membranes of organic carbon polymer, possibly partially halogenated, membranes of perfluorinated polymers, membranes of inorganic materials and hybrid membranes of inorganic ion-selective materials and organic polymers.

In the preparation of membranes, this material composition creates their characteristic structure, which can be homogeneous or heterogeneous, Fig. 4.2. In the homogeneous structure, the bound functional groups are statistically distributed in the polymer matrix, which is prepared by polymerization or polycondensation of a functionalized monomer or by polymer-analogous reaction (e.g. sulphonation) of a suitable polymer. The heterogeneous system contains macroscopic clusters (particles) with ion-exchange groups distributed in an inert non-charged polymer and it can be prepared by plastic technology from a polymer granulate or by casting from a dispersion of ion-exchange particles in a carrier polymer solution.

These two macroscopic structures are boundary variants because most membranes exhibit macroscopic heterogeneity occurring in systems of polymer blends of miscible or limitedly miscible polymers, Fig. 4.3.

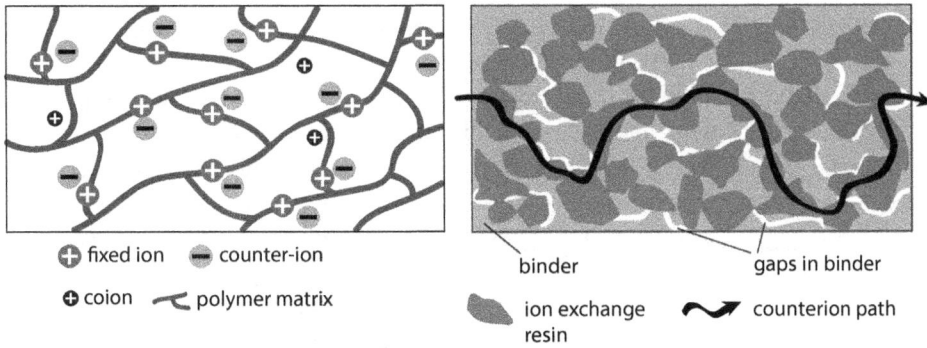

fixed ion  counter-ion

coion  polymer matrix

binder          gaps in binder

ion exchange      counterion path
resin

Fig. 4.2: Models of homogeneous (left) and heterogeneous (right) ion-selective membranes.

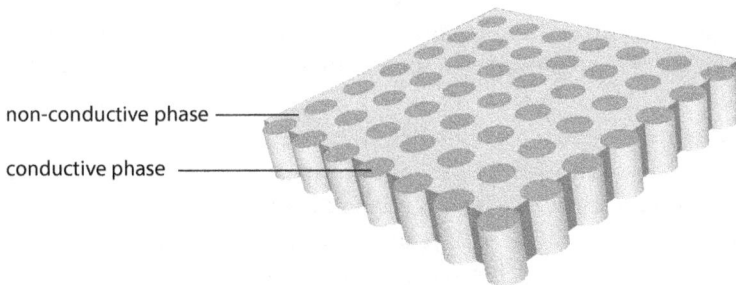

non-conductive phase —————

conductive phase —————

Fig. 4.3: Model of micro-heterogeneous ion-selective membrane.

From the structural point of view, membranes are divided into homogeneous ion-selective membranes, inter-polymer membranes, micro-heterogeneous grafted membranes, block copolymer membranes and heterogeneous membranes. These three transitional membrane variants belong in the problem of micro-heterogeneous polymer systems and polymer blends and are a promising area for development of new, higher-quality membranes. Inter-polymer membranes are prepared by casting from a blend of ionogenic and non-ionogenic polymer in a common solvent, that is, a homogeneous and macroscopically transparent solution. Block copolymers contain polymer domains from different monomers, which lead to their micro-heterogeneous arrangement in various geometric compositions. This type of membranes is subject to intensive research but is not on the market yet. If membranes are homogeneous or contain micro-heterogeneities smaller than 400 nm, they appear transparent; larger heterogeneities cause membrane opacity.

### 4.2.3 Preparation of membrane composites

All procedures for preparation and production of ion-selective membranes are based on the properties of ionogenic materials. These are high water sorption, depending on the pH and ionic strength of the solution, and fragility of the initial material. Both laboratory and industrial preparation of homogeneous membranes include conventional techniques of polymer processing, such as casting membranes from polymer solution and evaporation of the solvent, and hot working, mainly extrusion and pressing. Membranes of all degrees of heterogeneity (micro-heterogeneous, macro-heterogeneous) can be prepared in the laboratory as well as industrially similarly to homogeneous membranes by casting from polymer solution and evaporation. With respect to the low adhesion of the heterogeneous membrane phases and dimensional changes at the solvent evaporation from the polymer matrix, heterogeneous membranes are produced by thermoplastic methods of extrusion and pressing. The preparation of homogeneous and micro-heterogeneous materials of ion-selective membranes is done in two ways, either by direct polymerization of selected monomer or by introducing ion-exchange groups into the polymer film.

A typical example of a polymer cross-linked structure leading to both ion-exchange modifications is the styrene copolymer with divinylbenzene, the most widely used material for the preparation of membranes and column fillings. The preparation procedure is shown in Figs. 4.4 and 4.5. Radical copolymerization of styrene with divinylbenzene generates a spatial 3D matrix whose swelling capacity in a suitable solvent is controlled by the input ratio of reactants. Subsequent polymer-analogous reaction with acid (sulphuric acid, chlorosulphuric acid) creates a cation-exchange structure typical for cation-selective membranes. Another polymer-analogous reaction creates chloromethylene derivatives of the basic copolymer, and by reaction with a suitable amine (trimethylamine), anion-exchange structure of the anion-selective membrane is obtained. This scheme is generally applicable in various variants for the preparation of ion-selective materials from different polymer structures.

Functionalized monomers and their polymerization have undergone many years of development, and in recent years, systems have been created with the prospect of transfer to production technology. These include polymers of the type of modified polyethersulphones, polyetherketones or aromatic polymers of the structures shown in Fig. 4.6.

A typical method of the preparation of homogeneous membranes is the introduction of functional groups into suitable polymer films from polyolefins. The polymer-analogous reaction of the preparation of cation-selective membranes is the introduction of sulpho- group by reaction in the gaseous state with sulphur dioxide and chlorine, catalysed by UV radiation, and subsequent conversion to sulpho- group by reaction with sodium hydroxide. The sulphonation in solution is carried out by direct reaction with chlorosulphuric acid, in some cases also with concentrated sulphuric acid

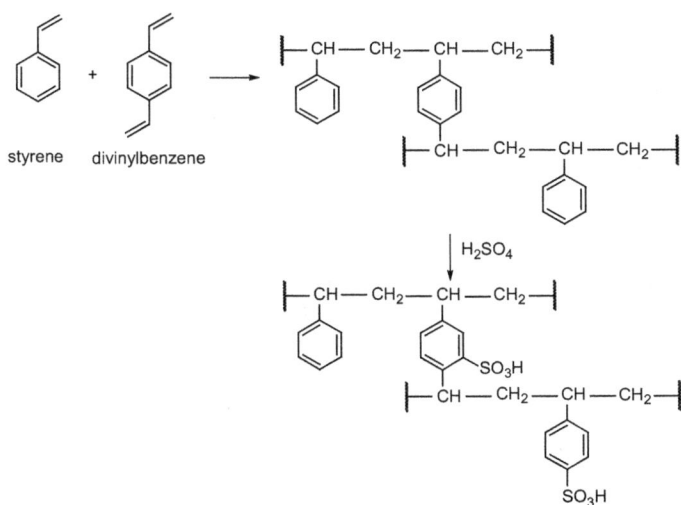

**Fig. 4.4:** Synthesis of cation-exchange material (cation exchanger) based on polystyrene cross-linked with divinylbenzene with subsequent sulphonation.

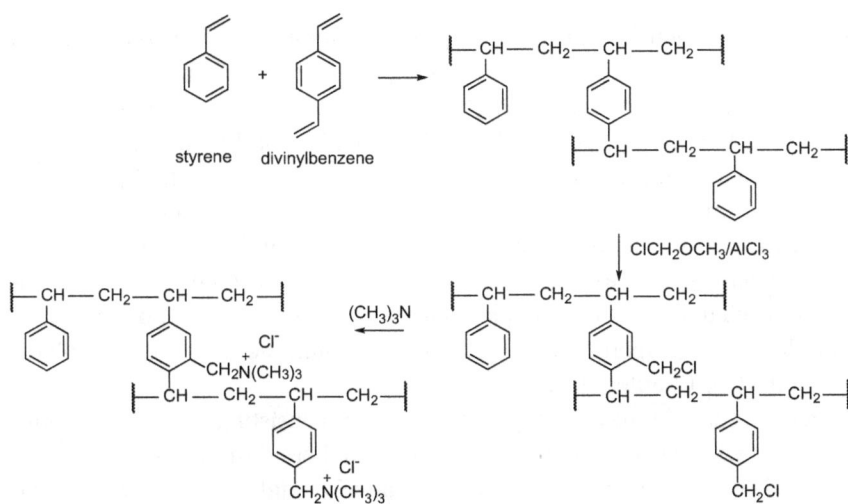

**Fig. 4.5:** Synthesis of anion-exchange material (anion exchanger) based on polystyrene cross-linked with divinylbenzene and subsequent chloromethylation and quarterization with trimethylamine.

(e.g. polyetheretherketones). Anion-selective membranes are prepared mainly from amine derivatives in the quaternized form. Amine bound by amidic bond is produced by polymer-analogous reaction on aliphatic polymer chains and it is quaternized with alkyl halide. The polymer chain with aromatic elements in the structure is usually subject to halogen methylation and subsequent quaternization with a tertiary amine.

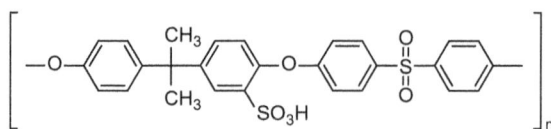

Structural unit of sulphonated poly(ether-sulphon)

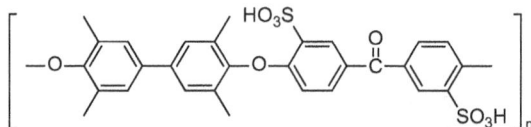

Structural unit of sulphonated poly(ether-ether-ketone)

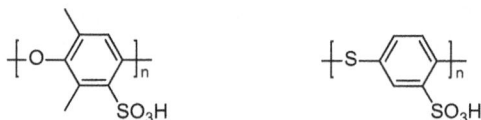

Structural units of sulphonated poly(2,6-dimethylphenyleneoxide) and poly(phenylene sulphide)

**Fig. 4.6:** Structure of selected ion-exchange materials.

The structure of homogeneous membranes produced from these materials respects their mechanical and chemical properties, requiring in some cases reinforcing textiles.

An independent group consists of highly resistant perfluorinated polymers and resistant cation-active membranes made from them with ion-functional groups of sulpho- or carboxylic acids on the basic polytetrafluorethylene. The first product of this kind was Nafion® membrane (Du Pont) and later Flemion® and Neosepta F® (Asahi Chemical, Tokuyama Soda). Membranes are used mainly in membrane electrolysers but sometimes also as the border membrane in electrodialysis stacks (separating electrode chambers), that is in places requiring high chemical and increased thermal stability. Such properties allow using these membranes as solid electrolytes in power systems (see Chapters 13 and 14).

Heterogeneous membranes are polymer composites consisting of particles of ion-selective polymer with the size usually less than 50 µm bound in an inert polyolefin, polyvinylchloride or phenolic resin. Grinding a relatively tough ion-selective material requires much energy, and experience also shows that ion exchanger can only be effectively ground in a certain ion form, which prevents corrosion of machinery as well. The amount of ion-selective polymer in the blend must be at least 50 mass% to form a conductive channel through the entire thickness of the membrane so that membranes have a sufficiently low working electrical resistance. It has been proven many times that finer distribution of the ion-selective polymer leads to membranes with lower electrical resistances. This fact also has its limitations, and very fine ion exchanger particles cannot be well dispersed into the polymer matrix. Such composite membranes have insufficient mechanical properties and their design often required

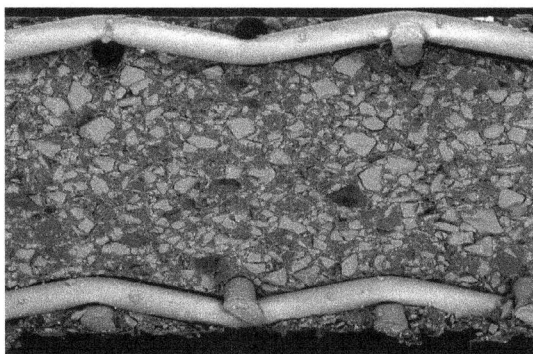

**Fig. 4.7:** Structure of reinforced heterogeneous ion-selective membrane (photo MemBrain).

reinforcing with a suitable textile, Fig. 4.7. The comparison of water content in a swollen heterogeneous membrane with a homogeneous membrane with the same ion-exchange capacity often shows higher swelling capacity of the heterogeneous system. Creating spaces between phases on the contact of the binder and ionic particles not only influences negatively the mechanical stability of the membrane but also decreases its permselectivity due to the transport of co-ions through these channels formed in the matrix. Repressing the membrane at a temperature above the melting point of the polymeric binder leads, in most cases, to improving the electrochemical properties of the membrane, but this operation demands much energy and time and it substantially influences the overall economy of the forming process of heterogeneous ion-selective membranes.

Experience with the discontinuous production of heterogeneous ion-selective membranes, applied mainly in the past, gradually led to the continuous production of heterogeneous membranes in the form of an endless belt. Continuous lamination made the process of membrane production significantly more effective. The production of heterogeneous membranes with a binder based on a thermoplastic polymer has a lesser environmental impact compared to homogeneous membranes, mainly due to less liquid toxic wastes and no contamination of the operating atmosphere.

Another advantage of heterogeneous ion-selective membranes is the possibility to store them in a dry inactive state for a relatively long time and their generally greater resistance to membrane poisons.

Membranes containing both types of ion-selective groups require special approach to their preparation. It can be said that the development of technologies using bipolar membranes is still limited by the unavailability of suitable membranes with high operational quality. They should exhibit low electrical resistance, high selectivity for counter-ions, that is, complete elimination of co-ions in monopolar layers, high stability at extreme pH values, high capacity of water dissociation and sufficient permeability of ionogenic layers for water. Several technological procedures are utilized to prepare bipolar membranes. The simplest is the lamination of cation- and anion-selective membranes at laboratory by pressure or by higher temperature with a catalytically active component attached to one of them or with an independent interlayer of catalytically active component. Casting technologies utilize a solution of one of the ion-selective components containing the catalytically active component onto a membrane underlay with an opposite charge. Direct technology is the co-extrusion of cation- and anion-active polymers, one of which contains a catalytically active component. The decisive parameter for quality membrane function is the contact area at the interface of monopolar membranes, that is, the place where water dissociation takes place. Weak acids or bases like carboxylic acid, phenolic and phosphorus derivatives, pyridine, amine derivatives and also heavy metal ion complexes of zirconium, chromium, indium or ruthenium have the best catalytic activity of the process. The search for suitable combinations of components and bipolar membrane designs is subject of many studies because the preparation of large-format products with long-time work stability and optimum parameters is still a problem. The solution could be hybrid bipolar membranes.

Hybrid (inorganic-organic) membranes gain importance, thanks to their extraordinary properties, primarily in unconventional applications of ion-selective membranes. These materials exhibit significantly modified mechanical, thermal, electrical and magnetic properties compared to the original components. Inorganic materials bring mechanical and thermal stability, new magnetic and dielectric properties, whereas organic components add structural flexibility, acceptable workability, possible photoconductivity and effective luminescence. The first hybrid membranes, prepared as physical composites by dispersing inorganic particles in an organic matrix, did not meet these advantages of hybrid membranes but at present, several methods are described leading to hybrid structures at the molecular level. The most common is to include the sol-gel process or liquid reaction procedure, using silane agents bound to the silica skeleton to create insoluble hybrid membranes. Similar procedures have been proposed for the synthesis of many organic precursors bound by condensation reactions to hydrolysed metal oxides. The morphology and resulting membrane properties of hybrids are a function of the precursors used and reaction conditions.

Amphoteric ion-selective membranes still do not belong among widespread and used types. The preparation of purely synthetic membranes based on copolymers of weak carboxylic acids and amine derivatives has been described, examples are copolymers of methacrylic acid, dimethylaminopropylacrylamide, butyl methacrylate and 2,3-epithiopropylmethacrylate. Other reported amphoteric membranes have derivatives of natural materials like chitosan or cysteine in their structure. A feature of such membranes is the fact that they act like cation-selective or anion-selective membranes depending on the pH of the processed solution. It has been found that the prevailing charge of the amphoteric membrane determines dissociation of ion-exchange groups, as well as ion sorption in the membrane [4, 5]. Several types of amphoteric ion-selective membranes have been synthetized on the basis of polyfluorinated ether-ketone, on which sulphone and quaternary ammonium groups were bound. These membranes were intended directly for flow-through vanadium batteries, in which they exhibited the selectivity higher by an order of magnitude than Nafion® and related suppression of the permeability of $VO^{2+}$ ions [6]. More applications for this type of membrane are likely to be found in the future in the fields of colloid chemistry, pharmacy and medicine, for example, for selective sorption of proteins depending on the pH of the solution [7].

Neither mosaic ion-selective membranes belong to the current manufacturing and application sphere. Membranes with well-defined structural domain of cation- and anion-exchange groups have been prepared from block copolymers of isoprene, styrene and their functionalized derivatives. Other possibilities are in the preparation of polymer blend based on polysulphone and polyphenyleneoxide or derivatives of styrene and acrylonitrile–styrene copolymer. At present, the technologies of 3D printing are also used for the development of these membranes. Separation procedures using these membranes are considered rather virtual and the applications most cited in literature are desalination of solutions containing both dissolved inorganic salts and dissolved organic substances [4].

The structural and functional properties of ion-selective membranes significantly influence their possible application. The main studied parameters are the ion selectivity and semi-permeability of the membrane, controlled by macro- and microstructure of the membrane matrix and also by the bound functional group. The architecture of the membrane polymer matrix utilizes a wide range of synthetic procedures to form channels with high permeability for transported ions. Polymers with functional groups on the main chain change to block copolymers whose components have different polarities and functional groups on branched network are replaced by atypical ions or groups.

## 4.3 Technology of membrane production

An ion-selective membrane produced for the use in electromembrane process must have parameters leading to its efficient operation. These are:

**High selective permeability**. As a result, some substances are excluded from the transfer, others are relatively slowed down and preferred substances should pass very easily. Therefore, the membrane should be very highly permeable for counter-ions but practically impermeable for co-ions. The decisive parameter for the production of a membrane close to this ideal state is its chemical functionality and transport characteristics defined by the transport number of the relevant ion. The conversion numbers of counter-ions stated for manufactured membranes under defined conditions are between 0.90 and 0.98.

**Low electrical resistance**. Since the electric charge is mainly transported by corresponding counter-ions, the membrane should be well conductive. High area resistance $R_A$ (unit: $\Omega$ m$^2$) or resistivity $R_S$ (unit: $\Omega$ m) causes a voltage loss which substantially influences the electrochemical yields of electromembrane processes. The resulting membrane resistance depends primarily on the concentration of the fixed ion-exchange groups and the thickness of the membrane. The area resistance of membranes intended for water desalination must be as low as possible to make the process economically viable and should be between 0.5 and 3 $\Omega$ cm$^2$. For desalination of low-conductive solutions, such as whey, membrane resistances could be higher, 6–10 $\Omega$ cm. Cataphoretic coating works with low-conductive solutions and membrane resistances for this process could be in the range between 15 and 20 $\Omega$ cm at high selectivity.

**Good mechanical properties**. The membrane should exhibit high mechanical resistance and dimensional stability which is characterized primarily by the cross-linking degree of the polymer structure, essentially the water content in the membrane, that is the swelling degree. In the case of heterogeneous membrane, the so-called reinforcing textile is also pressed in to improve mechanical properties. The values of the tensile strength and tear strength of the membrane are also studied in the wet, that is, swollen state.

**Chemical stability**. The membrane should be chemically stable in various aggressive environments (acidic, alkaline) and also as inert as possible to various surface-active substances and so-called membrane poisons. Degradation of the ion-selective membrane results in a change of electrical resistance, decrease of permselectivity and loss of mechanical properties. The polymer membrane can be degraded by various mechanisms, generally typical for polymer materials, such as thermo-oxidation, hydrolysis, biodegradation and chemical attack.

It must be noted that it is very difficult to produce an optimum ion-selective membrane with all the discussed properties because some requirements act in opposition. For example, a high degree of polymer cross-linking or greater membrane thickness significantly improves the mechanical strength of the membrane but at the same time they cause an increase of electrical resistance. Conversely, high concentration of fixed ion-exchange groups leads to low electric resistance but also causes a high degree of membrane swelling in combination with low dimensional stability and reduced permselectivity.

The current production of ion-selective membranes can be divided into two parts. Older types, including heterogeneous membranes, are produced and used by many manufacturers for applications where special electrochemical properties and high ion selectivity are not emphasized, which can be an advantage in some processed systems (less susceptibility to attacks by so-called membrane poisons, for example, in waste multicomponent systems, which limits their life cycle). On the contrary, higher resistance of these membranes in handling is required with respect to their greater thickness (0.2–0.5 mm), compared to homogeneous membranes, as well as using various reinforcing textiles [8].

Efficient membranes for qualified technological processes, especially water desalination in production of salts or pure water, are manufactured by Japanese companies Asahi Glass and ASTOM (formerly Tokuyama Soda), General Electric (formerly Ionics) in the USA and Fumatech (Germany). The common basis of these membranes is a cross-linked polymer skeleton on the basis of vinyl monomers, mainly ethylene, styrene, polyvinylchloride and divinylbenzene. The functional groups of these membranes are obtained either by direct polymer-analogous reactions (sulphonation) or by addition of functional or pre-functional co-monomers to polymerization blends and subsequent polymer-analogous modification. These membranes generally have a low area electrical resistance (usually at the thickness of 0.15–0.20 mm), high permselectivity (some of them are capable of separating monovalent ions from polyvalent ions), sufficient mechanical strength and long-time resistance against chemical action of acid solutions, alkalis and mild oxidizing agents. Since the 1980s, the company MEGA from the Czech Republic (formerly Czechoslovakia) has been one of the traditional producers, exporting products worldwide under the tradename of RALEX®. The others include, for example, Membranes International (USA) and recently also Chinese companies (Hangzhou Iontech). Heterogeneous ion-selective membranes are particularly useful in the food industry, especially in desalination of whey.

The basic technological procedure of the production of heterogeneous ion-selective membranes is either solvent-free or by casting technique from the blend of the basic matrix solution and ion-exchange component. The solvent-free system consists in the use of plastic-processing technologies, that is, mixing of melts and the extrusion technique. The membrane composite is produced by mixing the powder ion exchanger and particles of the polymer carrier in the melt. The resulting mixture passes through the extrusion line, either directly or in the granule form, and is extruded in the form of a foil, possibly reinforced with a suitable textile and protected by a separation layer. A typical product is a foil with the thickness of 200–500 μm, which is formatted to the required membrane size, and the functional groups are activated chemically.

Heterogeneous ion-selective membranes can also be produced by the method of casting from the polymer solution. In this procedure, the polymer forming the membrane matrix is dissolved first in a suitable solvent. A typical example is polyvinyl chloride dissolved in tetrahydrofuran. A powder ion exchanger of a suitable graininess resin is mixed into this solution. The resulting dispersion is cast into the membrane shape and the solvent is evaporated. The advantage of this process is less thermo-oxidative stress of the ion-exchange phase compared to melt processing; the disadvantage is the work with volatile solvents (explosive environment, waste management).

A similar method of the production of heterogeneous ion-selective membranes is the preparation of a suspension of finely ground ion exchanger in rubber or latex.

After shaping the suspension into a flat membrane, the rubber is vulcanized with a suitable agent. The rubber can be silicone or butadiene styrene. This production method is not used on an industrial scale at present but it is still frequent in the laboratory preparation of model ion-selective membranes.

The technology of casting and possible subsequent chemical reaction leading to the membrane foil formation is used in the production of heterogeneous membranes. The definition of membrane homogeneity is currently evaluated according to the micro-heterogeneity of the final membrane structure. It depends on the compatibility of the ion-exchange and inert parts in the membrane composition. A typical method to produce homogeneous or pseudo-homogeneous membranes is the casting technology. The components of the membrane polymer composite are dissolved in a strongly polar solvent and the resulting solution is cast onto a horizontal surface and then the membrane is formed as a foil, by evaporating the solvent and possibly adding reinforcement. An important factor influencing the application properties of membranes is the chemical composition of casting solutions. The simplest combination is a solution containing both functionalized and ion-inert component. Another type is a blend of two or more non-functionalized polymers, one of which contains a precursor of the functional group, generated after the formation of the membrane foil. Another variant is to use a solution of the functionalized monomer or polymer and to form the membrane composite during casting. This relatively wide range of material combinations of the resulting membrane composition regulate its application and mainly transport characteristics. The most important ones include the ion-exchange capacity and swelling capacity in the aqueous environment, others are chemical and mechanical stability of the membrane. The advantage of this technology is the variability of the material composition of the produced ion-selective membranes.

New approaches to optimize the separation properties of membranes by modifying their internal microstructure require new targeted production technologies. Solvent-free systems can be mentioned, based on monomer blends, capable of rapid polymerization reaction during the preparation of a membrane foil, or membranes from porous materials whose pores are filled with an ion-exchange component.

## 4.4 Development directions of membranes for targeted applications

Electromembrane processes are based on the fifty-year tradition of using ion-selective membranes and they prefer a minimum possible area resistance and high permselectivity. These requirements are motivated by the main field of the use in electrodialysis separation processes to process various types of water. The future utilization of membranes is directed to qualitatively higher separation processes in production technologies, combined use in separation and synthetic procedures including membrane reactors and also power applications in energy storage in batteries or conversion of electrical energy to chemical and vice versa. For these reasons, ion-exchange materials are synthetized today with properties required for particular applications or for creating a particular membrane microstructure [9]. There are many possibilities on how to design polymer ion-exchange material; the basic polymer chain can be linear, branched or grafted, functional groups can be bound to the basic chain randomly or in blocks, they can be on side chains only or they can be bound to the main chain, with side chains sterically protecting them. By combination of the above principles, a very varied architecture of ion-selective membranes [9, 10] can be achieved; membranes have been synthetized with hydrophilic main chain containing functional groups and grafted hydrophobic chains from fluorinated hydrocarbons [10], intended for fuel cells. Macroscopically, they are homogeneous membranes because they are made from one material. The different chemical nature of blocked or grafted copolymer causes self-arrangement of polymer chains and the resulting membrane has domains that lead or adsorb ions and domains that do not react with ions. This type of membranes is called micro-heterogeneous (Fig. 4.8).

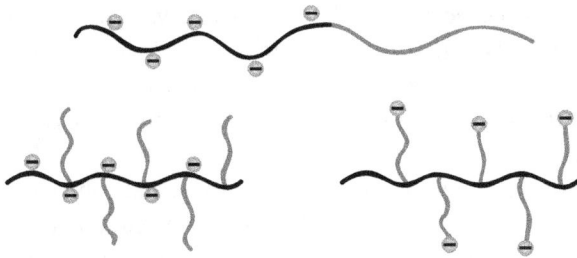

**Fig. 4.8:** Examples of different architecture of polymer material for cation-selective membranes.

In the last decade, the development of new materials and design modifications of membranes has been evident, based on the interdisciplinary approach and using new knowledge of material science. Research procedures are limited by the problems of long-time working stability of membranes, by the solution of the dependence of electrochemical properties on their structure and, last but not least, by the economic parameters of the new application processes. The study of inorganic–organic, so-called hybrid materials, prepared primarily by the sol–gel process at acceptable reaction conditions of the input materials has become a progressive direction. The wide range of silicone and silane derivatives allows extensive structural variations in the preparation of materials and their subsequent modifications to ion-selective membranes, exhibiting specific surface properties, especially reducing the risk of membrane fouling and increasing thermal and chemical resistance [10].

In recent years, much attention has been paid to the modification of both homogeneous and heterogeneous ion-selective membranes by nano- and microparticles. The motivation is often to improve mechanical properties or to influence physical properties, especially to reduce swelling. This can be achieved relatively easily by filling membranes with short inorganic fibres, carbon or glass [11, 12]. Much attention is also paid to nanoparticles of $TiO_2$ and silver due to their antibacterial effects [13–15]. Authors often highlight the advantageous properties of these membranes but the stability of these heterogeneous nanocomposite systems is less dealt with. In most cases, nanoparticles are not covalently bound to the membrane structure and they are eroded at least from the membrane surface layer by the flowing liquid. In general, the antibacterial effect of colloidal silver is overestimated in membrane technologies. Ion-selective membranes treated with small amount (1–8 mass%) of ferrous ferric oxide exhibit improved physical–chemical properties and improved selectivity for barium and lead ions [14]. The filler in the form of nanoparticles or fibres often positively influences the pursued properties, such as resistances, selectivity and swelling capacity up to certain content in the membrane only because the membrane properties deteriorate after exceeding the optimum amount.

In the field of the development of new materials, attention is still paid to the development of bipolar membranes, especially to their catalytic systems, which leads to decreasing the energy demands of their use and increasing chemical resistance in the preparation of acids and bases from their salts. The development is focused to the use of biopolymers from natural resources, whose structure contains weak basic and carboxyl groups, or catalysts from multivalent metal oxides.

A closely studied and large group of membranes is intended for use as solid electrolytes in fuel cells or separators in battery assemblies. Perfluorinated cation-active membranes do not meet the strict conditions for optimum activity of solid electrolyte, especially due to their low thermal stability, and therefore new materials are being developed on the basis of mixtures of chemically resistant polymers with ion-active oligomers (ionic liquids) or ion-conductive low-molecular compounds [16].

Besides completely new types of membranes, the properties of the existing composites are optimized to ensure their specific separation applications. Systems capable of separating monovalent from multivalent ions are studied, as well as modifications of the membrane surface with a defined surface hydrophilicity. Design modifications of profiled membranes exhibiting electrochemical properties applicable in separations processes with higher current efficiency or in power procedures of reverse electrodialysis are also described. These procedures, leading to optimized ion-selective membranes, require simultaneous study of their physical parameters and evaluation of the association between the membrane microstructure and its transport properties.

There are limiting factors that prevent the use of polymer ion-selective membranes in special technologies, such as processing of very acidic, strongly oxidizing or radioactive solutions. In such cases, the life cycle of the membrane can be very short and their frequent replacement increases the operating costs of electrodialysis significantly.

A logical step is to focus the membrane development on the ceramic materials with ion-exchange properties. Since at least the nineteenth century, it has been known that some sort of clays have ion-exchange properties. Natural aluminosilicate materials, such as montmorillonite or halloysite, are ion-exchangers, as well as natural or synthetic zeolites, and can be incorporated into a polymer or ceramic matrix and shaped into a flat membrane. There is a group of material called NASICON with the summary composition $Na_{1+x}Zr_2Si_xP_{3-x}O_{12}$, where $0 < x < 3$, which can also be shaped into thin flat desks and serve as membranes [17]. The material is highly selective for the transport of sodium ions and can be used as a solid electrolyte in a sulphur–sodium accumulator or as a membrane to remove sodium from highly alkaline radioactive aqueous solutions.

Acid salts of tetravalent metals like Zr, Ti and Ce exhibit ion-exchange properties, typically represented by α phase of zirconium hydrogen phosphate $Zr(HPO_4)_2 \cdot H_2O$, which has a layered structure. This compound is used to impregnate a porous ceramic carrier in the form of a thin plate, and after thermal fixation, a ceramic cation-selective membrane is formed [18]. An obvious property, greatly differentiating ceramic and polymer ion-selective membranes, is zero dimensional change with water absorption and ionic form change.

# References

[1]  M. Mulder, Basic Principles of Membrane Technology. 2nd Edition. Kluwer Academic Publishers, 1996. ISBN 079234247X.
[2]  T. Sata, Recent trends in ion-exchange membrane research. Pure Appl. Chem. 58: 1613–1625, 1986.
[3]  T. Sata, Ion Exchange Membranes. Preparation, Characterization, Modification and Application. RSC Advancing the Chemical Science, 2004.
[4]  T. Xu, Ion exchange membranes: State of their development and perspective. J. Membr. Sci. 263: 1–29, 2005.
[5]  R. Takagi and M. Nakagaki, Ionic dialysis through amphoteric membranes. Sep. Purif. Technol. 32: 65–71, 2003.
[6]  Y. Wang, et al. Science direct amphoteric ion exchange membrane synthesized by direct polymerization for vanadium redox flow battery application. Int. J. Hydrogen Energy 1–9, 2014.
[7]  H. Matsumoto, Y. Koyama, and A. Tanioka, Interaction of proteins with weak amphoteric charged membrane surfaces: Effect of pH. J. Colloid Interface Sci. 264: 82–88, 2003.
[8]  K. Scott, Ed., Handbook of Industrial Membranes. Elsevier Advance Technology, UK, 1995. ISBN 1856172333.
[9]  J. Ran, et al., Ion exchange membranes: New developments and applications. J. Membr. Sci. 522: 267–291, 2017.
[10]  M. Ingratta, E. P. Jutemar, and P. Jannasch, synthesis, nanostructures and properties of sulfonated Poly(phenylene oxide) bearing polyfluorostyrene side chains as proton conducting membranes. Macromolecules 44: 2074–2083, 2011.

[11]  R. Válek and J. Zachovalová, Cation-exchange membrane modified by inorganic short fibres. Desalin. Water Treat. 56(12): 3233–3237, 2015.

[12]  J. Křivčík, D. Neděla, and R. Válek, Ion-exchange membrane reinforcing. Desalin. Water Treat. 56(12): 3214–3219, 2015.

[13]  S. M. Hosseini, M. Nemati, F. Jeddi, E. Salehi, A. R. Khodabakhshi, and S. S. Madaeni, Fabrication of mixed matrix heterogeneous cation exchange membrane modified by titanium dioxide nanoparticles: Mono/Bivalent ionic transport property in desalination. DES 359: 167–175, 2015.

[14]  S. M. Hosseini, M. Askari, P. Koranian, S. S. Madaeni, and A. R. Moghadassi, Fabrication and electrochemical characterization of PVC based electrodialysis heterogeneous ion exchange membranes filled with $Fe_3O_4$ nanoparticles. J. Ind. Eng. Chem. 20(4): 2510–2520, 2014.

[15]  M. Zarrinkhameh, A. Zendehnam, and S. Hosseini, Preparation and characterization of nanocomposite heterogeneous cation exchange membranes modified by silver nanoparticles. Korean J. Chem. Eng. 31(4): 1187–1193, 2014.

[16]  C. Wu, T. Xu, and W. Yang, A new inorganic-organic negatively charged membrane: Membrane preparation and characterizations. J. Membr. Sci. 224: 117–125, 2003.

[17]  J. Fergus, Ion transport in sodium ion conducting solid electrolytes. Solid State Ion. 227: 102–112, 2012.

[18]  V. M. Linkov and V. N. Belyakov, Novel ceramic membranes for electrodialysis. Separ. Purif. Technol. 25: 57–63, 2001.

Miroslav Bleha, Milan Šípek, Dalimil Šnita

# 5 Transport processes

The theory of transport in solutions of electrolytes was introduced in Section 3.4.2. In this chapter, we describe the transport of particles in electrolyte solutions (Section 5.2) and the transport through membranes (Section 5.4).

## 5.1 Theory of transport

The process in which the value of a monitored variable at a certain place of the system changes with time is called a **transport process**. This process can also be defined as the transfer of system components in space; such system components can be, for example, the amount of substance, mass or energy [1].

The necessary condition for the transport process is an existence of the driving force, which is the gradient of a given intensive quantity. For example, transfer of substance occurs by diffusion, where the driving force is the chemical potential gradient, or by heat transfer (heat flux), where the driving force is the temperature gradient. The transport itself is usually an equilibrium process proceeding until equilibrium is established, or a non-equilibrium process until a stationary state is reached.

For mass transport, the mass flux per area unit is defined by the relation

$$J_m = \frac{1}{A}\left(\frac{dm}{d\tau}\right) \tag{5.1}$$

where $A$ is the area, $m$ is the mass passing through the area and $\tau$ is the time. The mass flow per area unit is called the **intensity of mass flux** and its dimension is $kg\,m^2\,s^{-1}$.

For the transport of the amount of substance per area unit, it similarly applies

$$J_n = \frac{1}{A}\left(\frac{dn}{d\tau}\right) \tag{5.2}$$

where $n$ is the amount of substance. The transport of the amount of the substance is often called the **intensity of molar flux**, and its dimension is $mol\,m^{-2}\,s^{-1}$. (Note: In physical chemistry, the flow of a given quantity per area unit is sometimes called density of molar flux.)

Transport can also be related to the substance volume. **Intensity of volume flux** $J_V$ describes the volume of substance passed through the unit membrane area per unit of time, and its dimension is $m^3\,m^{-2}\,s^{-1}$:

$$J_V = \frac{1}{A}\left(\frac{dV}{d\tau}\right) \tag{5.3}$$

https://doi.org/10.1515/9783110739466-006

In the case of gases and vapours, the volume of substance must be considered under standard conditions, that is, under standard temperature and standard pressure.

# 5.2 Transport of particles in electrolyte solutions

An essential condition limiting the transport of substances in electrolyte solutions is the requirement of **electroneutrality**, determining balanced positive and negative charges in the system. If cations and anions are considered as independent components of transport, their flow is conducted in a manner not allowing the accumulation of positive or negative charge in the macroscopic interpretation of the system.

The transport of particles in electrolyte solutions can be described in several ways, for example, by
- Nernst–Planck equation (it is a phenomenological equation),
- other phenomenological equations and
- equations based on the Maxwell–Stefan theory.

## 5.2.1 Nernst–Planck equation

This theory deals with **ion transport in electrolyte solutions**. In general, the rate of the transport process (ion diffusion, ion migration and ion convection) is proportional to the driving force of the transport process and the amount of particles on which the force acts.

The driving force in electrolyte solutions is the **electrochemical potential gradient**, defined by relation (2.55). With respect to relation (2.56), the intensity of molar flux is proportional to the amount of particles, expressed by their concentration $c_i$, their mobility $u_i$ and by the driving force in the direction of the decreasing gradient of electrochemical potential:

$$J_i = - u_i c_i \mathbf{R} T \, \nabla \ln a_i - u_i c_i z_i \, \mathbf{F} \, \nabla \varphi \quad [T,p] \tag{5.4}$$

In this equation, the mobility $u_i$ has the dimension of m mol s$^{-1}$ N$^{-1}$, the molar gas constant $\mathbf{R}$ has the value of 8.314 J K$^{-1}$ mol$^{-1}$, $T$ is the absolute (thermodynamic) temperature (dimension K), $z_i$ denotes the ion charge number (dimensionless), and $\varphi$ is the electrical potential (dimension V). The first term of the right side of this relation corresponds to the diffusion of ions, and the second to their migration.

Further in the text of this chapter, only the transport in the direction of $x$-axis is considered for the sake of simplicity. The one-dimensional version of eq. (5.4) is

$$J_i = - u_i c_i \mathbf{R} T \frac{\mathrm{d} \ln a_i}{\mathrm{d}x} - u_i c_i z_i \mathbf{F} \frac{\mathrm{d}\varphi}{\mathrm{d}x} \quad [T,p] \tag{5.5}$$

According to relations (2.54) and (5.5), the intensity of molar flux of the $i$-th component during diffusion is defined by the relation

$$J_{i,\text{dif}} = -u_i c_i RT \frac{d \ln a_i}{dx} = -u_i c_i \frac{RT}{a_i} \frac{da_i}{dx} \qquad [T,p] \qquad (5.6)$$

With respect to relation $a_i = y_i\, c_i$ – see eq. (2.54) – it applies that

$$J_{i,\text{dif}} = -u_i c_i \frac{RT}{c_i y_i} \left[ y_i \frac{dc_i}{dx} + c_i \frac{dy_i}{dx} \right] \qquad [T,p] \qquad (5.7)$$

Relation (5.7) can be modified in the final form

$$J_{i,\text{dif}} = -D_i \frac{dc_i}{dx} \left[ 1 + \frac{d \ln y_i}{d \ln c_i} \right] \qquad [T,p] \qquad (5.8)$$

where the thermodynamic diffusion coefficient $D_i$ is defined by the relation

$$D_i = RTu_i \qquad (5.9)$$

In this equation, $D_i$ has the dimension $m^2\, s^{-1}$. In the case of very dilute solutions, their ideal behaviour can be expected, with $a_i = c_i$ and $y_i$ approximately equal to 1. In this case, it is possible to write in correspondence with relation (5.8):

$$J_{i,\text{dif}} = -D_i \frac{dc_i}{dx} \qquad [T,p] \qquad (5.10)$$

which is **Fick's first law** for diffusion with a constant diffusion coefficient in the $x$-axis direction. It must be emphasized that in general, beyond ideal (i.e. very dilute) solution, the diffusion coefficient is a function of concentration because the concentration changes during diffusion. This fact is expressed by **Fick's second law of diffusion.**

The **electrolytic mobility**[1] of ions $U_i$ expresses the velocity of ions, which is related to the unit of electric field (V m$^{-1}$) and has the dimension $m^2\, V^{-1}\, s^{-1}$. This electrolytic mobility $U_i$ is related to the mobility $u_i$ (see Section 3.4.2) by the relation

$$U_i = z_i F u_i \qquad (5.11)$$

where $F$ is the Faraday constant, $F = 95{,}485$ C mol$^{-1}$. With respect to relation (5.9), we can write

$$D_i = \frac{RTU_i}{Fz_i} = \frac{RT}{Fz_i f_{r,i}} \qquad (5.12)$$

This relation for the diffusion coefficient (applied for charged particles) is called the **Nernst–Einstein equation.**

---

[1] Electrolytic mobility (also called electrophoretic mobility) determines the particle velocity in an electric field of the unitary intensity.

The reciprocal value of the electrolytic conductivity $U_i$ is called the **friction co-efficient** $f_{r,i}$. It represents certain resistance of the environment against the movement of the ion and it can be estimated, under simplified conditions, using the **Stokes–Einstein equation,** $f_{r,i} = 6\pi\eta r_i$, where $\eta$ is the viscosity of solution and $r_i$ is the hydraulic radius of the $i$-th ion (also including its solvation sheath).

The second term on the right side of eq. (5.5) describes ion transport by migration (transfer of charged particles in electric field). The driving force of this process is the electric energy gradient per the unit amount of substance. This quantity can be expressed as the product of the charge of 1 mol of the $i$-th ion ($z_iF$) and the electric field intensity as the electric potential gradient, $d\varphi/dx$. Thus, the intensity of molar flux by migration is expressed by the relation (again for simplicity in one-dimensional form, so all physical quantities can be considered as scalars)

$$J_{i,\text{migr}} = -u_i c_i z_i F \frac{d\varphi}{dx} \qquad [T,p] \qquad (5.13)$$

If we substitute relation (5.11) into relation (5.13), we get

$$J_{i,\text{migr}} = -U_i c_i \frac{d\varphi}{dx} \qquad [T,p] \qquad (5.14)$$

By substituting from eqs. (5.14) and (5.10) into relation (5.5), we get the form of the transport equation known as the **Nernst–Planck equation**

$$J_i = -D_i \frac{dc_i}{dx} - U_i c_i \frac{d\varphi}{dx} \qquad [T,p] \qquad (5.15)$$

where $D_i$ is the diffusion coefficient, $c_i$ is the mobile ion concentration, $U_i$ is the electrolytic mobility (dimension $m^2 V^{-1} s^{-1}$) defined by eq. (5.11) and $\varphi$ is the electric potential. By extending eq. (5.15) with the term $c_iv$ describing the convective electrolyte flow ($v$ is the flow velocity, dimension $m\ s^{-1}$, in the $x$-axis direction), we get the **extended Nernst–Planck equation:**

$$J_i = -D_i \frac{dc_i}{dx} - U_i c_i \frac{d\varphi}{dx} + c_i v \qquad [T,p] \qquad (5.16)$$

The extended Nernst–Planck equation is an example of a transport equation formed by the simple sum of contributions of individual transport mechanisms (diffusion, migration and convection) and these contributions and their transport coefficients of proportionality (diffusion coefficient, electrolytic mobility, viscosity, etc.) are linearly independent.

### 5.2.1.1 Application of the Nernst–Planck equation to charge transport

The application of the Nernst–Planck equation (5.15) for charge transport (the definitions of current density, transport number, electrolyte conductivity, etc.) was discussed in Section 3.4.2 and will be further explained in Chapter 6. Although this equation ranks among simpler transport relations, its solution is in many cases complicated and requires the tools of numerical mathematics. However, in simpler cases, an analytical solution of the problem can be obtained.

Let us consider a general **binary electrolyte** $C_{\nu_+}^{z+} A_{\nu_-}^{z-}$ and assume that the condition of electroneutrality applies; see eq. (3.37) and Section 6.3 Let us also consider that the convective electrolyte flow is negligible. In this case, the concentrations of cation and anion, $c_C$ and $c_A$, can be expressed using the electrolyte total concentration $c$ and flows of cation $J_{m,C}$ and anion $J_{m,A}$ by equations (for simplicity again in a one-dimensional form):

$$z^+c_+ + z^- c_- = 0, \quad c_+ = -cz_- = cv_+, \quad c_- = -cz_+ = cv_- \tag{5.17}$$

$$J_+ = +z_- D_+ \frac{dc}{dx} + D_+ z_+ z_- c \frac{F}{RT} \frac{d\varphi}{dx} \tag{5.18a}$$

$$J_- = -z_+ D_- \frac{dc}{dx} - D_- z_+ z_- c \frac{F}{RT} \frac{d\varphi}{dx} \tag{5.18b}$$

where $D_+$ and $D_-$ denote the cation and anion diffusion coefficients, respectively. If no current passes through the solution, that is, the local current density defined by relation (3.51) in the system is zero, then it applies for binary electrolyte

$$j_+ = F(z_+ J_+ + z_- J_-) = 0 \tag{5.19}$$

By substituting eqs. (5.18a) and (5.18b) into eq. (5.19) and by its mathematical modification, we obtain the relation for the electric potential gradient

$$\frac{d\varphi}{dx} = -\frac{D_+ - D_-}{z_+ D_+ - z_- D_-} \frac{F}{RT} \frac{1}{c} \frac{dc}{dx} = -D^* \frac{F}{RT} \frac{1}{c} \frac{dc}{dx} \tag{5.20}$$

From this relation, it is obvious that if $D^* \neq 0$ (for $D_+ \neq D_-$ or $z_+ \neq -z_-$), due to the ion diffusion ($dc/dx \neq 0$) an electric field is formed ($d\varphi/dx \neq 0$), referred to as the local **liquid-junction potential**. For a more general electrolyte, the liquid-junction potential is derived in Section 3.5.1 and its consequences are discussed in more detail for the case of a separation semi-permeable partition. By substituting for $d\varphi/dx$ in eq. (5.17) or (5.18) from eq. (5.20), we obtain, after modification, relations for the molar fluxes of individual ions:

$$J_+ = -\underbrace{\frac{(z_+ - z_-)D_- D_+}{z_+ D_+ - z_- D_-}}_{D_{eff}} \frac{dc_+}{dx} = J_- - \underbrace{\frac{(z_+ - z_-)D_- D_+}{z_+ D_+ - z_- D_-}}_{D_{eff}} \frac{dc_+}{dx} \tag{5.21}$$

Note that the above modifications led to the following changes:
a) transport equations (5.18a) and (5.18b) are simplified and the migration term is eliminated;

b) proportionality coefficient between the molar flux and the concentration gradient, $D_{\mathrm{eff}}$, is identical for the cation and the anion (resulting from the linear combination of $D_+$ and $D_-$);
c) to calculate the concentration distribution in the solution, it is not necessary to consider the electric field, which substantially simplifies the calculation (if the current is zero).

It results from points a) and b) that the cation and anion fluxes are linked, namely by electrostatic forces between ions of opposite charge. In short, the slower ion (with the lower diffusion coefficient) is pulled (accelerated) by the faster ion, while the faster ion (with the higher diffusion coefficient) is slowed down by the slower ion. This type of ion transport in electrolyte solutions is usually referred to as **coupled diffusion** and the transport coefficient $D_{\mathrm{eff}}$ is called the **effective coefficient of coupled diffusion**.

The theory of coupled diffusion is further developed for the current-loaded state in Section 5.3, where differences between symmetrical and unsymmetrical binary electrolytes are also discussed.

## 5.2.2 Phenomenological description of transport

In agreement with relation (2.35), every transport process can be described by appropriate phenomenological equations. In Fick's first law of diffusion, relation (5.10), the diffusion coefficient $D_i$ can be transformed to a phenomenological coefficient $L_{ii}$ and the concentration gradient $(-\mathrm{d}c_i/\mathrm{d}x)$ to the generalized force $X_i$. According to relation (2.35), we can write for the flow intensities $J_1$ and $J_2$,

$$J_1 = L_{11}X_1 + L_{12}X_2 \tag{5.22}$$

$$J_2 = L_{21}X_1 + L_{22}X_2 \tag{5.23}$$

An example of this case can be also thermal diffusion of ions whose driving forces are both the concentration gradient and the temperature gradient. Therefore, we can write

$$J_i = \sum_j L_{ij}\frac{\mathrm{d}c_j}{\mathrm{d}x} + L_{iQ}\frac{\mathrm{d}T}{\mathrm{d}x} \tag{5.24}$$

$$J_Q = \sum_j L_{Qj}\frac{\mathrm{d}c_j}{\mathrm{d}x} + L_{QQ}\frac{\mathrm{d}T}{\mathrm{d}x} \tag{5.25}$$

where $J_i$ and $J_Q$ are the intensity of molar flux of the $i$-th ion and the intensity of heat flux, respectively. In relations (5.23) and (5.24), the actual phenomenological coefficient $L_{ij}$ is the diffusion coefficient $D_i$, the actual phenomenological coefficient $L_{QQ} = \lambda_i$, where $\lambda_i$ is the **coefficient of thermal conductivity**, which is also present in the **Fourier's heat transfer law**:

$$J_{i,Q} = -\lambda_i \frac{\mathrm{d}T}{\mathrm{d}x} = L_{QQ}X_Q \tag{5.26}$$

For mutual coefficients $L_{DQ}$ and $L_{QD}$, which can be referred to as **thermal diffusion coefficients**, the Onsager reciprocity theorem applies; see relation (2.36).

If relations (5.22) and (5.23) for two fluxes and Onsager reciprocity theorem $L_{12} = L_{21}$ are valid, we can write, for $X_2 = 0$,

$$J_1 = L_{11}X_1 \tag{5.27}$$

$$J_2 = L_{21}X_1 \tag{5.28}$$

Therefore, it applies

$$\frac{J_2}{J_1} = \frac{L_{21}}{L_{11}} = \frac{L_{12}}{L_{11}} \qquad [X_2 = 0] \tag{5.29}$$

If $J_1 = 0$, it results from relation (5.22)

$$\frac{L_{12}}{L_{11}} = -\left(\frac{X_1}{X_2}\right)_{J_1=0} \tag{5.30}$$

Thus, in total

$$\left(\frac{J_2}{J_1}\right)_{X_2=0} = -\left(\frac{X_1}{X_2}\right)_{J_1=0} \tag{5.31}$$

Similarly, from relations (5.22) and (5.23) for $X_1 = 0$ we obtain

$$J_1 = L_{12}X_2 \tag{5.32}$$

$$J_2 = L_{22}X_2 \tag{5.33}$$

and

$$\frac{J_1}{J_2} = \frac{L_{12}}{L_{22}} = \frac{L_{21}}{L_{22}} \qquad [X_1 = 0] \tag{5.34}$$

For $J_2 = 0$, it applies

$$\frac{L_{21}}{L_{22}} = -\left(\frac{X_2}{X_1}\right)_{J_2=0} \tag{5.35}$$

and also

$$\left(\frac{J_1}{J_2}\right)_{X_1=0} = -\left(\frac{X_2}{X_1}\right)_{J_2=0} \tag{5.36}$$

From relations (5.22) and (5.23), it also results in

$$L_{12} = \left(\frac{J_1}{X_2}\right)_{X_1=0} \tag{5.37}$$

$$L_{21} = \left(\frac{J_2}{X_1}\right)_{X_2=0} \tag{5.38}$$

According to the Onsager reciprocity theorem, it must be valid

$$\left(\frac{J_1}{X_2}\right)_{X_1=0} = \left(\frac{J_2}{X_1}\right)_{X_2=0} \tag{5.39}$$

or

$$\left(\frac{\partial J_1}{\partial X_2}\right)_{X_1=0} = \left(\frac{\partial J_2}{\partial X_1}\right)_{X_2=0} \tag{5.40}$$

Relations (5.35), (5.36), (5.39) and (5.40) are called the **Saxen's relations** and they correspond to the **Maxwell's relations** of classical equilibrium thermodynamics.

Transport processes are irreversible processes and therefore the principles of linear irreversible mechanics can be applied to them, whose assumptions are briefly described in Section 2.4.


## 5.2.3 Maxwell–Stefan transport theory

This theory [2] describes the transport of uncharged particles in a multicomponent system. It is based on the assumption that there is a balance between the driving force acting on moving particles and the environment resistance, expressed by its frictional force (braking force due to mutual friction or other physical and chemical interactions between the moving particle and the environment). If the frictional force $f_{ij}$, by which molecule $j$ affects molecule $i$, is proportional to the difference in their velocities, then

$$f_{ij} = \frac{RT}{D_{ij}}\left(v_i - v_j\right), \qquad i = 1, 2, \dots, k \tag{5.41}$$

where $RT/D_{ij}$ are the friction coefficients, which are the reciprocal value of the mobility $u_i$, see relation (5.9), and $D_{ij}$ are the Maxwell–Stefan diffusion coefficients representing the resistance of the given component against its transport; they satisfy Onsager reciprocity theorem, $D_{ij} = D_{ji}$, see relation (2.36). In a system of $k$ components, the mean frictional force acting on component $i$ is the sum of the forces originated by the interactions of components with each other and is defined by the relation

$$\bar{f}_i = \sum_{j=1}^{k} \frac{RT}{D_{ij}} x_j (\bar{v}_i - \bar{v}_j), \qquad i = 1, 2, ..., k \qquad (5.42)$$

where $\bar{v}_i$ and $\bar{v}_j$ are the mean velocities of components $i$ and $j$ and $x_j$ is the local concentration of component $j$, expressed by its molar fraction. Since the frictional force acts against the driving force as the chemical potential gradient in the $x$-axis direction (see relation (2.52) or (2.53)), we can write

$$-\left(\frac{\partial \mu_i}{\partial x}\right)_{T,p} = \sum_{j=1}^{k} \frac{RT}{D_{ij}} x_j (\bar{v}_i - \bar{v}_j), \quad i = 1, 2, ..., k \qquad (5.43)$$

If for the intensity of total flux of component $i$ in stationary coordinates ($N_i$) it applies

$$N_i = c_i \bar{v}_i = c x_i \bar{v}_i \qquad (5.44)$$

where $c$ is the total concentration, and similarly for the intensity of total flux of component $j$ ($N_j$) in stationary coordinates

$$N_j = c_j \bar{v}_j = c x_j \bar{v}_j \qquad (5.45)$$

(where $c$ is again the total concentration) then relation (5.43) can be modified to the form

$$\left(\frac{\partial \mu_i}{\partial x}\right)_{T,p} = \sum_{j=1}^{k} \frac{RT}{D_{ij} c_i} (x_i N_j - x_j N_i), \quad i = 1, 2, ..., k; \ j \neq i \qquad (5.46a)$$

where $c_i$ is the concentration of component $i$, or

$$\left(\frac{\partial \mu_i}{\partial x}\right)_{T,p} = \sum_{j=1}^{k} \frac{RT}{D_{ij} c_i} (x_i J_j - x_j J_i), \quad i = 1, 2, ..., k; \ j \neq i \qquad (5.46b)$$

where $J_i$ and $J_j$ are molar fluxes of components $i$ and $j$, respectively.

These considerations apply in solution; in nanoporous environment (e.g. in the membrane), there are also forces between the membranes and components.

# 5.3 Ion-selective membranes

The basic phenomenon determining the separation process in an ion-selective membrane is transport of ions from one solution through a cation-selective membrane (CM) or anion-selective membrane (AM) to another solution. The decisive step of the separation process is the **mass transfer through the membrane**, consisting of three independent flows: **cation transfer, anion transfer** and **solvent**

**transfer**, controlled by kinetic or thermodynamic parameters. Kinetic parameters are expressed by the mobilities and diffusion coefficients of ions in the membrane and in the electrolyte solution (see Section 5.2) or by the electrical conductivity of the system; thermodynamic parameters are determined by the driving force needed to move the transported particles through the solution and membrane.

The transport properties of ion-selective membranes are determined by their composition and molecular structure and morphology (see Chapter 4 and Section 5.3.1). The solid phase of the membrane binds positive or negative electric charges that determine the preferential transport of ions, that is, cations in CM and anions in AM. This decides on the basic characteristics of mass and charge transport in ion-selective membranes.

Mass transport through an ion-selective membrane, that is, solid and mostly polymer electrolyte, is governed by basic physical and chemical principles applicable to electrolyte transport in general [3, 4], as described in Section 5.2. Ion-selective membranes have been produced and developed almost exclusively from polymer materials. Therefore, polymer chemistry is the starting discipline to prepare an appropriate substance composition of membrane matrix, its functionalization and formation of morphological structure. These three basic parameters of the membrane structure contribute to the final transport performance of the membrane [5], as was described in more detail in Chapter 4.

The first indicator of the nature of the membrane is its chemical disposition, that is, properties of the materials used. This indicator determines the thermal and corrosion resistance of the membrane, and with respect to the transport properties, it determines the physical dependences on the phase interface between the membrane and solution. The hydrophobic or hydrophilic nature of the polymer matrix of the membrane is manifested mainly at the passage of components through the membrane, and the ion-selective properties of the membrane are determined by the functional groups distributed in the polymer matrix. Besides the basic determination of the membrane polarity, that is, ion-selectivity type, their concentration and geometric arrangement in the polymer skeleton determines the separation process efficiency and its overall yield. The transport properties of the membrane are significantly influenced by its inner structure (morphology). From this point of view, it is possible to discuss the probability of transport paths in the membrane and functional manifestations of ion groups bound in the matrix.

## 5.3.1 Qualitative properties of ion-selective membranes

Ion-selective membranes are key components of electromembrane devices. They are selective partitions providing selective transport of the desired ions between the spaces (parts and chambers) of the electromembrane system and preventing unwanted mixing of other components. As mentioned earlier, ion-selective membranes are typically made of polymer materials, usually heterogeneous, consisting of at least two phases. The membranes contain a liquid phase (mostly an aqueous electrolyte solution) and porous (permeable) solid phase (an electrically non-conductive dielectric) that forms the more or less rigid skeleton of the membrane. The liquid phase is located in channels and cavities whose size is typically in nanometres. Their morphology can be very complicated and change with swelling, shrinking and mechanical stress.

If molecules with non-zero electric charge are bound to the phase interface, such membranes are called **homogeneous**. Although they are heterogeneous on the level of the characteristic diameter of nanopores (which is in nanometres), they appear homogeneous on the level of micrometres. In the case of **heterogeneous** membranes, there are also other solid phases (e.g. impermeable binder and reinforcing textile). These membranes are heterogeneous on the level of micrometres or higher. Still, they are sometimes described as pseudo-homogeneous on the level of membrane elements (Fig. 5.1).

According to the classification presented in Chapter 1, ion-selective membranes are either of the homogeneous (pseudo-homogeneous) or heterogeneous type. Typical homogeneous membranes produced from a single functionalized polymer are not used now so often and were replaced by membranes produced from copolymers and polymer blends, characterized by different morphology, primarily the level of micro-heterogeneities. Transport paths are determined by the presence of solvent – water – which can modify the membrane morphology depending on the polarity of the polymer material and distribution of functional groups in the polymer structure. Heterogeneous membranes are characterized by the distinctive phase separation of the ion-selective polymer and inert carrier. In these membranes, transport takes place through their ion-selective phase of the polymer composite, which must ensure transport across the membrane by its arrangement. Schematic variants of the first two described types were shown in Fig. 4.2.

Therefore, transport paths in the membrane profile are defined differently for each of these structures (see Fig. 4.2). In a homogeneous membrane, the transported particles pass through the entire profile of the membrane, and their movement is controlled by the statistical distribution of the functional groups in the membrane and by the concentration of these groups. In inter-polymer membranes, phase-separated domains are formed and transport is mainly conducted through functionalized channels created on the molecular level. Even in this system, the transport efficiency depends on the distribution and concentration of functional groups along the created transport lines. The material structures of heterogeneous membranes, totally separated on the macroscopic level, limit the transport possibility exclusively to their ion-functional part and the other component of the polymer composite can be considered inert as far as transport is concerned. Even in this case, the quality and concentration of functional groups and primarily their topography in the structure of the macroscopic polymer composite are decisive for the particle passage through the membrane.

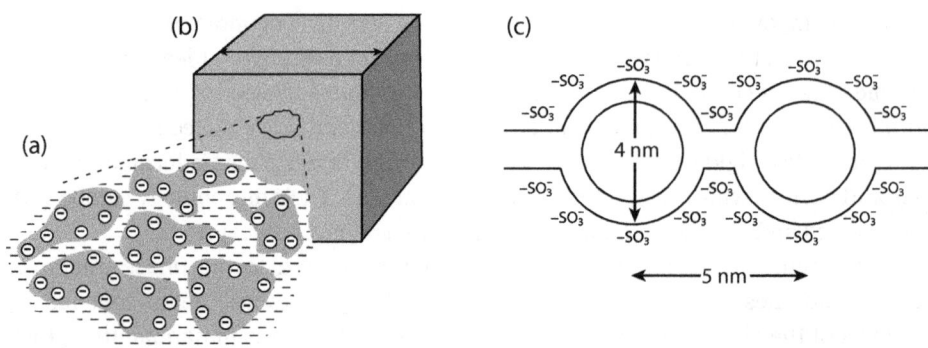

**Fig. 5.1:** (a) The zoomed area schematically displays a nanoporous structure of the membrane with nanometre scale pores forming a continuous ionic transport path. Charge fixed functional groups are indicated at the surface of domains of polymeric matrix or resin. (b) Pseudo-homogeneous interpretation of the ISM phase. (c) Model of the internal structure of Nafion®, considering cavities with a characteristic length of 4 nm connected by narrower channels 1 nm in diameter [3, 6].

Electric charge is chemically bound to the inner walls of the pores of the ion-selective membrane. Depending on its polarity, we can distinguish CM and AM. In CM, negative charges are bound, and in AM, positive charges are bound. Ions passing through the membrane are referred to as counter-ions (in CM they are positive cations, in AM negative anions), while ions repelled by the membrane are called co-ions (in CM they are negative anions, in AM positive cations).

## 5.3.1.1 Content of solvent – swelling capacity

In the case of ion-selective membranes, water or aqueous solutions are almost exclusively used as the polar solvent, ensuring the dissociation of the functional groups of the membrane and also solvating the present ions. Their presence in the membrane is a precondition of the transport process and their amount also affects the quality of membrane separation. The membrane polymer skeleton influences, to a certain degree, the amount of water in the membrane by its surface characteristics but the impact of the present bound ion-selective groups, determining the degree of the membrane swelling, is crucial. Their chemical nature and quantity determine the rate and degree of the membrane swelling. The swelling degree also increases with the growing ion-exchange capacity, but in limit cases, it can cause the loss of mechanical cohesion of the membrane or dissolution of the non-cross-linked polymer structure. Achieving the optimum degree of swelling is one of the key parameters of the preparation of a high-quality ion-selective membrane. Such a membrane transports counter-ions perfectly, and the transport of mass and charge exhibits high efficiency. The design of such a membrane is determined not only by the structural properties of the polymer matrix, and the mechanical cohesion can be increased by supporting reinforcement with a suitable textile.

The swelling capacity can be determined by several different physical methods. Similar to ion-exchange capacity, the swelling capacity values are directly related to the assessment of permeability and permselectivity of ion-selective membranes.

## 5.3.2 Quantitative characteristics of ion-selective membranes

Basic quantitative characteristics are void fraction $\varepsilon$, interphase area density $\alpha$ and ion-exchange capacity which is proportional to the amount of electric charge fixed on the interphase interface.

The **void fraction** $\varepsilon$ corresponds to the volume fraction of the liquid phase in the membrane

$$\varepsilon = \varphi_1 = V_1/V, \quad 1-\varepsilon = \varphi_s = V_s/V, \quad V = V_1 + V_s, \quad \varphi_1 + \varphi_s = 1 \tag{5.47}$$

where $V_1$ and $V_s$ are the volumes of the liquid and solid phases, $V$ is the total volume of the membrane, $\varphi_1$ is the volume fraction of the liquid phase (void fraction) and $\varphi_s$ is the volume fraction of the solid phase. The volume fractions depend on the conditions (temperature, pressure, stress, composition of liquid phase, swelling, shrinking, drying and ion exchange). The relaxation time of the change of the liquid volume in the membrane may be relatively long (hours). The void fraction of a dry membrane approaches zero.

The **interphase area density** $\alpha$ (unit: $\mathrm{m^2\,m^{-3}}$) is defined as

$$\alpha = A_{1-s}/V \tag{5.48}$$

where $A_{1-s}$ is the interphase area. The composition can be described by the apparent concentrations in membrane, $c_{i,M}$ (per unit of the membrane volume, unit: $\mathrm{mol\,m^{-3}}$)

$$c_{i,M} = n_i/V \tag{5.49}$$

or by the liquid phase concentrations, $c_{i,L}$ (per unit of the liquid phase volume, unit: $\mathrm{mol\,m^{-3}}$)

$$c_{i,L} = n_i/V_1 = n_i/\varphi_1 V = c_{i,M}/\varphi_1 = c_{i,M}/\varepsilon > c_{i,M} \tag{5.50}$$

The interphase concentrations of the fixed components, $c_{1-s,\mathrm{fix}}$ (the charged groups, unit: $\mathrm{mol\,m^{-2}}$), are

$$c_{1-s,\mathrm{fix}} = n_{\mathrm{fix}}/A_{1-s} = c_{M,\mathrm{fix}} V/aV = c_{M,\mathrm{fix}}/a \tag{5.51}$$

The mass of the dry solid, $m_s$, can be expressed as

$$m_s = V_s\rho_s = V\varphi_s\rho_s \tag{5.52}$$

where $\rho_s$ is the density of the solid phase.

### 5.3.2.1 Permselectivity of ion-selective membranes

Permselectivity is the ability of the membrane to prefer the transport of some component or components of the mass penetrating the membrane under the influence of the driving force. The permeability coefficient determines the amount of substance passed through a unit area of the membrane per time unit at a unit driving force. Both of these transport quantities are influenced by the properties of the ion-selective membrane and are indirectly characterized by basic material characteristics.

The permselectivity of the ion-selective membrane is an important parameter determining the membrane performance in the membrane separation process [7]. It determines how the membrane carries ions of a certain charge and retains ions of the opposite charge and is defined by the ion transport number:

$$\psi_{CM} = \frac{t_{+,CM} - t_+}{t_-} \tag{5.53}$$

$$\psi_{AM} = \frac{t_{-,AM} - t_-}{t_+} \tag{5.54}$$

where $\psi$ is the membrane permselectivity and $t$ represents the transport numbers of ions in the membrane or in solution. The transport number (see Chapters 3 and 6) characterizes the proportion of the electric charge transferred by the respective ion in the total charge transferred by all ions in the membrane. The transport number of the ion in the membrane is proportional to its concentration in the membrane, which is also dependent on its equilibrium concentration in the solution. This relation follows the principle of the Donnan exclusion, which means that the selectivity of an ion-selective membrane is the result of the exclusion of the relevant co-ion from the membrane phase [8]. See also Section 5.3.2.3.

### 5.3.2.2 Ion-exchange capacity

The ion-exchange capacity (IEC) is an important parameter of ion-selective membrane, determining the amount of bound cation- or anion-active centres in the membrane. This parameter has no direct relation to the transport of substances through the membrane but characterizes the available amount of free-moving counter-ions in the membrane, and indicates the amount of the electric charge transferred in the membrane–solution system. Similarly to the previous section, the chemical nature, concentration and form of the distribution of functional centres in the membrane structure are decisive for monitoring the influence of these centres on the transport process.

Based on their chemical composition, the functional groups are strong or weak acids or bases, CMs having sulpho- or carboxyl-groups are fixed on a polymer skeleton and AMs mostly use quaternized amine derivatives, whose structure limits the achievement of large ion-exchange capacity. Functional groups are distributed on polymer carriers statistically by polymer-analogous reactions and can form an anion-conductive space of counter-ions for charge transport in the electric field. The chemical quality and the amount of bound ion-selective groups thus indirectly influence the concentration of free-moving charge carriers in the membrane and subsequently its basic transport characteristic.

The ion-exchange capacity can be expressed per volume of the membrane, IECV $(\text{mol m}^{-3})$, or per mass of the dry membrane, IECM $(\text{mol kg}^{-1})$:

$$\text{IECV} = ac_{\text{l-s, fix}}z_{\text{fix}} = c_{\text{M, fix}}z_{\text{fix}} \tag{5.55}$$

$$\text{IECM} = ac_{\text{l-s, fix}}z_{\text{fix}}/m_{\text{s}} = ac_{\text{l-s, fix}}z_{\text{fix}}/V_{\text{s}}\rho_{\text{s}} = ac_{\text{l-s, fix}}z_{\text{fix}}/V\varphi_{\text{s}}\rho_{\text{s}} = c_{\text{l-s, fix}}z_{\text{fix}}/\varphi_{\text{s}}\rho_{\text{s}} \tag{5.56}$$

For these reasons, the ion-exchange capacity is one of the characteristic parameters of the quality of ion-selective membranes and its determination is usually one of the typical procedures of chemical analysis, showing methodological differences and therefore the need for standardization to compare the results.

### 5.3.2.3 Limiting current

The most important characteristic of an ion-selective membrane is the dependence of the electric current density on the inserted voltage (difference of electric potentials). A typical dependence is shown in Fig. 5.2 and is discussed in Section 5.4.2.

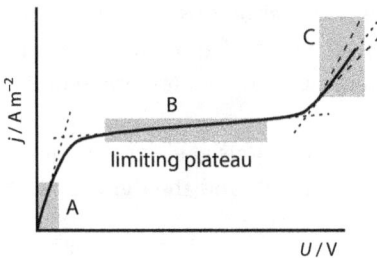

Fig. 5.2: Schematic drawing of current–voltage curve including three typical regions: A, ohmic region; B, limiting region; C, over-limiting region. The relative extent and slope of the individual regions vary significantly depending on the type of electromembrane process.

If we immerse the ion-selective membrane in an electrolyte solution for a suffi-ciently long time, a concentration of counter-ions and co-ions in the membrane is established, corresponding to the equilibrium between the external solutions and solution inside the membrane, sometimes called the **Donnan equilibrium** [9]. It is defined by the equality of the electrochemical potentials of electrically charged moving particles present in the considered membrane–electrolyte system. To calcu-late the concentration of ions in the membrane phase, the model of membrane (Fig. 5.3) can be used, separating two solutions of the uni-univalent electrolyte with different concentrations.

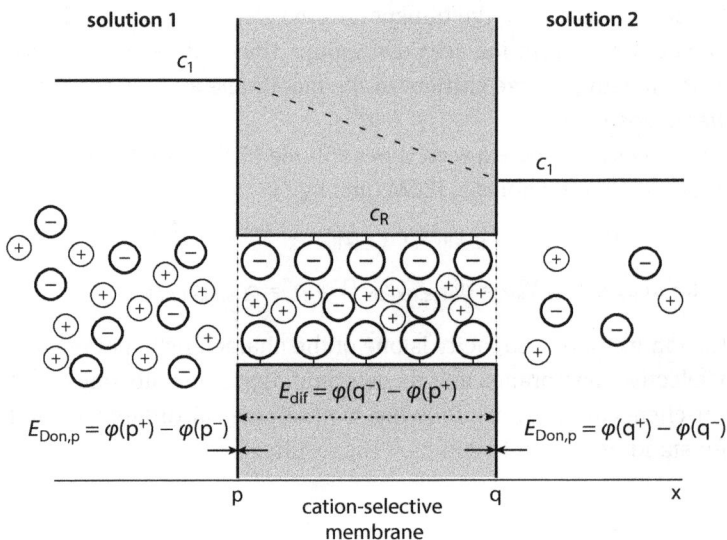

**Fig. 5.3:** Schematic representation of ion distribution inside and outside of the pore in cation-selective membranes.

The membrane consists of a system of pores with ion-exchange groups bound on its inner walls and through which counter-ions and co-ions move. Since the dimension of a membrane pore is many times greater than the diameter of ions, it is possible to assume the equilibrium between both phases for all kinds of moving (unbound) ions are established on the electrolyte phase interface outside the membrane and in the membrane pores.

Concentrations of ions and electric potential are discontinuous at electrolyte–mem-brane interfaces in a macroscopic scale. Let us denote the left and the right limits of a function $f(x)$ as follows:

$$f(x^-) = \lim_{\delta \to 0} f(x-\delta), \quad f(x^+) = \lim_{\delta \to 0} f(x+\delta)$$

Donnan potential (Section 3.5.3) differences on the left and right interfaces can then be expressed for $z_i \neq 0$ as follows:

$$E_{\text{Don,p}} = \varphi(\text{p}^+) - \varphi(\text{p}^-) = -\frac{RT}{z_i F} \ln \frac{c_i(\text{p}^+)}{c_i(\text{p}^-)} \tag{5.54a}$$

$$E_{\text{Don,q}} = \varphi(\text{q}^+) - \varphi(\text{q}^-) = -\frac{RT}{z_i F} \ln \frac{c_i(\text{q}^+)}{c_i(\text{q}^-)} \tag{5.54b}$$

and the diffusion potential on the membrane as follows:

$$E_{\text{dif}} = \varphi(\text{q}^-) - \varphi(\text{p}^+) \tag{5.55}$$

The near to equilibrium membrane potential can be expressed by the following equation:

$$E_{\text{M}} = \varphi(\text{q}^+) - \varphi(\text{p}^-) = \left( \overbrace{\varphi(\text{p}^+) - \varphi(\text{p}^-)}^{E_{\text{Don},p}} \right) + \left( \overbrace{\varphi(\text{q}^-) - \varphi(\text{p}^+)}^{E_{\text{dif}}} \right) + \left( \overbrace{\varphi(\text{q}^+) - \varphi(\text{q}^-)}^{E_{\text{Don},q}} \right) \tag{5.56}$$

We can define the dimensionless electric potential as follows:

$$\bar{\varphi} = \frac{\varphi F}{10RT} \doteq \frac{\varphi F}{2.303RT} \tag{5.57}$$

One dimensionless unit represents about 59.2 mV of the dimensional potential. Equations (5.54)–(5.56) can then be rewritten into dimensionless form (here for an uni-univalent electrolyte $C^+A^-$):

$$\bar{E}_{\text{Don},p} = \bar{\varphi}(\text{p}^+) - \bar{\varphi}(\text{p}^-) = -\log \frac{c_C(\text{p}^+)}{c_C(\text{p}^-)} = \log \frac{c_A(\text{p}^+)}{c_A(\text{p}^-)}$$

$$\bar{E}_{\text{Don},q} = \bar{\varphi}(\text{q}^+) - \bar{\varphi}(\text{q}^-) = -\log \frac{c_C(\text{q}^+)}{c_C(\text{q}^-)} = \log \frac{c_A(\text{q}^+)}{c_A(\text{q}^-)}$$

$$\bar{E}_{\text{dif}} = \bar{\varphi}(\text{q}^-) - \bar{\varphi}(\text{p}^+)$$

$$\bar{E}_{\text{M}} = \bar{\varphi}(\text{q}^+) - \bar{\varphi}(\text{p}^-) = \bar{E}_{\text{Don},p} + \bar{E}_{\text{dif}} + \bar{E}_{\text{Don},q} \tag{5.58}$$

For concentrations and dimensionless potentials in the electrolyte layer at the left side of the membrane (p), inside the membrane and at the right side of the membrane, see Figs. 5.3 and 5.4. The concentration of bound ion-exchange groups is $c_R$ (this value depends on the degree of swelling of the membrane, pH and concentration of the external electrolyte) and the functional groups have the charge $z_R$.

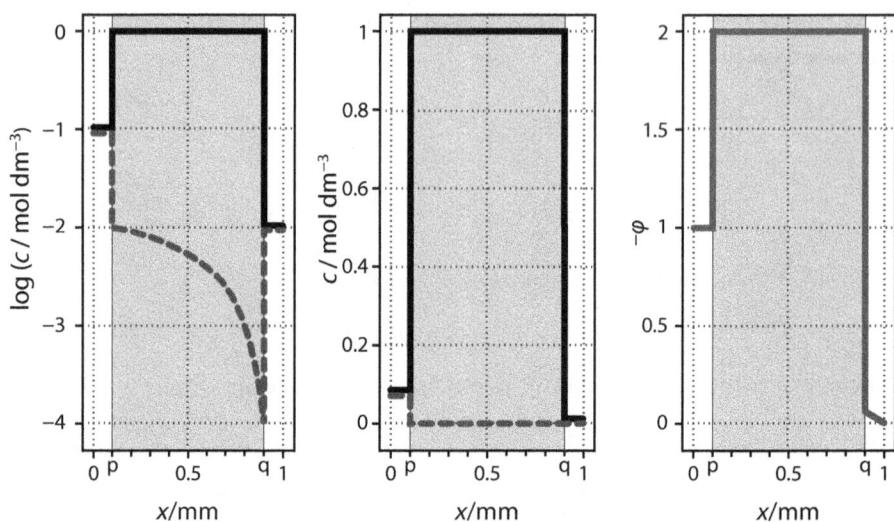

**Fig. 5.4: Example of space profiles of concentrations and dimensionless electric potential in the cation-selective membrane and its surroundings in current-less state.** (a) Concentrations of cation ($c_C$) and anion ($c_A$) in logarithmic scale; (b) concentrations in normal scale; (c) dimensionless potential changes at membrane boundaries. $c_{KCl} = 0.1$ mol dm$^{-3}$ in solution 1 and $c_{KCl} = 0.01$ mol dm$^{-3}$ in solution 2; $c_R = 1$ mol dm$^{-3}$. Thickness of the membrane is 0.8 mm and boundary layer of the electrolyte is 0.1 mm.

Thus, the membrane potential is the electric voltage established on the ion-selective membrane separating two solutions with different concentrations. It can be defined as the difference between electric potentials on the left and right sides of the membrane and consists of two Donnan potentials and diffusion potential. Individual electric potentials in the membrane and on the phase interfaces can be separated only in theoretical considerations, and experiments determine the resulting membrane potential, which is their sum.

Potential change (dimensionless Donnan potential) on the left boundary, $\Delta\varphi F/2.303\,RT = 1$, corresponds to the change in the concentration of one decimal order and on the right boundary, $\Delta\varphi F/\log RT = 2$, to the change of the concentration of two decimal orders. No current flows through the membrane and the Donnan potential difference is blocked by the applied voltage $[\varphi(p) - \varphi(q)]F/\log RT = 1$. At the same time, the slow diffusion of both ions from left to right occurs.

Further, it can be deduced that with the equilibrium of electrochemical potentials on the phase interface of electrolyte–membrane, the equilibrium applies to any kind of ions on this phase interface resulting from the ratio of ion activities (concentrations), called the **Donnan distribution coefficients**.

The concentrations of solutions outside the membrane are determined by the experiment and the concentration of groups bound in the membrane pores can be determined by ion-exchanger titration and pore volume determination. Thus, it can be deduced that in an equilibrium system, the concentrations of counter-ions and co-ions vary in magnitude. The result shows that the concentration of counter-ions in the membrane is only slightly higher than the concentrations of the corresponding bound ions. The concentration of co-ions in the membrane is very low. This effect is called the **Donnan exclusion** (of co-ions). The Donnan exclusion is caused by the increased content of ions with the opposite sign than that of the bound functional groups (counter-ions) in the membrane and at a greater concentration difference between the membrane and solution also by almost zero content of identical ions (co-ions). These concentration differences are the source of the chemical potential gradient on the membrane–solution interface, resulting in the Donnan potential acting against the concentration diffusion. The contact between the membrane and solutions of different concentrations on both sides of the membrane creates a potential difference, and the mentioned (Donnan) membrane potential appears on the membrane, which is an experimentally measurable characteristic of the permselectivity of the ion-selective membrane.

## 5.4 Transport fluxes in ion-selective membranes

The transport of components in electromembrane devices consists of several steps:
- transport of components between the bulk solution and the very narrow layer near the membrane surface (electric double layer); these processes are affected by hydrodynamics up to hundreds of micrometres;
- transport of solution components (including solvent) in the double layer (in the distance of nanometres from the membrane interface);
- transport of components within the membrane.

Driving forces are determined by gradients (local forces) or differences (integral forces) of relevant intensive quantities. To describe transport through membranes, the following driving forces are important:
a) difference (gradient) of electric potential,
b) difference (gradient) of concentrations of components,
c) difference (gradient) of hydrostatic pressure,
d) difference (gradient) of osmotic pressure.

With the exception of hydrostatic pressure, all these differences (gradients) can be included in the difference (gradient) of the electrochemical potentials of components. **In the linear region** (for not very high driving forces), **the transport fluxes are directly proportional to the driving forces and inversely proportional to resistances against transport**. Inside the membrane and its close vicinity, intensive variables describing local composition and local electric potential are not uniform in space (see Fig. 5.4) but form spatial fields described by functions for spatial coordinate functions perpendicular to the membrane.

**Differential (local) description** (Section 5.4.1) is based on the assumption that the membrane can be imaginarily divided into elementary layers whose thickness is considerably smaller than the thickness of the membrane, but at the same time large enough to contain a substantial number of repeating morphological structures of the membrane. (Similarly, we can divide the layers of electrolytes adjacent to the membrane.) In mathematical abstraction, we consider the thickness of these layers as infinitesimally small. The **gradients** of convenient intensive quantities are considered as **local (differential) driving forces**, and these elementary layers are described using **differential equations**.

By **integrating** these equations within the given limits (for the given boundary conditions), we obtain dependences of fluxes on intensive quantities in boundary conditions. As **global (integral) driving forces**, we consider the **differences** of the given intensive quantities between boundaries. In most cases, we do not know the analytical solution of this integration and have to solve it using numerical methods (Section 5.4.3). Numerical solutions can be approximated using algebraic equations (functions) that explicitly describe the dependence of component flows on global (integral) driving forces.

The basic postulate is that zero driving forces imply zero fluxes – it is **equilibrium**. At a sufficiently small difference from equilibrium, we can approximate these dependences by **linear functions** (Section 5.4.2).

## 5.4.1 Differential (local) description

Depending on how small or large morphological structures we take into account (by which microscope magnification we observe them), we can construct **pseudo-homogeneous models** and **heterogeneous models**.

### 5.4.1.1 Pseudo-one-phase (pseudo-homogeneous) models

In this case, we assume that the elementary volume is much smaller than the membrane volume and at the same time large enough to contain a large number of repeating parts of the membrane morphology. If the membrane thickness is in millimetres and the characteristic diameter of channels in nanometres, then a reasonable size of the elementary volume should be in micrometres. All pseudo-local quantities have then an average value over the elementary volume.

However, the pseudo-one-phase model is not convenient to describe processes in the membrane–electrolyte interface at the nanometre spatial scale, especially for heterogeneous membranes. It is still often used and assumed to provide meaningful results. The comparison of the modelling results with experimental findings confirms this assumption but it is necessary to assess these results critically. Some phenomena, for example, electrokinetic flows near the channel outlet from the membrane to electrolyte, cannot be described by this type of model in principle.

Pseudo-homogeneous models describe the processes using conveniently mean, effective quantities. The **effective** (apparent, pseudo-homogeneous) material properties as the diffusion coefficient, $D_{i,\text{eff}}$, and the effective permittivity, $\varepsilon_{\text{eff}}$, differ from the ones in the free liquid ($D_{i,\text{eff}} < D_{i,\text{free}}$; $\varepsilon_1 \sim \varepsilon_{H_2O} > \varepsilon_s$, and therefore $\varepsilon_{\text{eff}} < \varepsilon_{\text{free}}$).

The effective concentration, $c_{i,\text{eff}}$, the charge density, $q_{\text{eff}} = F \sum_i z_i \, c_{i,\text{eff}}$, and electric potential $\varphi_{\text{eff}}$ represent the mean quantities (averaged per elementary volume of the membrane or per volume of the liquid phase inside the elementary volume of the membrane). They essentially differ from local (on nanometre scale) ones in the liquid phase inside of the pores.

Then, the **Nernst–Planck equation**

$$J_{i,\text{eff}} = -D_{i,\text{eff}} \frac{dc_{i,\text{eff}}}{dx} - c_{i,\text{eff}} \frac{D_{i,\text{eff}}}{RT} z_i F \frac{d\varphi_{\text{eff}}}{dx} \tag{5.58}$$

and the **Poisson equation**

$$\frac{d^2 \varphi_{\text{eff}}}{dx^2}(x) = \frac{q_{\text{eff}}}{\varepsilon_{\text{eff}}} \tag{5.59}$$

must include these effective parameters and variables. The independent variable $x$ in eqs. (5.58) and (5.59) is the coordinate perpendicular to the membrane, and the transport in the directions parallel to the membrane is neglected. The pseudo-homogeneous models combine these equations with the local balances of the components, which can be written in the form

$$\frac{dJ_{i,\text{eff}}}{dx} = 0 \tag{5.60}$$

Here, no chemical reactions occur and the stationary state is considered.

## 5.4.1.2 Two-phase (heterogeneous) models

The complex morphology of the membranes can be described, for example, by the model of the **equivalent parallel channels** [4], or by the models of the packed beds, or by computer-aided 3D reconstruction method. Such models describe, among other things, the detailed flow of the liquid in the membrane structure, inside channels or in gaps between particles. The model of equivalent parallel channels shows deviations from linear transport processes (Ohm's law and Smoluchowski's equation for electro-osmotic flow [4]).

## 5.4.2 Local and global equilibria

The important question arises whether the establishment of local equilibria is sufficiently fast, even if the overall macroscopic system is not in equilibrium. One useful guide is the dependence on terminal voltage [5, 10], which indicates how much the system differs from macroscopic equilibrium. In this respect, three distinct regions can be identified (see Section 5.3.2.3 and Fig. 5.2):

- At low voltage values, the current increases almost linearly with increasing voltage (**ohmic region**). All reversible processes are locally in a near-equilibrium state since the characteristic time for approaching a local equilibrium is significantly lower in comparison to the rate of mass and charge transfer across the electromembrane unit. The mass and charge transport processes behave according to the principles of linear irreversible thermodynamics.
- At higher values of applied voltage, the current tends to increase non-linearly due to hindrance by the mass transport of ions in the membrane or solution phase, and/or the rate of approaching local equilibria. The current then inclines to a **limiting value** plateau characterized by only moderate dependence on the increasing voltage (**limiting region**).
- For even higher voltage, the current starts to increase significantly (**over-limiting region**). In this region, the overall process in the system is driven by phenomena taking place in very narrow zones at the phase interfaces, which are characterized by a very low concentration of ions, that is, by a very low ionic conductivity. These zones are distinguished by high local intensity of the electric field. The rate of movement of individual ions is significantly higher compared to the rate of approaching local dissociation or phase equilibrium. Complex phenomena responsible for the over-limiting current take place [5, 10, 11]: the electric field enhances water splitting, giving rise to the exaltation effect (enhancement of counter-ion flux) and instability of the current values due to electro-convection. Additionally, a concentration decrease of fixed charge in an AM (membrane discharge) at elevated pH inside the membrane is discussed as a possible explanation for the over-limiting current [12]. However, it is usually difficult or even impossible to experimentally characterize the highly developed over-limiting region reproducibly. An understanding of the phenomena in this region is on the horizon of the current state of technical and scientific knowledge.

In the **locally equilibrated description**, we assume electroneutrality in the membrane and the Donnan potential in the adjacent layers (for details, see Chapter 6). In the **locally non-equilibrated description**, we assume that the condition of electroneutrality is not met. In the adjacent spaces, we solve local balances of components and Poisson's equation. Transport can be described again by the Nernst–Planck equation.

### 5.4.3 Integral (global) description

Integral (global) description consists of functions (or algebraic equations) that explicitly (or implicitly) describe the dependence of component flows on global (integral) driving forces. The integral procedures include the phenomenological procedure (Section 5.2.2) and empirical procedure. For example, the Fourier's heat transfer law and Fick's first law of diffusion can be reformulated into forms in agreement with relation (2.35). For details, see Chapter 7.

Empirical description is usually based on the knowledge of **transport numbers,** either taken as empirical constants or estimated using detailed models. If we know the total electric current, we also know the flows of individual components carrying the electric charge.

The membrane selectivity for the $i$th ion can be quantified by the mean charge transfer number

$$t_i = I_i/I; \quad I = \sum I_i \tag{5.61}$$

where $I_i$ is the electric current transferred solely by the $i$th ion. In the case of an ideal CM, the sum of the charge transfer numbers of cations is $t_+ = 1$ and for anions $t_- = 0$, whereas for an ideal AM one, it is $t_- = 1$ and $t_+ = 0$. According to Faraday's law, the molar flux $J_i$ of the $i$th ion across the membrane is related to the current:

$$J_i = \frac{t_i}{z_i F} I \tag{5.62}$$

A theoretical evaluation of the charge transfer numbers can be difficult and requires more detailed mathematical modelling analysis. In practice, an alternative approach is an experimental measurement.

### 5.4.4 Solvent flow

We can distinguish several solvent transport mechanisms: flow by solvent entrainment in **solvation shell** of transported ions, **hydrodynamic flow** (pressure gradient as the driving force, Darcy's law), **osmotic flow** (composition gradient as the driving force, van't Hoff's law) and **electro-osmotic flow** (potential gradient as the driving force, Helmholtz–Smoluchowski equation).

## 5.5 Osmotic equilibrium

The equality of chemical potentials of components is a criterion of equilibrium in multi-component and multi-phase systems.

If we consider a case where two solutions (phase I) and (phase II) are separated by a semi-permeable membrane which only permits the solvent molecules, component (1), and does not permit the solute molecules, component (2), then the chemical potentials of component (1) in both phases are equal in equilibrium

$$\mu_{1,I} = \mu_{1,II} \tag{5.63}$$

According to relation (2.51), which expresses the dependence of the chemical potential of a component on temperature, pressure and composition, we can write for a dilute solution (with $a_i = x_i$) at constant temperature

$$\mu_{1,I} = \mu_1^\circ + RT \ln x_{1,I} + \bar{V}_{1,I} (p_I - p_{st}) \qquad [T] \tag{5.64}$$

$$\mu_{1,II} = \mu_1^\circ + RT \ln x_{1,II} + \bar{V}_{1,II} (p_{II} - p_{st}) \qquad [T] \tag{5.65}$$

where $\mu_1^\circ$ is the chemical potential of the pure solvent at the given temperature and standard pressure $p_{st}$, $x_{1,I}$ and $x_{1,II}$ are the molar fractions of the solvent in phases I and II, $\bar{V}_{1,I}$ and $\bar{V}_{1,II}$ are the partial molar volumes of the solvent in phases I and II, and $p_I$ and $p_{II}$ are pressures in respective phases.

In very dilute solutions, the partial volumes of the solvent are identical on both sides of the membrane and equal to the molar volume of the pure solvent, $V_{n,1}^0$ ($\bar{V}_{1,I} = \bar{V}_{1,II} = V_{n,1}^0$); then after achieving equilibrium, according to relations (5.64) and (5.65) we can write

$$RT(\ln x_{1,I} - \ln x_{1,II}) = V_{n,1}^0(p_{II} - p_I) \tag{5.66}$$

If $x_{1,I} > x_{1,II}$, then at the same temperature and pressure, the chemical potential of the solvent in phase I will be higher and the chemical potential of the solute will be lower, $\mu_{1,I} > \mu_{1,II}$, $\mu_{2,I} < \mu_{2,II}$. In consequence of the difference between chemical potentials, the solvent flows from the dilute phase I to the more concentrated phase II, till achieving equilibrium, that is, balancing the chemical potentials of components in both phases. This situation is shown in Fig. 5.5.

After the arrangement of relation (5.66), we get

$$\Delta p = p_{II} - p_I = \frac{RT}{V_{n,1}^0} (\ln x_{1,I} - \ln x_{1,II}) = \Delta \pi \tag{5.67}$$

The difference in hydrostatic pressures $\Delta p$ on both sides of the osmotic membrane is equal to the difference in osmotic pressures $\Delta \pi$, and the osmotic pressure is a quantity defined by the relation

$$\pi = -\frac{RT}{V_{n,1}^0} \ln x_1 \tag{5.68}$$

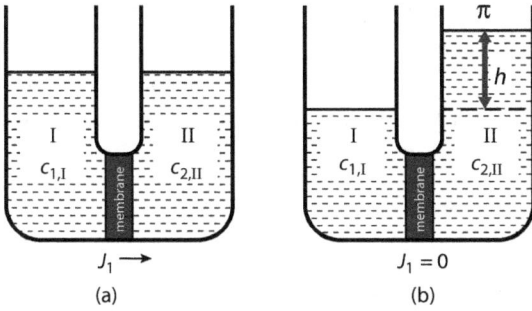

**Fig. 5.5:** Osmotic equilibrium: (a) initial condition: $\mu_{1,I} > \mu_{1,II}$, $c_{2,II} > c_{2,I}$, $p_I = p_{II}$; (b) equilibrium condition: $\mu_{1,I} = \mu_{1,II}$, $c_{2,II} > c_{2,I}$, $p_{II} > p_I$; $p_{II} - p_I = \pi_{II} - \pi_I = \Delta\pi$.
*Note*: The difference of osmotic pressures is also given by the relation $\Delta\pi = h\rho g$, where $h$ is the difference of surface-level heights, $\rho$ is the density of the solution and $g$ is the gravitational acceleration.

where $V_{n,1}^0$ is the molar volume of the pure solvent and $x_1$ is its molar fraction. For the dilute binary solution ($x_2 \to 0$), it applies according to Euler's series

$$\ln x_1 = \ln(1 - x_2) \cong -x_2 \tag{5.69}$$

therefore

$$\Delta\pi = \frac{RT}{V_{n,1}^0}(x_{2,II} - x_{2,I}) \tag{5.70}$$

where $x_{2,I}$ and $x_{2,II}$ are the molar fractions of the solute in phases I and II.

In very dilute solutions, the molar fraction of the solute $x_2$ can be replaced with the ratio of the amount of substance of the solute and solvent ($x_2 \cong n_2/n_1$) to obtain the **Morse equation**:

$$\Delta\pi = \frac{RT}{V_{n,1}^0}\left[\left(\frac{n_2}{n_1}\right)_{II} - \left(\frac{n_2}{n_1}\right)_{I}\right] \tag{5.71}$$

The product of the amount of substance of solvent $n_1$ and molar volume of pure solvent $V_{n,1}^0$ is the volume of pure solvent $V_1^0$, which can be approximated, in a very dilute solution, by the total volume of solution $V$ in respective phases. From relation (5.73), it follows that

$$\Delta\pi = RT\left[\left(\frac{n_2}{V}\right)_{II} - \left(\frac{n_2}{V}\right)_{I}\right] = RT(c_{2,II} - c_{2,I}) \tag{5.72}$$

This relation is called the **van't Hoff equation** for osmotic pressure and allows the calculation of the difference of osmotic pressures on both sides of the membrane

$$\Delta\pi = RT\,\Delta c_2 \quad [T, c_2 \to 0] \tag{5.73}$$

where $\Delta c_2$ is the difference in the molar concentrations of the solute on both sides of the membrane.

Phase I is very often a pure solvent (e.g. water). In this case, according to relation (5.68), $\pi_I = 0$, the difference $\Delta \pi = \pi_{II} = \pi$ and the difference in osmotic pressures are equal to the osmotic pressure in phase II, which is proportional to the concentration of the solute in the solution

$$\pi_{II} = \pi = \boldsymbol{R}T\, c_2 \quad [T, c_2 \rightarrow 0] \tag{5.74}$$

Since $c_2 = (m_2/M_2/V)$, it applies for the molar mass of solute $M_2$

$$M_2 = \frac{RTm_2}{\pi V} \tag{5.75}$$

where $m_2$ is the mass of the solute and $V$ is the solution volume. The equation suggests a possible method to determine the molar masses of high-molecular substances (e.g. polymers and proteins) by measuring osmotic pressure – this method is called **osmometry**. In determining the osmotic pressures of aqueous electrolyte solutions, the experimental values substantially exceed the values calculated on the basis of derived relations. The reason is that the ions produced by dissociation considerably increase the number of particles in the system and thus decrease the molar fraction of the solvent. After the dissociation equilibrium is established, the total concentration of particles in the solution equals $c_2[1 + \alpha(v_C + v_A - 1)]$, where $c_2$ is the molar concentration of the electrolyte, $\alpha$ is the dissociation degree, $v_C$ is the number of cations and $v_A$ is the number of anions produced by salt dissociation.

According to relation (5.75), the osmotic pressure is then given by the relation

$$\pi = \boldsymbol{R}Tc_2[1 + \alpha(v_C + v_A - 1)] \tag{5.76}$$

For example, NaCl is fully dissociated, so $\alpha = 1$, $v_C = 1$ and $v_A = 1$; therefore, for the osmotic pressure of 0.5-molar solution at the temperature of 25 °C, we calculate the incredibly high value as follows:

$$\pi = 2 \times 0.5 \times 10^3 \times 8.314 \times 298.15 = 2,478.82 \times 10^3 \text{Pa} \tag{5.77}$$

Note: In human physiology, osmotic pressure is expressed by the so-called **tonicity**. A solution with the same osmotic pressure as the given standard medium (blood plasma) is **isotonic,** a solution with higher osmotic pressure is **hypertonic** and with lower $\pi$ is **hypotonic**. An isotonic solution corresponds to a 0.155-molar solution of NaCl. The osmotic pressure of blood at 36.6 °C is then $\pi = 2 \times 0.155 \times 10^3 \times 309.7 = 798.2$ kPa. We realize how high this value is if we know that average values of atmospheric pressure are around 100 kPa.

## 5.6 Reverse osmosis

For the difference in the chemical potentials of the solvent in phases I and II, we can write on the basis of relation (5.65):

$$\Delta\mu_1 = V^0_{n,1}(p_{II}-p_I) + RT(\ln x_{1,II} - \ln x_{1,I}) \tag{5.78}$$

or

$$\Delta\mu_1 = V^0_{n,1}(\Delta p - \Delta\pi) \tag{5.79}$$

The flow of the solvent (usually water) $J(H_2O)$ from phase I to phase II will cease at $\Delta\mu_1 = 0$. This condition is satisfied if the difference in hydrostatic pressures $\Delta p$ is equal to the difference in osmotic pressures $\Delta\pi$. If the osmotic semi-permeable membranes separate pure water from the aqueous salt solution, then the water flow $J(H_2O)$ is zero at $\Delta p = \pi$, as evident from Fig. 5.6. When $\Delta p < \pi$, water flows through the membrane to the salt solution; when $\Delta p > \pi$, water flows in the opposite direction.

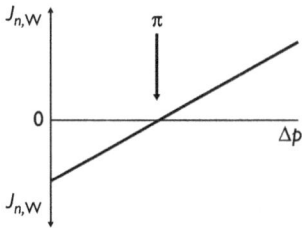

**Fig. 5.6:** Reverse osmosis. (a) $\Delta p < \pi$, $J(H_2O) < 0$, (b) $\Delta p = \pi$, $J(H_2O) = 0$, (c) $\Delta p > \pi$, $J(H_2O) > 0$.

This phenomenon is technologically used in water purification by the method of reverse osmosis. It applies for the water flow

$$J(H_2O) = A(\Delta p - \pi) \tag{5.80}$$

where $A$ is a constant, which is a characteristic for the particular membrane.

## References

[1]  R. B. Bird, W. E. Stewart, and E. N. Lightfoot, Transport Phenomena. 2nd Edition. Wiley, New York, 2002. ISBN 0-471-07392-X.
[2]  J. A. Wesslinger and R. Krishna, Mass Transfer. Ellis Horwood, New York, 1990.
[3]  R. Kodym, D. Snita, and K. Bouzek, Mathematical Modelling of Electromembrane Processes. In Current Trends and Future Developments on (Bio-)Membranes. Membrane Desalination Systems: The Next Generation. (A. Basile, C. Curcio, and D. Inamuddin, Eds.). Elsevier, Amsterdam, 2019, Chapter 12, 285–326. ISBN 978-0-12-813551-8.

[4]   T. Postler, et al, Parametrical studies of electroosmotic transport characteristics in sub-micrometer channels. J. Colloid Interface Sci. 320(1): 321–332, 2008.

[5]   V. V. Nikonenko, N. D. Pismenskaya, E. I. Belova, P. Sistat, P. Huguet, G. Pourcelly, and C. Larchet, Intensive current transfer in membrane systems: Modelling, mechanisms and application in electrodialysis. Adv. Colloid Interface Sci. 160(1–2): 101–123, 2010.

[6]   K. A. Mauritz and R. B. Moore, State of understanding of Nafion. Chem. Rev. 104(10): 4535–4586, 2004.

[7]   H. Strathmann, Ion-Exchange Membrane Separation Processes. Elsevier, UK, 2004.

[8]   K. S. Spiegler, Transport process in ionic membranes. Trans. Faraday Soc. 54: 1408–1428, 1958.

[9]   A. Černín, Aplikace elektrodialýzy pro zpracování oplachových vod z galvanického zinkování. PhD Thesis (in Czech), VŠCHT Praha, 2001.

[10]  H. Strathmann, A. Grabowski, and G. Eigenberger, Ion-exchange membranes in the chemical process industry. Ind. Eng. Chem. Res. 52(31): 10364–10379, 2013.

[11]  V. V. Nikonenko, A. V. Kovalenko, K. Mahamet, K. Urtenov, N. D. Pismenskaya, J. Han, P. Sistat, and G. Pourcelly, Desalination at overlimiting currents: State-of-the-art and perspectives. Desalination 342: 85–106, 2014.

[12]  M. B. Andersen, M. van Soestbergen, A. Mani, H. Bruus, P. M. Biesheuvel, and M. Z. Bazant, Current-induced membrane discharge. Phys. Rev. Lett. 109: 108301, 2012.

Dalimil Šnita, Roman Kodym

# 6 Mathematical modelling of electromembrane processes

## 6.1 Introduction

Chapter 3 offers readers the basics of electrochemistry, useful for understanding electromembrane processes. This chapter contains less general information leading to the **mathematical description of the behaviour of electrolyte solutions near ion-selective membranes** (the electrolyte solution will often be abbreviated, not entirely correctly, as the **electrolyte**) and outlines the methodology of the mathematical modelling of partial processes occurring in electromembrane devices. The description and modelling of electromembrane devices as a whole can be found in Chapter 7.

Mathematical modelling is a widely recognized method enabling to analyse, understand and use experimental and practical knowledge. **A mathematical model** is a set of mathematical relations describing the processes and principles of the real world. A part of this world can be represented, for example, by an industrial unit consisting of several partial operations or just by a chemical reactor representing a single manufacturing operation. Wider use of mathematical models dates back to the mid-1960s. Not by chance, computer technology had been simultaneously developed to a sufficient level by that time. Since that time, the key components of industrial design have no longer been solved by the empirical approach and with personal experience of chemical engineers and complex mathematical tasks enabled to estimate the dependence of basic operating parameters and the dimensions of the device only with the use of computers. However, the first computers had limited abilities and it was problematic to follow.

One of the first overviews of the quantitative mathematical description of electrochemical engineering processes (and also electromembrane processes), including mass, charge and heat transfer, was introduced by Newman et al. [1]. The work of these authors is still very relevant. The need to analyse systematically and critically the use of the methods of numerical mathematics to solve the problems of electrochemical engineering was also understood by Roušar et al. [2] – it was the first systematic study focused on the field at that time. The problems of transport and transfer phenomena are summarized in detail by Bird et al. [3] and Taylor et al. [4] in their books. An important aspect is also the problem of fluid dynamics. From this perspective, the book by Wilkes et al. [5] can be recommended, emphasizing applications in chemical engineering. The mathematical fundamentals necessary to resolve general chemical engineering tasks can be found in Kreyszig et al. [6].

https://doi.org/10.1515/9783110739466-007

The high computing power of today's computers allows to design complex mathematical models encompassing more physical processes – and these models are closer and closer to reality. Besides, they can be resolved in hours or days, which is particularly important from the practical point of view. At present, there is a wide selection of simulation programs with intuitive user environment. They have one or more numerical methods included in their source code to resolve sets of partial differential equations, so the time-consuming part of designing models, that is the design of the program code, is largely transferred to the software itself and the engineer's work is easier. However, it does not diminish the need to understand the mathematical background of proposed models and the physical and chemical nature of studied processes. Some examples of commercial simulation programs are COMSOL Multiphysics$^{TM}$, ANSYS (CFX, FLUENT) or the ACE + software package.

Mathematical models allow simulating the behaviour of a device or an interconnected system of more devices. This particularly contributes to the following goals:

- **Basic analysis of the system and quantitative description of its behaviour**. Without basic understanding of the system, successful resolution of operational problems cannot be expected.
- **Optimization of the system** (e.g. modification of geometry or operating conditions) **or scaling-up and scaling-down** of the system dimensions with the goal of increasing/decreasing its performance. The model results serve as a supportive argument in the difficult strategic decision-making of designers, because the empiric approach based on the experience of the engineers themselves or on experimental results may be insufficient and in the case of large industrial devices, it is very expensive and time consuming.
- **Study of local phenomena occurring at the level of millimetres or less**. At the sub-millimetre level, it is especially difficult to measure quantities like concentration, temperature and voltage.
- **Study of the behaviour of the device in limit to extreme cases at the edge of its destruction**, which allows establishing boundary operating conditions. Similar experiments cannot be performed on a real device under semi-operational and industrial conditions due to extreme financial and time demands. Sometimes they happen accidentally, for example due to a failure of control mechanisms. The analysis of these accidents can be used to verify predictions of mathematical models.

If any mathematical model is to become a useful and effective tool for engineers, its use must be verified by experimental or operational measurements. The reliability of the model then depends on adopted simplification assumptions, that is on the necessary simplifications included in the model in comparison with reality. The systems or devices subject to mathematical modelling are de facto removed from the infinite space that surrounds them. The inner space of the device is then separated from the surroundings by boundaries on which the solution is known and more or less corresponds to reality.

The following text presents an overview of the basic types of mathematical models, model relationships and principles, and last but not least, their practical applications.

## 6.1.1 Subject of mathematical modelling of electromembrane processes

The most important process in electromembrane devices is mass transfer, in particular the transfer of the mass of components. What is crucial is that at least two components have ionic character, that is they carry a non-zero electric charge. This allows influencing the behaviour of the system using an external electric field (usually with a voltage source) which is induced by at least two electrodes (**anode** and **cathode**, see Chapter 3).

Devices for electromembrane processes are **heterogeneous**; their space contains different components with different properties. The behaviour on the interface of these components (e.g. **membrane–electrolyte**) is also important.

**The transport of components is selective** based on not only the differences in their electric charge but also in the size and shape of molecules and the degree of their interaction with the environment, both chemical (e.g. dissociation degrees) and physical (e.g. adsorption). The degree of transport selectivity in various parts of the system can be influenced by the **choice of materials.** Thanks to our knowledge of many creations of nature (cells of living organisms could not exist without the selectivity of the transport of components through biological membranes), we can now design artificial materials with convenient selective properties on the basis of scientific, practical, technical and technological information. The core is a **rigid skeleton**, consisting of inorganic or more often organic **polymers** modified physically or by chemical bonding of other substances (especially those carrying electric charge). These are **ion-active, ion-selective, ion-exchange** (on a more general level, these terms are synonymous) materials in the form of, for example, **particles** or **membranes.**

## 6.1.2 Methods of mathematical modelling of electromembrane processes

We will use the methods of chemical and electrochemical engineering that can be summarized in the scheme

balance and general laws + equilibrium + rate of processes.

The mass and energy balances represent the solid ground of any model. Actually, they can be considered as the accounting of natural and technical sciences that can be written in the form

accumulation = input – output + source

We balance some extensive quantity, for example the amount of substance of a component. Depending on the selection of the balancing system, we can distinguish macroscopic balances, **integral in space, concerning the entire system**, for example, some device as a whole. Such balances cannot provide information about detailed behaviour within the system. If we want to analyse phenomena at different places within the system, we must divide it into subsystems – spatial domains, that is, larger or smaller parts – study these parts and express the total behaviour by means of a sum, by **integration**. If we work with relatively large spatial domains, this is integration in the general sense, that is, creation of a whole from particulars (particulars are, e.g. **nodes** and **fluxes**), forming together the whole, **balance scheme**. If spatial domains are very small, infinitely small in mathematical abstraction, the balances of these volumes (called **elementary volumes**) are formed by **differential equations** and we obtain the behaviour of the entire system by **mathematical integration in space**. The location of an infinitely small part is precisely determined, localized in space, so balances are **spatially local**.

Similarly, according to the selection of the balance period, time interval in which we are balancing, we can distinguish macroscopic balances, **integral in time, concerning the entire balance period**, for example some batch process as a whole. Such balances cannot provide information about detailed behaviour at different moments within the balance period. If we want to analyse what happens in different periods or moments of the balance period, we must divide the balance period into large or smaller parts, **time intervals**, study these intervals and express the total behaviour by means of a sum, by **integration**. If time intervals are relatively long, it is a sum, integration in the general sense, that is creation of a whole from particulars (particulars are for example months forming a whole, year). If time intervals are very short, infinitely small in mathematical abstraction, the balances are formed by **differential equations** and we obtain the behaviour of the entire system by **mathematical integration in time**. The location of an infinitely short time interval is precisely determined, localized in time, so balances are **time local**.

Very useful are balances that are local both in space $(x, y, z)$ and in time $(\tau)$:

$$\frac{\partial \rho_Y}{\partial \tau} = -\nabla \cdot \Phi_Y + \sigma_Y \tag{6.1}$$

where individual terms represent (from left to right) accumulation, input and source of the quantity $Y$, all in volume unit per time unit, $\rho_Y$ is the density of the quantity $Y$, $\Phi_Y$ is the flux (intensity, surface density) of the quantity $Y$ and $\sigma_Y$ is the source of the quantity $Y$ (see Section 6.9). In this book, we will mainly use **material balances of components**:

$$\frac{\partial c_i}{\partial \tau} = -\nabla \cdot \boldsymbol{J}_i + \sum_j v_{i,j} r_j \qquad (6.2)$$

where $c_i$ and $\boldsymbol{J}_i$ are the molar concentration and intensity of flux of the $i$-th component (unit: mol s$^{-1}$ m$^{-2}$), $v_{i,j}$ is the stoichiometric coefficient and $r_j$ is the rate of the $j$-th chemical reaction (unit: mol s$^{-1}$ m$^{-3}$). The linear combination of the balances of components represents the **local balance of the electric charge**

$$\frac{\partial q}{\partial \tau} = -\nabla \cdot \boldsymbol{j}, \qquad q = F \sum z_i c_i, \qquad \boldsymbol{j} = F \sum z_i \boldsymbol{J}_i \qquad (6.3)$$

where $q$ (unit: C m$^{-3}$) is the density of the electric charge, $\boldsymbol{j}$ (unit: A m$^{-2}$) is current density (intensity of the charge flux) and $F$ is the Faraday constant. In examples given in this book, we will mainly deal with systems in a steady state, that is with zero accumulation ($\partial c_i / \partial \tau = 0$). In a steady state without chemical reactions, it applies for spatially one-dimensional (1D) systems

$$0 = -\frac{\mathrm{d}J_i}{\mathrm{d}x} + 0, \qquad 0 = -\frac{\mathrm{d}j}{\mathrm{d}x} \qquad (6.4)$$

In such case, the fluxes of components $J_i$ and electric charge $j$ are spatially uniform (constant). Spatially 1D approximation is typically a very useful simplification in the description of a very thin layer (e.g. membrane) where the component transfer in the direction parallel to the layer is negligible in comparison with the transfer in the direction perpendicular to the layer.

Out of general laws, in this book, we will use mainly **Faraday's law, Coulomb's law** and **Gauss laws of electrostatics** (see Chapter 3), which will be expressed in the form of **Poisson's equation** for the practical modelling purposes:

$$\nabla^2 \varphi = -\frac{q}{\varepsilon} \qquad (6.5)$$

where $\varphi$ is the electrostatic potential, $q$ is the charge density and $\varepsilon$ is the permittivity (unit: C V$^{-1}$ m$^{-1}$ = F m$^{-1}$; in the entire chapter, we assume that the permittivity is constant). If inputs and outputs are zero, we call the system **isolated**. The steady state of an isolated system is called the **equilibrium state**. The key issues of mathematical modelling (and not only mathematical one) include:
- Is the system sufficiently steady?
- Is the system sufficiently isolated?
- Is the system sufficiently close to the equilibrium?
- If not, how far is it from the equilibrium?

These issues should be studied by the analysis of experimental results and mathematical models (analytically and/or numerically). If the system is far from the equilibrium, the equations describing the equilibrium state probably cannot be used to describe the system. If the system is closed to the equilibrium (it is only slightly non-equilibrium), we can use the principle of local equilibrium, although the system is in an overall global non-equilibrium. **Electric current flows through electromembrane devices. Therefore, the system is not isolated and it is out of equilibrium.**

Two types of extremely fast processes occur in aqueous electrolyte solutions: chemical reactions connected with dissociation of electrolytes (molecule loses or gains protons) and transport of ions caused by a local surplus or lack of the electric charge. The first type of processes approaches the state of the system to dissociation equilibria. In aqueous solutions, for example, the dissociation equilibrium is achieved, given by the ionic product of water:

$$c(H^+)c(OH^-) = K_w \tag{6.6a}$$

The second type of processes approaches the state of the system to the local electroneutrality, where

$$q = 0 \tag{6.6b}$$

(see Section 6.8.3). If the system is not far from equilibrium (the intensity of the electric current and applied electric voltage are not very high), **equilibrium equations** are not completely **accurate** but they can be a very good **approximation** of non-equilibrium models and probably also the reality. Understanding the conditions of electroneutrality and possible deviations from electroneutrality is one of the most important conditions for understanding electromembrane and generally electrochemical processes. For describing electromembrane processes, **phase equilibria** are also very important (see Chapter 3).

In this book, we will use two types of models: **non-equilibrium models**, based on **Poisson's equation and the kinetics of dissociation reactions**, which will be marked with the acronym **PNP** (Poisson–Nernst–Planck equation), and mainly **local equilibrium models**, based on the **electroneutrality condition and the dissociation and phase equilibria.**

## 6.1.3 Transport and transformation processes

The processes that take place in electromembrane devices (and probably any processes) can be divided into two groups: **transport processes** in which something moves, and **transformation processes**, especially **chemical reactions**, in which something changes into something else. The result of these processes can be an **accumulation**, which means that something somewhere is being gathered or diminished. Actually, this is an expression of the balance definitions above.

Generally, it is assumed that **near equilibrium, the rate of processes is proportional to the deviation from equilibrium,** which is the basis of **linear irreversible thermodynamics**. In equilibrium, the rate of processes is defined as zero. Far from equilibrium, almost nothing can be ruled out and in hydrodynamics, **turbulent** behaviour occurs. An extreme and probably the most important manifestation of phenomena far from equilibrium is life itself.

The equations describing **transport processes, transport relations,** are historically referred to as laws, even though they have not the character of generally valid laws but they are constitutive equations, more or less precisely valid for different environments and different conditions. Yet transport relations are extremely useful because often nothing better is available. They include, for example, Newton's law of viscosity, Fourier's law of thermal conduction, Fick's law of diffusion, Ohm's law of the charge flow, Darcy's law for the flow of a fluid through a porous medium and law describing the relationship between polarization and intensity of the electric field. In the context of this book, perhaps the most important transport relation is **Nernst–Planck equation** (see Chapter 3), given here in the form for ideal behaviour:

$$J_i = -D_i \nabla c_i - c_i \frac{D_i}{RT} z_i F \nabla \varphi + \boldsymbol{v} c_i = -c_i \frac{D_i}{RT} \nabla \tilde{\mu}_i + \boldsymbol{v} c_i \tag{6.7}$$

where $J_i$ is the intensity of the molar flux of component $i$, $\boldsymbol{v}$ is the vector of the flow velocity (unit: m s$^{-1}$), $c_i$, $D_i$, $z_i$ and $\tilde{\mu}_i$ are molar concentration, diffusion coefficient (diffusivity), charge number and electrochemical potential of the component, respectively. The transport rate of any quantity is a vector (tensor) quantity because transport of any scalar (vector) quantity in a given point of space always occurs through some oriented surface in a particular direction.

**Transformation (chemical) processes** are described by **kinetic equations,** which are not general laws either. The situation is considerably complicated by the fact that transformation (chemical) processes do not happen in elementary steps but in more or less complicated sequences controlled by mechanisms about which we know very little and we only have more or less verified hypotheses even in the case of important processes. The change of a certain substance (component, molecule, ion etc.) into another substance is a scalar quantity because time only has one direction, from the past to the future.

## 6.1.4 Flow

To understand electromembrane processes, it is often very important to describe the flow of electrolytes in the working chambers or channels of the device. In most cases, we can assume that the flow is largely caused by a pressure gradient, and it is not affected by the composition of flowing solutions, electric field and transport of components. Then we can solve the problem of hydrodynamics **in advance** (independently on other physical and chemical processes in the system), whether by an analytical or numerical solution of the flow equations or by a reasonable hypothesis about the distribution of the rate field within the device. **Subsequently, we solve the material and transport equations together with the equations describing the electric field.** In special cases, however, we have to solve the equations of flow (hydrodynamics) **together** with the substance balances and field equations because the spatial distribution of concentrations and electric field can influence the flow (e.g. in the case of electroosmotic flow) and vice versa.

### 6.1.5 Distribution of the electric potential at zero fluxes of components

In equilibrium, component fluxes are zero. For electrically neutral components, the condition of their zero flux requires a spatially uniform (identical everywhere) field of their concentrations (see Fick's law). To describe the fluxes of electrically charged components, we can use Nernst–Planck equation without any convective term (we assume a stationary environment). However, the spatial field of concentrations need not be uniform if the diffusion flux is compensated by the same migration flux in the electric field in the opposite direction. The electrical potential ($\varphi$) difference then brings us information on the concentration fields of components in equilibrium (near equilibrium). The analytical solution leads to the **Boltzmann's distribution**:

$$\boldsymbol{J}_i = -c_i \frac{D_i}{RT} \nabla \tilde{\mu}_i = -D_i \nabla c_i - c_i \frac{D_i}{RT} z_i \boldsymbol{F} \nabla \varphi = 0 \quad \Rightarrow c_i = c_{i,0} \exp\left(-\frac{RT}{z_i F}(\varphi - \varphi_0)\right) \quad (6.8)$$

The same fact can be expressed by means of electrochemical potentials that are constant in space under given conditions, that is

$$\tilde{\mu}_i = RT \ln c_i + z_i \boldsymbol{F}\varphi = RT \ln c_{i,0} + z_i \boldsymbol{F}\varphi_0 = \tilde{\mu}_{i,0} \quad (6.9)$$

## 6.2 Transport numbers

Even though transport numbers are important electrochemical quantities, they have various definitions and interpretations. According to the basic definition, the mean value of the transport number of a component is the ratio between the electric current carried by the component $I_i$ and the total current $I$:

$$\widehat{t}_i = \frac{I_i}{I} = \frac{I_i}{\sum_k I_k} \quad (6.10)$$

To determine these currents, we must define the macroscopic area $A$ through which the currents flow, because even the partial currents $I_i$ are the surface integrals of the current densities $\boldsymbol{j}_i$ ($\boldsymbol{s}$ marks the trajectory):

$$I_i = \int_A \boldsymbol{j}_i \cdot d\boldsymbol{s}, \qquad I = \int_A \boldsymbol{j} \cdot d\boldsymbol{s}, \qquad \boldsymbol{j}_i = Fz_iJ_i, \qquad \boldsymbol{j} = \sum_k \boldsymbol{j}_k \tag{6.11}$$

The values of transport numbers defined in this way generally depend on the selection of area $A$, but under certain conditions, they have a practical interpretation. If this surface is planar, we will only consider the vector components of current densities perpendicular to this plane. Let us also assume that the field of current densities on this surface is constant (uniform), which applies well to very small surfaces. Then

$$\hat{t}_i = \frac{\int_A \boldsymbol{j}_i \cdot d\boldsymbol{s}}{\int_A \boldsymbol{j} \cdot d\boldsymbol{s}} = t'_i = \frac{j_i \oint_A d\boldsymbol{s}}{j \oint_A d\boldsymbol{s}} = \frac{j_i}{j} = \frac{j_i}{\sum_k j_k} = \frac{Fz_iJ_i}{\sum_k Fz_kJ_k} \tag{6.12}$$

Quantities $\hat{t}_i$ can be considered as global (integral) transport numbers, while $t'_i$ are local (differential) transport numbers. If we do not simplify our consideration to a 1D system and work with a spatially multidimensional system (2D, 3D), the local transport number $t'_i$ would have the tensor nature: $\boldsymbol{j}_i = t'_i \cdot \boldsymbol{j}$ (a formal ratio between two vectors is a tensor).

For the surface intensities of the molar fluxes of components $J_i$, we substitute from Nernst–Planck equation:

$$t'_i = \frac{j_{i,\text{dif}} + j_{i,\text{mig}} + j_{i,\text{conv}}}{\sum_k \left(j_{i,\text{dif}} + j_{i,\text{mig}} + j_{i,\text{conv}}\right)} = \frac{j_{i,\text{dif}} + j_{i,\text{mig}} + j_{i,\text{conv}}}{j_{\text{dif}} + j_{\text{mig}} + j_{\text{conv}}} \tag{6.13a}$$

$$J_{i,\text{dif}} = -z_iFD_i\frac{dc_i}{dx}, \qquad J_{i,\text{mig}} = -\frac{c_iz_i^2D_iF^2}{RT}\frac{d\varphi}{dx}, \qquad J_{i,\text{conv}} = vz_iFc_i \tag{6.13b}$$

$$j_{\text{dif}} = -F\sum_k z_ic_iD_i\frac{dc_i}{dx}, \quad j_{\text{mig}} = -\overbrace{\sum_k \frac{c_kz_k^2D_kF^2}{RT}}^{\kappa}\frac{d\varphi}{dx}, \quad j_{\text{conv}} = v\overbrace{F\sum_k z_ic_i}^{q} \tag{6.13c}$$

Here, $j_{i,\text{diff}}$, $j_{i,\text{mig}}$ and $j_{i,\text{conv}}$ are the convective, diffusion and migration components of current densities, $q = F\sum_k z_ic_i$ is the charge density (unit: C m$^{-3}$) and

$$\kappa = \sum_k \frac{c_kz_k^2D_kF^2}{FT}$$

is the conductivity (unit: S m$^{-1}$).

Assuming that **convection and diffusion are negligible**, then

$$t'_i = t_i = \frac{j_{i,\text{mig}}}{j_{\text{mig}}} = \frac{D_ic_iz_i^2}{\sum_k D_kc_kz_k^2} \tag{6.14}$$

This assumption is very well met in **well-mixed macroscopic** (electroneutral) systems (without any composition gradient, i.e. without diffusion), in which the transport numbers are mostly **determined experimentally**. Therefore, we have three definitions of transport numbers, global $\hat{t}_i$, local $t'_i$ and local without diffusion (migration) $t_i$:

$$\hat{t}_i = \frac{I_i}{I} \qquad\qquad t'_i = \frac{j_i}{j} \qquad\qquad t_i = \frac{j_{i,\mathrm{mig}}}{j_{\mathrm{mig}}} \qquad\qquad (6.15)$$

which only give the same result under special conditions. Since the last of these definitions is **most frequently used in electrochemical practice**, if we talk about transport numbers without any further specification in the following text, we will mean migration (non-diffusion, experimental) numbers. In the interpretation of the meaning of transport numbers, **binary electrolytes** have a completely exceptional status (see Section 6.6.2), because due to electroneutrality, migration transport numbers are independent of concentration.

## 6.3 Electroneutral binary electrolytes

We consider the **binary electrolyte** $(C^{z+})_{\nu_+}(A^{z-})_{\nu_-}$ **in condition of electroneutrality**. We express the concentrations of cation and anion by the total salt concentration $c$:

$$c_+ = \nu_+ \, c, \quad c_- = \nu_- \, c \quad z_+ c_+ + z_- c_- = z_+ \, c_+ \, c + z_- \, c_- \, c = 0$$

$$z_+ c_+ + z_- c_- = 0 \quad\Rightarrow\quad -\frac{\nu_+}{z_-} = \frac{\nu_-}{z_+} = n, \quad \nu_+ = -nz_-, \quad \nu_- = nz_+ \qquad (6.16)$$

where $z_i$, $\nu_i$ and $c_i$ are charge numbers, stoichiometric coefficients and molar concentrations of cation and anion (for transcript simplification, we use convention $i = +$ for cation and $i = -$ for anion). For example, for $Fe_2(SO_4)_3 \to 2\,Fe^{3+} + 3\,SO_4^{2-}$, it applies that

$$z_+ = 3, \quad \nu_+ = 2, \quad z_- = -2, \quad \nu_- = 3 \qquad\qquad (6.17)$$

By substituting into Nernst–Planck equation, we get for the intensities of molar fluxes:

$$\boldsymbol{J}_+ = -D_+ \nu_+ \nabla c - D_+ z_+ \nu_+ c \frac{F}{RT} \nabla \varphi = +D_+ nz_- \nabla c + D_+ z_+ nz_- c \frac{F}{RT} \nabla \varphi \qquad (6.18\mathrm{a})$$

$$\boldsymbol{J}_- = -D_- \nu_- \nabla c - D_- z_- \nu_- c \frac{F}{RT} \nabla \varphi = -D_- nz_+ \nabla c - D_- z_- nz_+ c \frac{F}{RT} \nabla \varphi \qquad (6.18\mathrm{b})$$

where $D_i$ (unit: $m^2\,s^{-1}$) are the diffusion coefficients (to simplify the notation, we use again for cation $i = +$ and for anion $i = -$). Then we express the densities of partial electric currents $j_{\pm}$ and of the total electric current $j$ (unit: A m$^{-2}$) and the electric potential gradient $\nabla\varphi$ (unit: V m$^{-1}$):

$$j_+ = Fz_+ J_+, \qquad\qquad\qquad j_- = Fz_- J_- \qquad\qquad\qquad (6.19)$$

$$j = j_+ + j_- = Fz_+ J_+ + Fz_- J_- =$$

$$= nF(D_+z_+ z_- - D_-z_+ z_-)\nabla c + nF(D_+z_+ z_- z_+ - D_-z_-z_- z_+)c\frac{F}{RT}\nabla\varphi$$

$$\frac{F}{RT}\nabla\varphi = \frac{j - nF(D_+z_+ z_- - D_-z_+ z_-)\nabla c}{nF(D_+z_+ z_- z_+ - D_-z_-z_- z_+)}$$

$$= \frac{j}{nF(D_+z_+ z_- z_+ - D_-z_-z_- z_+)c} - \frac{(D_+ - D_-)\nabla c}{(D_+z_+ - D_-z_-)c} \qquad (6.20)$$

Then we substitute the electric potential gradient expressed in this manner into the relations (6.18a) and (6.18b), and we get:

$$J_+ = +D_+nz_- \nabla c + D_+z_+ nz_- c\left(\frac{j}{nF(D_+z_+ z_- z_+ - D_-z_-z_- z_+)c}\right) - \frac{(D_+ - D_-)\nabla c}{(D_+z_+ - D_-z_-)c}$$

$$J_+ = -\overbrace{\frac{D_+D_-(z_+ - z_-)}{D_+z_+ - D_-z_-}}^{D_{\text{eff}}} nz_- \nabla c_- - \overbrace{\frac{D_+z_+}{(D_+z_+ - D_-z_-)z_+}}^{t_+}\frac{j}{F} = -D_{\text{eff}}\nabla c + \frac{t_+}{z_+}\frac{j}{F} \qquad (6.21a)$$

and similarly

$$J_- = -D_{\text{eff}}\nabla c + \frac{t_-}{z_-}\frac{j}{F} \qquad\qquad (6.21b)$$

$$j_+ = t'_+ j = -F z_+ D_{\text{eff}}\nabla c_+ + t_+ j \qquad\qquad (6.22a)$$

$$j_- = t'_- j = -F z_- D_{\text{eff}}\nabla c_+ + t_- j \qquad\qquad (6.22b)$$

Here $D_{\text{eff}}$ is the effective diffusion coefficient and $t_i$ are the migration (non-diffusion) transport numbers of cation and anion:

$$D_{\text{eff}} = \frac{D_+D_-(z_+ - z_-)}{D_+z_+ - D_-z_-} \qquad t_+ = \frac{D_+z_+}{(D_+z_+ - D_-z_-)} \qquad t_- = \frac{-D_-z_-}{(D_+z_+ - D_-z_-)} \qquad (6.23a)$$

In the special case when $z_+ = -z_- = |z|$, it applies:

$$D_{\text{eff}} = \frac{2D_+D_-}{D_+ + D_-} \qquad t_+ = \frac{D_+}{(D_+ + D_-)}, \qquad t_- = \frac{D_-}{(D_+ + D_-)} \qquad (6.23b)$$

and if the diffusion coefficients of both components are identical, $D_+ = D_- = D$, it applies:

$$D_{\text{eff}} = D, \qquad t_+ = \frac{z_+}{(z_+ - z_-)}, \qquad t_- = \frac{-z_-}{(z_+ - z_-)} \qquad (z_- < 0) \qquad (6.23c)$$

In the special case of the symmetrical binary electrolyte $z_+ = -z_- = |z|$ and simultaneously $D_+ = D_- = D$, it applies

$$D_{\text{eff}} = D, \quad t_+ = t_- = 0.5 \qquad (6.23d)$$

A currentless state is typical for the transport of components through membranes without any external electric field, without electrodes (e.g. in dialysis). By integrating eq. (6.20) on the assumption of **zero electric current**, we obtain the difference of electric potentials between the general point and reference point 0, the **diffusion potential**:

$$\varphi - \varphi_0 = -\frac{RT}{F}\frac{D_+ - D_-}{D_+ z_+ - D_- z_-}\ln\frac{c}{c_0} \qquad (6.24)$$

# 6.4 Limiting current

It is well known that by increasing the driving force of electromembrane processes, that is the difference of electric potential, the electric current is not increased linearly, but reaches the **maximum value** (approaches it asymptotically), **limiting current** (unless the device gets damaged sooner by overheating). This phenomenon can have two substantially different causes:

- **Global:** All charge carriers, ions, that could be transported in the device have already been transported, for example for electrodialysis the state approaches complete desalination of the diluate.
- **Local:** In a certain thin layer, through which the electric current has to pass, there will be local depletion of charge carriers, that is **local reduction of the electric conductivity**. The total voltage difference is then inefficiently concentrated on this layer (these layers). In the following text, we will discuss this second case. These layers typically occur in the electrolyte very close to the phase interface (electrolyte–membrane, electrolyte–electrode).

## 6.4.1 Symmetrical binary electrolyte

Here we will study the stabilized behaviour of the fixed (Nernst diffusion) layer of thickness $\delta$, containing the univalent symmetrical electrolyte ($D_+ = D_- = D$, $z_+ = -z_- = 1$), while the condition of electroneutrality is observed, that is $c_+(x) = c_-(x) = c(x)$. (Note: In some parts of this book, the axis perpendicular to membranes is marked as $z$ because the $x$-axis denotes the predominant flow direction.) The local balance of components in a spatially 1D system can be written for a steady state and no chemical reaction occurring in the form:

$$\frac{\partial c_\pm}{\partial \tau} = -\frac{\partial J_\pm}{\partial x} = 0 \quad \Rightarrow \quad J_\pm(x) = J_\pm \neq f_\pm(x) \tag{6.25}$$

In a steady state, the intensities of flux $J_\pm$ thus do not depend on the spatial coordinate $x$. Suppose there is the left electrode – cathode on the left outside the layer and the right electrode – anode on the right. This arrangement of electrodes corresponds to a negative electric current (i.e. flowing from right to left) and the gradient of the electric potential is positive (i.e. decreasing from right to left). (If the anode is on the left and cathode on the right, the electric current is positive and the gradient potential negative.) We assume that on one side $(x = x_0)$ the layer adheres to an ideally mixed environment and the concentration on this interface is equal to the concentration in this environment, $c(x_0) = c_{bulk}$. On the other side $(x = x_0 - \delta$ or $x = x_0 + \delta)$, the layer adheres to an ion-selective membrane, see Fig. 6.1a, c (for other possible arrangements see Fig. 6.1b, d, the electric current is in all cases oriented from right to left). We assume that the membrane is ideal, so it is permeable for only one type of ions. Through cation-selective membranes (CM) only cations can pass, while anion-selective membranes (AM) is permeable only for anions only. The current density

$$j = j_+ + j_- \tag{6.26}$$

is the sum of the current densities carried by individual components:

$$j_+ = +FJ_+ = t'_+ j, \quad t'_+ = \frac{j_+}{j} = \frac{J_+}{J_+ - J_-}, \quad t'_+ + t'_- = 1 \tag{6.27a}$$

$$j_- = -FJ_- = t'_- j, \quad t'_- = \frac{j_-}{j} = \frac{J_+}{J_+ - J_-}, \quad t'_+ - t'_- = \alpha \tag{6.27b}$$

where $\alpha = t'_+ - t'_-$ is the difference between transport numbers. For ideal membranes, it applies:

cation–selective membrane:     $j = j_+, t'_+ = 1, t'_- = 0, \alpha = +1$ \hfill (6.28a)

anion–selective membrane:      $j = j_+, t'_+ = 0, t'_- = 1, \alpha = -1$ \hfill (6.28b)

The intensities of the molar fluxes will be expressed using Nernst–Planck equation:

$$J_+ = -D\frac{dc}{dx} - D\frac{F}{RT}c\frac{d\varphi}{dx} \tag{6.29a}$$

$$J_- = -D\frac{dc}{dx} + D\frac{F}{RT}c\frac{d\varphi}{dx} \tag{6.29b}$$

If we express the current densities carried by individual components,

$$j_+ = +FJ_+ = -DF\frac{dc}{dx} - D\frac{F^2}{RT}c\frac{d\varphi}{dx} = t'_+ j \tag{6.30a}$$

$$j_- = -FJ_- = DF\frac{dc}{dx} - D\frac{F^2}{RT}c\frac{d\varphi}{dx} = t'_- j \tag{6.30b}$$

the difference and sum of these equations have the form:

$$j_+ - j_- = -2DF\frac{dc}{dx} = (t'_+ - t'_-)j = \alpha j \tag{6.31a}$$

$$j_+ + j_- = -2Dc\frac{F^2}{RT}\frac{d\varphi}{dx} = (t'_+ + t'_-)j = j \tag{6.31b}$$

For generalization, it is convenient to transcribe eq. (6.31) into a dimensionless form:

$$\frac{d\tilde{c}}{d\tilde{x}} = -\alpha\tilde{j} \qquad\qquad \frac{d\tilde{\Phi}}{d\tilde{x}} = -\frac{\tilde{j}}{\tilde{c}} \tag{6.32}$$

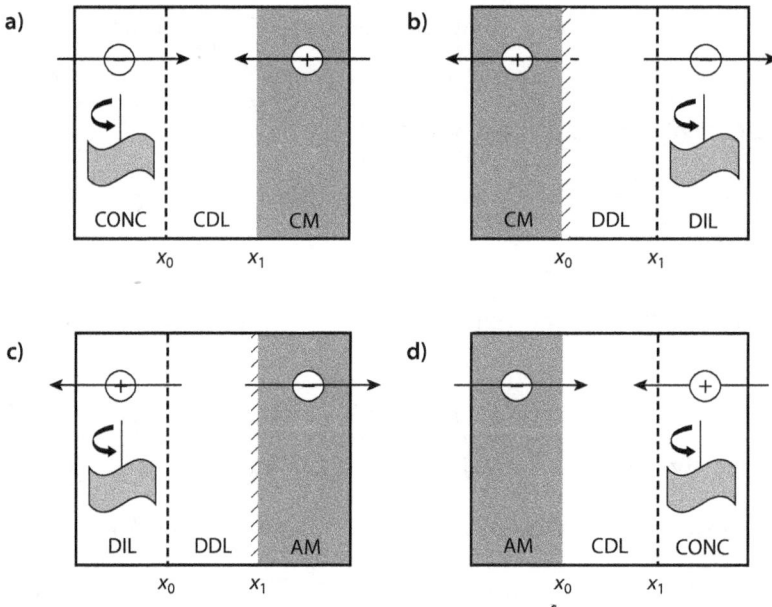

**Fig. 6.1:** Order of layers: (a) ideally mixed concentrate (CONC), concentrate diffusion layer (CDL), cation-selective membrane (CM), zone of low conductivity does not occur, eq. (6.39a) applies; (b) CM, diluate diffusion layer (DDL), ideally mixed diluate (DIL), zone of low conductivity occurs (marked by shading) near the membrane, eq. (6.39b) applies; (c) DIL, DDL, anion-selective membrane (AM), zone of low conductivity occurs near the membrane, eq. (6.39c) applies; (d) AM, CDL, CONC, zone of low conductivity does not occur, eq. (6.39d) applies. In all cases, the electric current is negative (from right to left, the anode is on the left).

where

$$\tilde{x} = \frac{x}{l_0}, \qquad \tilde{\varphi} = \varphi \frac{F}{RT}, \qquad \tilde{c} = \frac{c}{c_0}, \qquad \tilde{j} = \frac{j}{j_0}, \qquad j_0 = 2DF \frac{c_{bulk}}{l_0}$$

If we analyse one layer only, it is natural to select the characteristic length and concentration:

$$l_0 = \delta, \quad c_0 = c_{bulk} \tag{6.33}$$

By integration of eq. (6.32)

$$\int_{\tilde{c}_0}^{\tilde{c}} d\tilde{c} = -\alpha \tilde{j} \int_{\tilde{x}_0}^{\tilde{x}} d\tilde{x} \qquad \qquad \int_{\tilde{\varphi}_0}^{\tilde{\varphi}} d\tilde{\varphi} = -\tilde{j} \int_{\tilde{x}_0}^{\tilde{x}} \frac{1}{\tilde{c}(\tilde{x})} d\tilde{x} \tag{6.34}$$

we get:

$$\tilde{c}(\tilde{x}) = \tilde{c}(\tilde{x}_0) - \alpha \tilde{j} (\tilde{x} - \tilde{x}_0) \tag{6.35a}$$

$$\tilde{\varphi}(\tilde{x}) = \tilde{\varphi}(\tilde{x}_0) + \frac{1}{\alpha} \ln \frac{\tilde{c}(\tilde{x}_0) - \alpha \tilde{j}(\tilde{x} - \tilde{x}_0)}{\tilde{c}(\tilde{x}_0)} \tag{6.35b}$$

The function $\tilde{c}(\tilde{x})$ is linear, while the function $\tilde{\varphi}(\tilde{x})$ is non-linear. These analytical solutions are shown in Fig. 6.2. The limit for $\alpha \to 0$ (transport numbers of both components in neutral membrane are identical) has the form:

$$\tilde{\varphi}(\tilde{x}) = \tilde{\varphi}(\tilde{x}_0) - \frac{\tilde{j}}{\tilde{c}(\tilde{x}_0)} (\tilde{x} - \tilde{x}_0) \tag{6.36}$$

The concentration may not be negative, just in the limit case it can be zero, and then the current density approaches the limit value $\tilde{j}_{lim}$:

$$\frac{\tilde{c}(\tilde{x})}{\tilde{c}(\tilde{x}_0)} = 1 - \frac{\alpha \tilde{j} (\tilde{x} - \tilde{x}_0)}{\tilde{c}(\tilde{x}_0)} = 0 \quad \Rightarrow \quad \tilde{j} = \frac{\tilde{c}(\tilde{x}_0)}{\alpha (\tilde{x} - \tilde{x}_0)} = \tilde{j}_{lim} = \frac{j_{lim}}{j_0}$$

$$j_{lim} = \frac{\tilde{c}(\tilde{x}_0)}{\alpha (\tilde{x} - \tilde{x}_0)} 2DF \frac{c_{bulk}}{\delta} = \frac{\tilde{c}(\tilde{x}_0)}{(t_+ - t_-) (\tilde{x} - \tilde{x}_0)} 2DF \frac{c_{bulk}}{\delta} \tag{6.37}$$

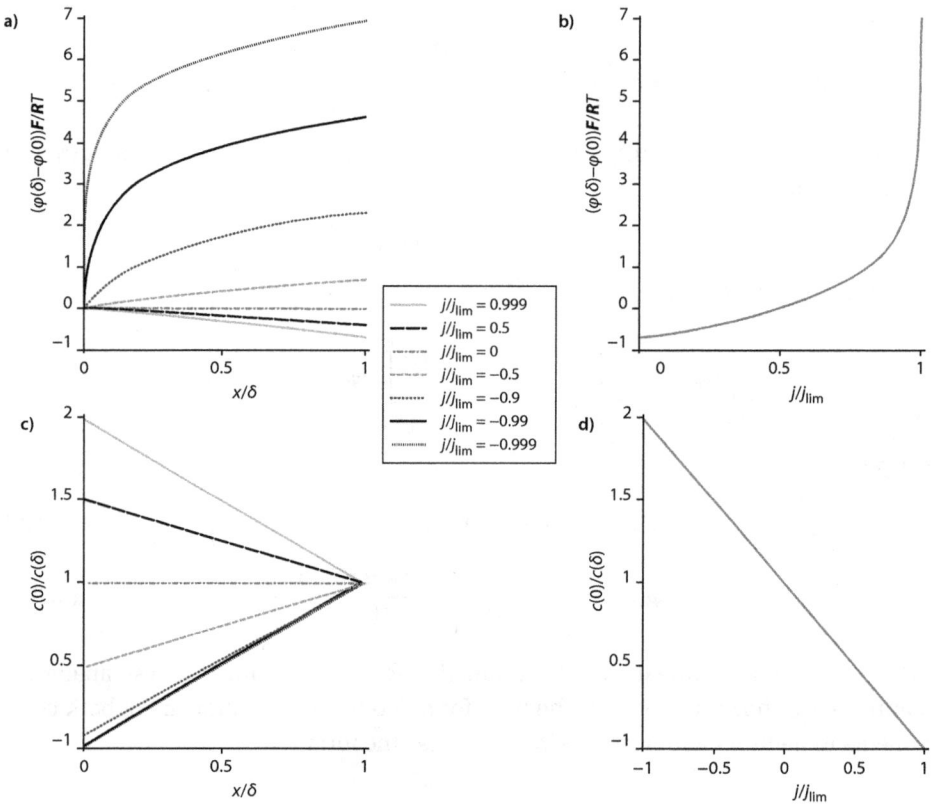

**Fig. 6.2:** Representation of eq. (6.30) for the case (c), see eq. (6.34c). (a) spatial field of the electric potential for different values of the current density, (b) dependence of the potential difference on the current density, (c) concentration field for different values of the current density, (d) dependence of the concentration on the membrane–electrolyte interface on the current density.

and the absolute value of the difference of the electric potentials $\tilde{\varphi}(\tilde{x}) - \tilde{\varphi}(\tilde{x}_0)$ grows beyond all limits. For example:

$$(\tilde{x} - \tilde{x}_0) = 1, \quad \tilde{c}(\tilde{x}_0) = 1, \quad j_{\lim} = \frac{2DF}{(t_+ - t_-)} \frac{c_{\text{bulk}}}{\delta}$$

$$\tilde{\varphi}(\tilde{x}) - \tilde{\varphi}(\tilde{x}_0) = \frac{1}{\alpha} \ln\left(1 - \frac{\alpha \, \tilde{j}(\tilde{x} - \tilde{x}_0)}{\tilde{c}(\tilde{x}_0)}\right) = \frac{1}{\alpha} \ln(1-1) = \frac{-\infty}{\alpha} \tag{6.38}$$

Figure 6.1 shows four possible situations that can occur in electrodialysis (with the anode placed on the left). In cases (b) and (c), thin layers of very low conductivity can occur in the diluate near the membrane, due to the removal of ions from this layer. This example is completely symmetrical (ideal membranes, symmetrical binary electrolyte) so both layers can exist simultaneously. In general cases, layers with very low conductivity only occur either by CM, or by AM, because the layer that is formed as first prevents increasing the current to the limit value and creating the other layer.

**Case (a):** Concentrate is on the left, CM is on the right (Fig. 6.1a):

$$\alpha = 1, \; \tilde{x}_0 = -1, \; \tilde{c}(\tilde{x}_0) = c_{CONC}, \; \tilde{c}(0) = \tilde{c}_{CONC} - \tilde{j}, \; \tilde{\varphi}(0) - \tilde{\varphi}(-1) = \ln\left(\frac{\tilde{c}_{CONC} - \tilde{j}}{\tilde{c}_{CONC}}\right) \quad (6.39a)$$

**Case (b):** CM is on the left, diluate is on the right (Fig. 6.1b):

$$\alpha = 1, \; \tilde{x}_0 = 1, \; \tilde{c}(\tilde{x}_0) = c_{DIL}, \; \tilde{c}(0) = \tilde{c}_{DIL} + \tilde{j}, \; \tilde{\varphi}(1) - \tilde{\varphi}(0) = -\ln\left(\frac{\tilde{c}_{DIL} + \tilde{j}}{\tilde{c}_{DIL}}\right) \quad (6.39b)$$

**Case (c):** Diluate is on the left, AM is on the right (Fig. 6.1c):

$$\alpha = -1, \; \tilde{x}_0 = -1, \; \tilde{c}(\tilde{x}_0) = c_{CONC}, \; \tilde{c}(0) = \tilde{c}_{DIL} + \tilde{j}, \; \tilde{\varphi}(0) - \tilde{\varphi}(-1) = -\ln\left(\frac{\tilde{c}_{DIL} + \tilde{j}}{\tilde{c}_{DIL}}\right) \quad (6.39c)$$

**Case (d):** AM is on the left, concentrate is on the right (Fig. 6.1d):

$$\alpha = -1, \; \tilde{z}_0 = 1, \; \tilde{c}(\tilde{x}_0) = c_{CONC}, \; \tilde{c}(0) = \tilde{c}_{CONC} - \tilde{j}, \; \tilde{\varphi}(1) - \tilde{\varphi}(0) = \ln\left(\frac{\tilde{c}_{CONC} - \tilde{j}}{\tilde{c}_{CONC}}\right) \quad (6.39d)$$

An example of a particular value of the limiting current:

$$D = 2 \times 10^{-9} \text{ m}^2\text{s}^{-1}, \quad F \cong 10^8 \text{ C kmol}^{-1}, \quad c_{bulk} = 10^{-2} \text{ kmol m}^{-3}, \quad \delta = 10^{-5} \text{ m},$$

$$j_{lim} = 2 \times 10^{-9} \times 10^8 \times \frac{10^{-2}}{10^{-5}} = 400 \text{ A m}^{-2} \quad (6.40)$$

## 6.4.2 Non-symmetrical binary electrolyte

In most cases, neither $D_+ = D_-$, nor $z_+ = -z_-$ applies. We will express the concentration of components

$$c_+ = -z_- \, c, \quad c_- = z_+ \, c \quad (6.41)$$

as though they automatically fulfil the electroneutrality condition

$$z_+ c_+ + z_- c_- = -z_+ z_- c + z_- z_+ c = 0 \quad (6.42)$$

We will express the flux intensities using Nernst–Planck equation:

$$J_+ = -D_+ \frac{dc_+}{dx} - D_+ \frac{z_+ F}{RT} c_+ \frac{d\varphi}{dx} = +D_+ z_- \frac{dc}{dx} + D_+ \frac{z_+ z_- F}{RT} c \frac{d\varphi}{dx} \quad (6.43a)$$

$$J_- = -D_-\frac{dc_-}{dx} - D_-\frac{z_-F}{RT}c_-\frac{d\varphi}{dx} = -D_-z_+\frac{dc}{dx} - D_-\frac{z_-z_+F}{RT}c\frac{d\varphi}{dx} \qquad (6.43b)$$

We will divide these equations by the diffusion coefficients and add them up:

$$\frac{J_+}{D_+} + \frac{J_-}{D_-} = (z_- - z_+)\frac{dc}{dx} = \left(\frac{j_+}{z_+D_+} + \frac{j_-}{z_-D_-}\right)\frac{1}{F} = \left(\frac{t'_+}{z_+D_+} + \frac{t'_-}{z_-D_-}\right)\frac{j}{F} \qquad (6.44)$$

$$\frac{dc}{dx} = -\left(\frac{t'_+}{z_+D_+} + \frac{t'_-}{z_-D_-}\right)\left(\frac{1}{z_+ - z_-}\right)\frac{j}{F} \qquad (6.45)$$

Thus, the term $c(d\varphi/dx)$ is removed from the equations. Then we express the current densities corresponding to individual components:

$$j_+ = Fz_+J_+ = +D_+z_+z_-F\frac{dc}{dx} + D_+\frac{z_+^2z_-F^2}{RT}c\frac{d\varphi}{dx} = t'_+j \qquad (6.46a)$$

$$j_- = Fz_-J_- = -D_-z_-z_+F\frac{dc}{dx} - D_-\frac{z_-^2z_+F^2}{RT}c\frac{d\varphi}{dx} = t'_-j \qquad (6.46b)$$

We will divide the equations by the diffusion coefficients and add them up,

$$\frac{j_+}{D_+} + \frac{j_-}{D_-} = (z_+^2z_- - z_-^2z_+)c\frac{d\varphi}{dx}\frac{F^2}{RT} = \left(\frac{t'_+}{D_-} + \frac{t'_-}{D_-}\right)j \qquad (6.47a)$$

$$c\frac{d\varphi}{dx}\frac{F}{RT} = \left(\frac{t'_+}{D_+} + \frac{t'_-}{D_-}\right)\left(\frac{1}{z_+z_-}\right)\left(\frac{1}{z_+ - z_-}\right)\frac{j}{F} \qquad (6.47b)$$

to remove the term $d\varphi/dx$ from the equations. Then we transcribe them into the dimensionless form:

$$\tilde{x} = \frac{x}{\delta}, \tilde{c} = \frac{c}{c_0}, \tilde{\varphi} = \frac{\varphi F}{RT}, \tilde{D}_\pm = \frac{D_\pm}{D_0}, \tilde{j} = \frac{j}{j_0}, j_0 = \frac{(z_+ - z_-)D_0c_0F}{\delta} \qquad (6.48a)$$

$$\frac{d\tilde{c}}{d\tilde{x}} = -\left(\frac{t'_+}{z_+\tilde{D}_+} + \frac{t'_-}{z_-\tilde{D}_-}\right)\left(\frac{1}{z_+ - z_-}\right)\frac{j\delta}{D_0c_0F} = -\overbrace{\left(\frac{t'_+}{z_+\tilde{D}_+} + \frac{t'_-}{z_-\tilde{D}_-}\right)}^{\alpha} = -\alpha\tilde{j} \qquad (6.48b)$$

$$\frac{d\tilde{\varphi}}{d\tilde{x}} = \frac{1}{z_+z_-}\left(\frac{t'_+}{\tilde{D}_+} + \frac{t'_-}{\tilde{D}_-}\right)\left(\frac{1}{z_+ - z_-}\right)\frac{j\delta}{D_0c_0F} = -\frac{1}{z_+z_-}\overbrace{\left(-\frac{t'_+}{\tilde{D}_+} - \frac{t'_-}{\tilde{D}_-}\right)}^{\beta}\frac{\tilde{j}}{\tilde{c}} \qquad (6.48c)$$

We get analytically integrable equations:

$$\frac{d\tilde{c}}{d\tilde{x}} = -\alpha\tilde{j} \qquad (6.49a)$$

$$\frac{d\tilde{\varphi}}{d\tilde{x}} = -\beta \frac{\tilde{j}}{\tilde{c}} \tag{6.49b}$$

that differ from eq. (6.32) for the symmetrical electrolyte ($D_+ = D_- = D_0$, $z_+ = -z_- = 1$) only by the definition of parameter $\alpha$ and parameter $\beta$, which equals to 1 for the symmetrical electrolyte. Here

$$\alpha = \frac{t_+}{z_+ \tilde{D}_+} + \frac{t_-}{z_- \tilde{D}_-} \overset{\tilde{D}_+ = \tilde{D}_- = 1}{=} \frac{t_+}{z_+} - \frac{t_-}{z_-} \overset{z_+ = -z_- = 1}{=} t_+ - t_- \tag{6.49c}$$

$$\beta = -\frac{1}{z_+ z_-} \left( \frac{t_+}{\tilde{D}_+} + \frac{t_-}{\tilde{D}_-} \right) \overset{\tilde{D}_+ = \tilde{D}_- = 1}{=} -\frac{1}{z_+ z_-} (t_+ + t_-) \overset{z_+ = -z_- = 1}{=} t_+ + t_- = 1 \tag{6.49d}$$

Equation (6.32) is a special case of eqs. (6.49a, b).

Examples:

KCl:
$$\alpha = \frac{t_+}{\tilde{D}_+} - \frac{t_-}{\tilde{D}_-} \qquad \beta = \left( \frac{t_+}{\tilde{D}_+} + \frac{t_-}{\tilde{D}_-} \right) \tag{6.50a}$$

Na$_2$SO$_4$:
$$\alpha = \frac{t_+}{\tilde{D}_+} - \frac{t_-}{2\tilde{D}_-} \qquad \beta = \frac{1}{2} \left( \frac{t_+}{\tilde{D}_+} + \frac{t_-}{\tilde{D}_-} \right) \tag{6.50b}$$

FeCl$_3$:
$$\alpha = \frac{t_+}{3\tilde{D}_+} - \frac{t_-}{\tilde{D}_-} \qquad \beta = \frac{1}{3} \left( \frac{t_+}{\tilde{D}_+} + \frac{t_-}{\tilde{D}_-} \right) \tag{6.50c}$$

Fe$_2$(SO$_4$)$_3$:
$$\alpha = \frac{t_+}{3\tilde{D}_+} - \frac{t_-}{2\tilde{D}_-} \qquad \beta = \frac{1}{6} \left( \frac{t_+}{\tilde{D}_+} + \frac{t_-}{\tilde{D}_-} \right) \tag{6.50d}$$

Another method of deducing the limiting current is based on the transport numbers in the solution $t_{+,S}$ and in the membrane $t_{+,M}$. We assume that diffusion in the membrane is negligible, so local transport numbers $t'_{i,M}$ here are equal to migration transport numbers $t_{i,M}$. We assume that component fluxes are identical in the solution and in the membrane (1D system, steady state, without chemical reactions). It is sufficient to analyse the cation flux $J_+$. The analysis based on the anion flux $J_-$ would be complementary

$$\overbrace{J_{+,S} = -D_{eff}v_+ \frac{dc}{dx} + \frac{t_{+,S}}{z_+} \frac{j}{F}}^{\text{in the solution}} = \overbrace{J_{+,M} \frac{t_{+,M}}{z_+} \frac{j}{F}}^{\text{in the membrane}} \tag{6.51a}$$

$$j = \frac{\overbrace{z_+ FD_{eff}v_+}^{K}}{t_{+,S} - t_{+,M}} \frac{dc}{dx} = K \frac{dc}{dx} \tag{6.51b}$$

$$K = \frac{z_+ F D_{\text{eff}} v_+}{t_{+,S} - t_{+,M}} = -\frac{F D_{\text{eff}} z_+ z_-}{t_{+,S} - t_{+,M}} = -\frac{F \dfrac{D_+ D_- (z_+ - z_-)}{D_+ z_+ - D_- z_-} z_+ z_-}{\dfrac{D_+ z_+}{(D_+ z_+ - D_- z_-)} - t_{+,M}} = -\frac{F D_+ D_- (z_+ - z_-) z_+ z_-}{D_+ z_+ - t_{+,M}(D_+ z_+ - D_- z_-)}$$

$$(6.51c)$$

$$j = -\overbrace{\frac{F D_+ D_- (z_+ - z_-) z_+ z_-}{D_+ z_+ - t_{+,M}(D_+ z_+ - D_- z_-)}}^{K} \frac{\mathrm{d}c}{\mathrm{d}x} \qquad (6.51d)$$

Here, we assume that the interface of the membrane and non-flowing electrolyte (diffusion layer) lies at the coordinate $x = 0$ and the interface of the diffusion layer and ideally mixed electrolyte core lies at the coordinate $x = \delta$. The concentration is always non-negative, therefore

$$c(0) \ge 0 \quad \Rightarrow \quad \frac{\mathrm{d}c}{\mathrm{d}x} \le \frac{c(\delta)}{\delta} \qquad (6.52)$$

For an ideal CM, it applies:

$$t_{+,M} = 1 \quad \Rightarrow \quad j = -\overbrace{F D_+ (z_+ - z_-) z_+}^{\text{positive number}} \frac{\mathrm{d}c}{\mathrm{d}x} \ge -F D_+ (z_+ - z_-) z_+ \frac{c(\delta)}{\delta} = j_{\text{lim}} < 0 \quad (6.53)$$

$$\text{NaCl:} \; z_+ = 1, \; z_- = -1 \; \Rightarrow \; j = -2 F D_+ \frac{\mathrm{d}c}{\mathrm{d}x} \ge -2 F D_+ \frac{c(\delta)}{\delta} = j_{\text{lim}} < 0 \qquad (6.54a)$$

$$\text{Na}_2\text{SO}_4: \; z_+ = 1, \; z_- = -2 \; \Rightarrow \; j = -3 F D_+ \frac{\mathrm{d}c}{\mathrm{d}x} \ge -3 F D_+ \frac{c(\delta)}{\delta} = j_{\text{lim}} < 0 \qquad (6.54b)$$

$$\text{FeCl}_3: \; z_+ = 3, \; z_- = -1 \; \Rightarrow \; j = -12 F D_+ \frac{\mathrm{d}c}{\mathrm{d}x} \ge -12 F D_+ \frac{c(\delta)}{\delta} = j_{\text{lim}} < 0 \qquad (6.54c)$$

$$\text{Fe}_2(\text{SO}_4)_3: \; z_+ = 3, \; z_- = -2 \; \Rightarrow \; j = -15 F D_- \frac{\mathrm{d}c}{\mathrm{d}x} \ge -15 F D_+ \frac{c(\delta)}{\delta} = j_{\text{lim}} < 0 \qquad (6.54d)$$

For an ideal AM, it applies:

$$t_{+,M} = 0 \quad \Rightarrow \quad j = -\overbrace{F D_- (z_+ - z_-) z_-}^{\text{negative number}} \frac{\mathrm{d}c}{\mathrm{d}x} \le -2 F D_- (z_+ - z_-) z_- \frac{c(\delta)}{\delta} = j_{\text{lim}} > 0 \quad (6.55)$$

$$\text{NaCl:} \; z_+ = 1, \; z_- = -1 \; \Rightarrow \; j = 2 F D_- \frac{\mathrm{d}c}{\mathrm{d}x} \le 2 F D_+ \frac{c(\delta)}{\delta} = j_{\text{lim}} > 0 \qquad (6.56a)$$

$$\text{Na}_2\text{SO}_4: \; z_+ = 1, \; z_- = -2 \; \Rightarrow \; j = 6 F D_- \frac{\mathrm{d}c}{\mathrm{d}x} \le 6 F D_+ \frac{c(\delta)}{\delta} = j_{\text{lim}} > 0 \qquad (6.56b)$$

$$\text{FeCl}_3: \; z_+ = 3, \; z_- = -1 \; \Rightarrow \; j = 4 F D_- \frac{\mathrm{d}c}{\mathrm{d}x} \le 4 F D_+ \frac{c(\delta)}{\delta} = j_{\text{lim}} > 0 \qquad (6.56c)$$

$$\text{Fe}_2(\text{SO}_4)_3: \quad z_+ = 3, \quad z_- = -2 \quad \Rightarrow \quad j = 10FD_- \frac{dc}{dx} \leq 10FD_+ \frac{c(\delta)}{\delta} = j_{\lim} > 0 \quad (6.56d)$$

Next, we will show that both these methods lead to the same limiting current because the equations obtained by the first method (6.49a) and by the second method (6.51d) are equivalent. The first method is more general since it also produces eq. (6.49b) that allows the calculation of the spatial distribution of the electric potential. From eq. (6.49a), it follows that

$$\frac{j}{j_0} = \tilde{j} = -\frac{1}{\alpha}\frac{d\tilde{c}}{d\tilde{x}} = -\frac{1}{\alpha}\frac{\delta}{c_0}\frac{dc}{dx} \qquad (6.57)$$

By gradual substitution, we get:

$$j = -\frac{j_0}{\alpha}\frac{\delta}{c_0}\frac{dc}{dx} = -\frac{(z_+ - z_-)D_0 c_0 F}{\alpha\delta}\frac{\delta}{c_0}\frac{dc}{dx} = -\frac{(z_+ - z_-)D_0 F}{\frac{t_+}{z_+ \tilde{D}_+} + \frac{t_-}{z_- \tilde{D}_-}}\frac{dc}{dx} =$$

$$= -\frac{(z_+ - z_-)F}{\frac{t_+}{z_+ D_+} + \frac{t_-}{z_- D_-}}\frac{dc}{dx} = -\frac{z_+ D_+ z_- D_- (z_+ - z_-)F}{t_+ z_- D_- + t_- z_+ D_+}\frac{dc}{dx} = \qquad (6.58)$$

$$= -\frac{z_+ D_+ z_- D_- (z_+ - z_-)F}{t_+ z_- D_- + (1 - t_+)z_+ D_+}\frac{dc}{dx} = \overbrace{-\frac{z_+ D_+ z_- D_- (z_+ - z_-)F}{z_+ D_+ - t_+(z_+ D_+ - z_- D_-)}}^{K}\frac{dc}{dx} = K\frac{dc}{dx}$$

# 6.5 Electric potential field near the membrane

## 6.5.1 *N*-component electrolyte

Let us consider a stationary layer of a solution of strong electrolytes, perpendicular to the $x$ axis and having the thickness of $\delta$. To describe the component fluxes, we will use Nernst–Planck equation, here for a 1D case and ideal behaviour:

$$\frac{J_i}{D_i} = -\frac{dc_i(x)}{dx} - c_i(x)z_i\frac{d\tilde{\varphi}(x)}{dx}, \quad \tilde{\varphi}(x) = \varphi(x)\frac{F}{RT}, \quad x \in (0, \delta) \qquad (6.59)$$

where $\tilde{\varphi}(x)$ is the dimensionless electric potential. We assume the validity of the local electroneutrality condition,

$$\sum_i z_i c_i(x) = 0 \qquad (6.60)$$

By summations

$$\sum_i \frac{J_i}{D_i} = -\frac{d}{dx}\overbrace{\sum_i c_i}^{f} - \overbrace{\sum_i c_i z_i}^{0}\frac{d\varphi}{dx} \tag{6.61a}$$

$$\sum_i \frac{z_i J_i}{D_i} = -\frac{d}{dx}\overbrace{\sum_i z_i c_i}^{0} - \overbrace{\sum_i c_i z_i^2}^{g}\frac{d\varphi}{dx} \tag{6.61b}$$

we obtain the ordinary linear differential equation (6.62a) and generally non-linear ordinary differential equation (6.62b):

$$\frac{df(x)}{dx} = A = -\sum_i \frac{J_i}{D_i}, \quad f(x) = \sum_i c_i(x) \tag{6.62a}$$

$$\frac{d\varphi(x)}{dx} = \frac{B}{g(x)}, \quad B = -\sum_i \frac{z_i J_i}{D_i}, \quad g(x) = \sum_i z_i^2 c_i(x) \tag{6.62b}$$

The following description of the electrolyte layer is based on the analysis of the functions $f(x)$ and $g(x)$. From eq. (6.62a), it is obvious that the function $f(x)$ is linear. For the following boundary conditions

$$c_i(0) = c_{i,0}, \quad \sum_i c_{i,0} = f(0) = f_0$$

$$c_i(\delta) = c_{i,\delta}, \quad \sum_i c_{i,\delta} = f(0) = f_\delta, \quad \varphi(\delta) = \varphi_\delta 0 \tag{6.63}$$

we obtain the solution of eq. (6.40) in the form:

$$f(x) = f_0 + \frac{f_\delta - f_0}{\delta}x = f_0 + Ax = A(x_0 + x), \quad z_0 = \frac{f_0}{A}$$

$$f(x) = A(x_0 + x) \tag{6.64}$$

In special cases of multicomponent electrolytes, we can easily determine the relationship between the functions $f$ and $g$ – see eq. (6.62a, b). For a uni-uni-uni- . . . -valent electrolyte (the absolute value of the charge numbers of all ions is equal to one), it applies:

$$|z_i| = 1 \quad \Rightarrow g = f \tag{6.65a}$$

For a bi-bi-bi- . . . -valent electrolyte, it applies:

$$|z_i| = 2 \quad \Rightarrow g = 4f \tag{6.65b}$$

For a n-n-n- . . . -valent electrolyte, it applies:

$$|z_i| = n \quad \Rightarrow g = n^2 f \tag{6.65c}$$

In general case, however, the relations between functions $f$ and $g$ can be very complicated. Here, we will analytically solve this problem for a binary electrolyte (Section 6.5.2) and indicate the solution for a ternary electrolyte (Section 6.5.4). For general multicomponent electrolytes, analytical solutions are not known and numerical solutions must be used.

## 6.5.2 Binary electrolyte

For a binary (two components) electrolyte, it applies, see eqs. (6.62a), (6.60) and (6.62b), summation for $i = 1, 2$:

$$c_1 + c_2 = f \tag{6.66a}$$

$$z_1 c_1 + z_2 c_2 = 0 \quad \Rightarrow \quad c_2 = -c_1 \frac{z_1}{z_2} \quad \Rightarrow \quad c_1 \left(1 - \frac{z_1}{z_2}\right) = f \tag{6.66b}$$

$$z_1^2 c_1 + z_2^2 c_2 = g = z_1^2 \overbrace{\frac{f}{1 - \frac{z_1}{z_2}}}^{c_1} - z_2^2 \overbrace{\frac{f}{1 - \frac{z_1}{z_2}} \frac{z_1}{z_2}}^{c_2} = f \frac{z_1^2 z_2 - z_2^2 z_1}{z_2 - z_1} = z_1 z_2 \frac{z_1 - z_2}{z_2 - z_1} f = -z_1 z_2 f \tag{6.66c}$$

Possible variations of binary electrolytes and corresponding relations between the functions $f(x)$ and $g(x)$ are shown in Tab. 6.1.

We will find the solution of eq. (6.62b) in the form:

$$\frac{d\varphi(x)}{dx} = \frac{B}{g(x)} = -\frac{B}{z_1 z_2 f(x)} = -\frac{B}{z_1 z_2 A(x_0 + x)} = \frac{K}{(x_0 + x)}, \quad K = -\frac{B}{z_1 z_2 A}$$

$$K \int_x^\delta \frac{dx}{x_0 + x} = \int_{\varphi(x)}^{\varphi(\delta)} d\varphi, \quad K \ln \frac{x_0 + \delta}{x_0 + x} = \varphi(\delta) - \varphi(x) \tag{6.67}$$

$$\varphi(x) = K \ln \frac{x_0 + x}{x_0 + \delta}$$

An analogous procedure can also be used for multicomponent mixtures, provided that all cations have the same charge number and all anions also have the same charge number but generally with a different absolute value than the charge number

**Tab. 6.1:** Examples of binary electrolytes and corresponding relations between $f(x)$ and $g(x)$.

| NaCl | Na$_2$SO$_4$ | Na$_3$PO$_4$ | CuSO$_4$ | Fe$_2$(SO$_4$)$_3$ | FeC$_6$H$_8$O$_7$ |
|---|---|---|---|---|---|
| $z_1 = 1$ | $z_1 = 1$ | $z_1 = 1$ | $z_1 = 2$ | $z_1 = 3$ | $z_1 = 3$ |
| $z_2 = -1$ | $z_2 = -2$ | $z_2 = -3$ | $z_2 = -2$ | $z_2 = -2$ | $z_2 = -3$ |
| $g = f$ | $g = 2f$ | $g = 3f$ | $g = 4f$ | $g = 6f$ | $g = 9f$ |

of cations. It is a so-called pseudobinary electrolyte (e.g. a solution containing sodium sulphate and potassium carbonate).

## 6.5.3 Pseudobinary electrolyte

Here, we assume that all anions have the same charge number and all cations have the same charge number (other than anions), that is

$$z_i > 0 \Rightarrow z_i = z_+ \quad z_i < 0 \Rightarrow z_i = z_- \tag{6.68}$$

We define

$$c_+ = \sum_{i,\, z_i > 0} c_i, \quad c_- = \sum_{i,\, z_i < 0} c_i \tag{6.69}$$

From the analogy with the binary electrolyte, it results

$$\sum_i c_i = c_+ + c_- = f \tag{6.70a}$$

$$\sum_i z_i c_i = z_+ c_+ + z_- c_- = 0 \tag{6.70b}$$

$$\sum_i z_i^2 c_i = z_+^2 c_+ + z_-^2 c_- = g = -z_+ z_- f \tag{6.70c}$$

$$\frac{d\varphi(x)}{dx} = \frac{B}{g(x)} = -\frac{B}{z_+ z_- f(x)} = -\frac{B}{z_+ z_- A(x_0 + x)} = \frac{K}{(x_0 + x)} \tag{6.71}$$

$$K = -\frac{B}{z_+ z_- A} \qquad \varphi(x) = K \ln \frac{x_0 + x}{x_0 + \delta}$$

## 6.5.4 Ternary electrolyte

$$c_1 + c_2 + c_3 = f \tag{6.72a}$$

$$z_1 c_1 + z_2 c_2 + z_3 c_3 = 0 \tag{6.72b}$$

$$z_1^2 c_1 + z_2^2 c_2 + z_3^2 c_3 = g \tag{6.72c}$$

We look for a solution for the electric potential field in the form

$$\varphi(x) = -\ln \zeta(x) \quad \Rightarrow \quad \frac{d\varphi(x)}{dx} = -\frac{1}{\zeta(x)} \frac{d\zeta}{dx} \tag{6.73}$$

For components with a non-zero flux, it applies:

$$0 = \frac{J_i}{D_i} = -\frac{dc_i(x)}{dx} - c_i(x)z_i\frac{d\varphi(x)}{dx} = -\frac{dc_i(x)}{dx} + c_i(x)z_i\frac{1}{\zeta(x)}\frac{d\zeta}{dx} \tag{6.74a}$$

$$\frac{dc_i}{c_i} = z_i\frac{d\zeta}{\zeta} \quad\Rightarrow\quad d\ln c_i = z_i d\ln\zeta = d\ln\zeta^{z_i} = z_i\zeta^{z_i-1}\frac{1}{\zeta^{z_i}}d\ln\zeta = z_i\frac{d\zeta}{\zeta} \tag{6.74b}$$

$$c_i = c_{i,\delta}\zeta^{z_i} \quad\Rightarrow\quad \frac{dc_i}{dx} = c_{i,\delta}z_i\zeta^{(z_i-1)} \tag{6.74c}$$

$$0 = -z_i\zeta^{(z_i-1)} + z_i\zeta^{z_i}\frac{1}{\zeta(x)}\frac{d\zeta}{dx} \tag{6.74d}$$

$$c_i(\delta) = c_{i,\delta} \quad\Rightarrow\quad \zeta(\delta) = 1, \quad \varphi(\delta) = 0 \tag{6.75}$$

Now, we will consider a special case when **only one (first) component passes through the layer** (it has a non-zero transport number):

$$c_1 + \sum_{i=2}^{N} c_{i,\delta}\zeta^{z_i} = f \quad\Rightarrow\quad c_1 = f - \sum_{i=2}^{N} c_{i,\delta}\zeta^{z_i} \tag{6.76a}$$

$$z_1 c_1 + \sum_{i=2}^{N} z_i c_{i,\delta}\zeta^{z_i} = 0 \quad\Rightarrow\quad c_1 = -\frac{1}{z_1}\sum_{i=2}^{N} z_i c_{i,\delta}\zeta^{z_i} \tag{6.76b}$$

$$g = z_1^2 c_1 + \sum_{i=2}^{N} z_i^2 c_{i,\delta}\zeta^{z_i} = -\frac{z_1^2}{z_1}\sum_{i=2}^{N} z_i c_{i,\delta}\zeta^{z_i} + \sum_{i=2}^{N} z_i^2 c_{i,\delta}\zeta^{z_i} = \sum_{i=2}^{N} (z_i - z_1) z_i c_{i,\delta}\zeta^{z_i} \tag{6.76c}$$

$$\frac{d\varphi}{dx} = -\frac{1}{\zeta}\frac{d\zeta}{dx} = \frac{B}{g} = \frac{B}{\sum_{i=2}^{N} (z_i - z_1) z_i c_{i,\delta}\zeta^{z_i}} \tag{6.76d}$$

$$\sum_{i=2}^{N} (z_i - z_1) z_i c_{i,\delta}\zeta^{z_i-1}d\zeta = -B\,dx \tag{6.77}$$

Another method of deduction is:

$$z_1 c_1 + \sum_{i=2}^{N} z_i c_{i,\delta}\zeta^{z_i} = z_1\left(f - \sum_{i=2}^{N} c_{i,\delta}\zeta^{z_i}\right) + \sum_{i=2}^{N} z_i c_{i,\delta}\zeta^{z_i} = 0 \tag{6.78}$$

Equation (6.76) can be analytically integrated for particular $z_1$, $z_2$ and $z_3$. For a given value of the parameter

$$B = -\sum_i \frac{z_i J_i}{D_i} = -\frac{j}{FD_1}$$

the equation has just one solution for boundary conditions (6.75). Equation (6.78) is a solution of eq. (6.77). The explicit expression of $\zeta(x)$ can be a problem because we determine the roots of a polynomial, which can be performed analytically in a practicable manner up to the third order.

Now, let us consider another special case when **only the first two components can permeate the layer** (they have a non-zero transport number):

$$c_1 + c_2 + \sum_{i=3}^{N} c_{i,\delta}\zeta^{z_i} = f(x) \quad \Rightarrow \quad c_1 + c_2 = f(x) - \sum_{i=3}^{N} c_{i,\delta}\zeta^{z_i}$$

$$z_1 c_1 + z_2 c_2 + \sum_{i=3}^{N} z_i c_{i,\delta}\zeta^{z_i} = 0 \quad \Rightarrow \quad z_1 c_1 + z_2 c_2 = -\sum_{i=3}^{N} z_i c_{i,\delta}\zeta^{z_i}$$

$$\begin{pmatrix} c_1(x,\zeta) \\ c_2(x,\zeta) \end{pmatrix} = \begin{pmatrix} 1 & 1 \\ z_1 & z_2 \end{pmatrix}^{-1} \begin{pmatrix} f(z) - \sum_{i=3}^{N} c_{i,\delta}\zeta^{z_i} \\ -\sum_{i=3}^{N} z_i c_{i,\delta}\zeta^{z_i} \end{pmatrix}$$

$$\begin{pmatrix} 1 & 1 \\ z_1 & z_2 \end{pmatrix}^{-1} = \frac{1}{z_2 - z_1}\begin{pmatrix} z_2 & -1 \\ -z_1 & 1 \end{pmatrix}$$

$$c_1(x,\zeta) = \frac{1}{z_2 - z_1}\left( z_2\left( f(x) - \sum_{i=3}^{N} c_{i,\delta}\zeta^{z_i}\right) + \sum_{i=3}^{N} z_i c_{i,\delta}\zeta^{z_i}\right)$$

$$c_2(x,\zeta) = \frac{1}{z_2 - z_1}\left( -z_1\left( f(x) - \sum_{i=3}^{N} c_{i,\delta}\zeta^{z_i}\right) - \sum_{i=3}^{N} z_i c_{i,\delta}\zeta^{z_i}\right)$$

$$g(x,\zeta) = z_1^2 c_1(x,\zeta) + z_2^2 c_2(x,\zeta) + \sum_{i=3}^{N} z_i^2 c_{i,\delta}\zeta^{z_i} \tag{6.79}$$

$$\frac{d\zeta}{dx} = -B\frac{\zeta}{g(x,\zeta)} \tag{6.80}$$

Equation (6.80) can be analytically integrated for particular $z_1$, $z_2$ and $z_3$. For a given value of the parameter

$$B = -\sum_i \frac{z_i J_i}{D_i} = -\left(\frac{t_1}{D_1} + \frac{t_2}{D_2}\right)\frac{j}{F}$$

the equation has an infinite number of solutions (for different choices of one non-zero transport number, the second is calculated up to one, the third is zero). We can also find a limit solution, corresponding to the maximum possible (limiting) current. However, necessary mathematical procedures often are not trivial.

In a general case when all components pass through the layer (they have non-zero transport numbers), the utility of the above analytical procedures is not proven. It is necessary to supplement boundary conditions and solve the equations numerically.

# 6.6 Non-ideal membrane surrounded by diffusion layers

Here, we will discuss a system consisting of three layers, see Fig. 6.3 and Section 6.8. We assume that the membrane is oriented in the $x$–$y$ plane and it is perpendicular to the $z$-axis. The $z$-axis is oriented from left to right.

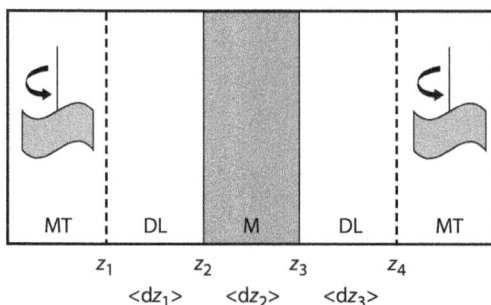

**Fig. 6.3: Scheme of a three-layer arrangement consisting of membrane M surrounded by diffusion layers DL.** Outside the diffusion layers, there are chambers with ideally mixed electrolytes (ideally mixed tank MT).

A key part of the formulation of the mathematical model is the definition of boundary conditions at the interfaces between the system and environment, as well as the conditions of the transition on internal boundaries between subsystems. The considered (possible) boundary conditions are:

Left environment ($z < z_1$):     ideally mixed tank, known composition $c_{i,L}$, $i = 1, \ldots, N$.

Interface ($z_1$):     known composition $c_{i,L}(z_1)$, $i = 1, \ldots, N$, and potential $\varphi_L(z_1) = U$.

First layer:     stationary diffusion (Nernst) layer, $c_{fix} = 0$.

Interface ($z_2$):     continuous fluxes, continuous composition and potential in the case of the PNP model, non-continuous and potential (Donnan) but continuous electrochemical potential for the equilibrium model.

Second layer:     membrane, $c_{fix} \neq 0$.

Interface ($z_3$):     qualitatively same conditions as on interface $z_2$.

Third layer:     stationary diffusion (Nernst) layer, $c_{fix} = 0$.

Interface ($z_4$):     known composition $c_{i,R}$, $i = 1, \ldots, N$, and potential $\varphi_R = 0$.

Right environment ($z > z_4$):     ideally mixed tank, known composition $c_{i,R}$, $i = 1, \ldots, N$.

Under these conditions, $N$ is the number of components and $c_{fix}$ is the concentration of fixed ions (non-zero in the membrane). Mixed tanks represent ideally mixed cores of chambers in which the electroneutrality condition is met. We also assume that there are no chemical reactions, the system is uniform in the direction of the $x$- and $y$-axes, and its state is steady.

The surface density of the electric current passing through the membrane from the left chamber to the right chamber is determined by the equation

$$j = \sum j_i = F \sum z_i J_i \qquad (6.81)$$

The molar flux of the $i$-th components passing through the membrane from the left chamber to the right chamber is described by Nernst–Planck equation, through the relation

$$J_i = -D_i \frac{dc_i}{dz} - c_i \frac{D_i}{RT} z_i F \frac{d\varphi}{dz}, \quad i = 1, \ldots, N \qquad (6.82)$$

Under these assumptions, the local balance of components can be expressed in the form

$$\frac{dJ_i}{dz} = 0, \quad i = 1, \ldots, N \qquad (6.83)$$

To describe the electric potential field in the case of the PNP model, we will use Poisson's equation

$$\frac{d^2\varphi}{dz^2} = -\frac{q}{\varepsilon} \qquad (6.84)$$

where $q = q_{mob} + q_{fix} = F \sum z_i c_i + F z_{fix} c_{fix}$ is the electric charge density consisting of the sum of the contributions of moving (mobile) ions and ions bound (fixed) in the membrane. The equations form a set of $N+1$ differential equations of the second order for $N+1$ variables $c_i(z)$, $i = 1, \ldots, N$. These can be solved as a boundary value problem for the following boundary conditions

$$c_i(z_0) = c_{i,L}, \quad \varphi(z_0) = U, \quad c_i(z_3) = c_{i,R}, \quad \varphi(z_3) = 0 \qquad (6.85)$$

Let us remind that at interfaces $z_1$ and $z_4$, we assume that the electroneutrality condition is met

$$\sum z_i c_{i,L} = \sum z_i c_{i,R} = 0 \qquad (6.86)$$

In the case of the local equilibrium model, we replace Poisson's equation with the electroneutrality condition. The equations then form a set of $N$ differential equations of the second order and one algebraic equation for $N+1$ variables $c_i(z)$, $i = 1, \ldots, N$, $\varphi(z)$, $z \in \Omega$.

In the following examples, the equations were solved using the COMSOL Multiphyscs® programme in the environment of the MATLAB programme. In the simplest example, we assume a uni-univalent symmetrical electrolyte (e.g. KCl):

$$N = 2; \quad i \in (1, 2) \equiv (A, B) \equiv (K^+, Cl^-), \ z_1 = 1; \quad z_2 = -1; \tag{6.87}$$

$$D_1 = D_2 = 2 \times 10^{-9} \, \text{m}^2 \, \text{s}^{-1}$$

Selected geometric quantities are

$$dz_1 = 0.1 \, \text{mm}; \quad dz_2 = 1 \, \text{mm}; \quad dz_3 = 0.1 \, \text{mm} \tag{6.88}$$

For an AM, the fixed charge is positive. Here, we select

$$z_{\text{fix}} = 1, \quad c_{\text{fix}} = \begin{cases} 0 & , \quad z \in (z_0, z_1) \\ 1 \, \text{kmol m}^{-3} & , \quad z \in (z_1, z_2) \\ 0 & , \quad z \in (z_2, z_3) \end{cases} \tag{6.89}$$

The selected boundary conditions are, for example A, B, C and D

$$c_{1,L} = c_{2,L} = c_{1,R} = c_{2,R} = c = \begin{cases} 0.001 \, \text{kmol m}^{-3} < \ < c_{\text{fix}} & \text{A} \\ 0.01 \, \text{kmol m}^{-3} < \ < c_{\text{fix}} & \text{B} \\ 0.1 \, \text{kmol m}^{-3} < c_{\text{fix}} & \text{C} \\ 1 \, \text{kmol m}^{-3} = c_{\text{fix}} & \text{D} \end{cases} \tag{6.90}$$

## 6.6.1 Selectivity loss for high concentration in the concentrate chamber

Equations (6.60) and (6.83) were solved numerically. The behaviour in concentrated solutions differs from the case of ideal (hypothetical) membranes, which are not permeable for co-ions (ions having the same charge as charges fixed in the membrane), regardless of how high their concentration is in the outer electrolyte. For non-ideal membranes, concentration of co-ions must be significantly lower than the concentration of ions bound in the membrane. If this condition is not met, there is a dramatic decrease of selectivity (Fig. 6.4).

## 6.6.2 Influence of the different composition in the left and right chambers

In the previous paragraph, the symmetrical case was discussed where the electrolyte compositions on both sides of the membrane were the same. In the non-symmetrical case (see Fig. 6.5), the dependence of the current density on the voltage, $j(\Delta\varphi)$, does not go through the beginning (of short-circuited electrode, $j(0) \neq 0$) and the non-zero voltage difference (of disconnected electrode, $j(\Delta\varphi \neq 0) = 0$) corresponds to the currentless state.

**a)** $c_L = 0.001$ kmol m$^{-3}$, $c_R = 0.001$ kmol m$^{-3}$    **b)** $c_L = 0.01$ kmol m$^{-3}$, $c_R = 0.01$ kmol m$^{-3}$

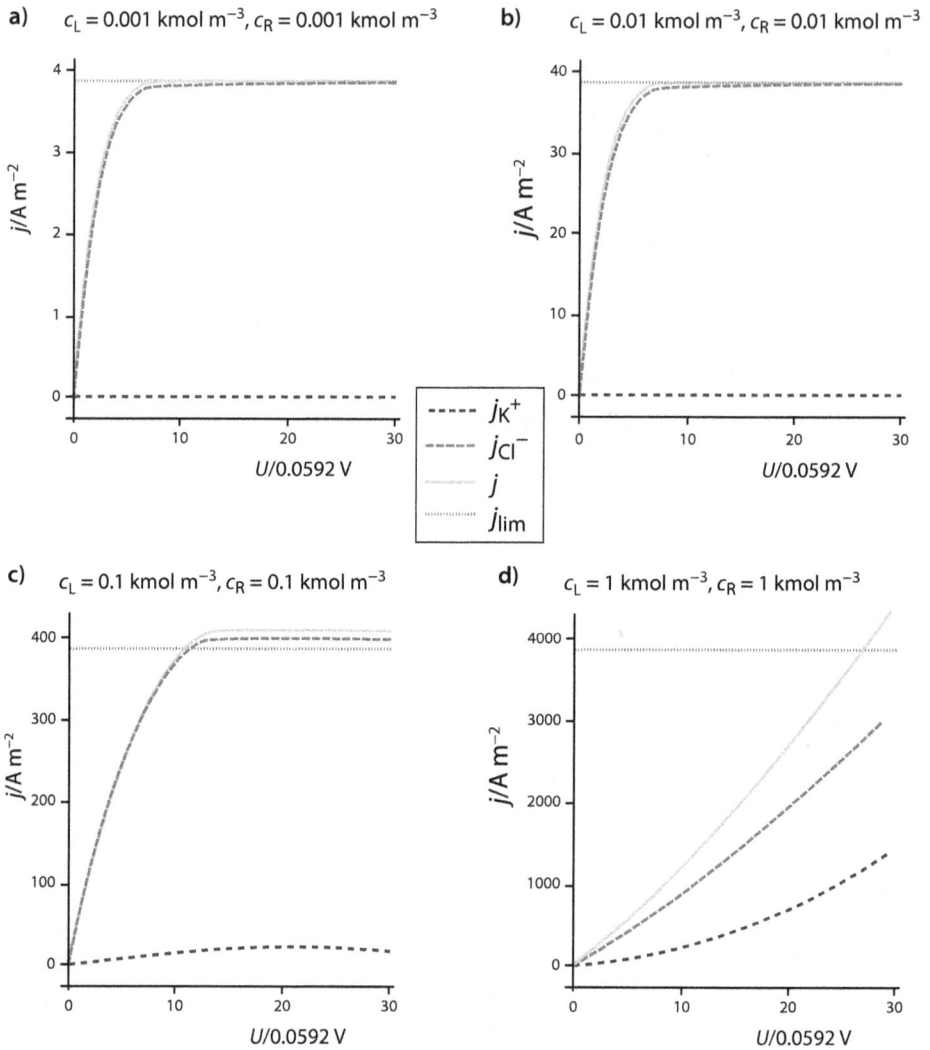

**c)** $c_L = 0.1$ kmol m$^{-3}$, $c_R = 0.1$ kmol m$^{-3}$    **d)** $c_L = 1$ kmol m$^{-3}$, $c_R = 1$ kmol m$^{-3}$

Legend:
- - - - $j_{K^+}$
-- -- -- $j_{Cl^-}$
——— $j$
·········· $j_{lim}$

**Fig. 6.4: The dependence of the partial current densities and the limiting current on the applied voltage for anion-selective membrane.** (a) and (b) The concentration of mobile ions in the chambers are considerably lower than the assumed concentration of fixed cations in the membrane and almost no cations pass through the anion-selective membrane. We observe the asymptotic dependence towards the limit value. (c) The concentrations of mobile ions in the chambers are lower that the assumed concentration of cations fixed in the membrane. Cations also permeate the anion-selective membrane but in a much lesser extent than anions. We observe the asymptotic dependence towards the limit value. (d) The concentrations of mobile ions in the chambers are the same as the assumed concentration of cations fixed in the membrane; cations also permeate the anion-selective membrane in a comparable extent as anions and the membrane loses its selectivity. The asymptotic development of the current towards the limit value is not observed.

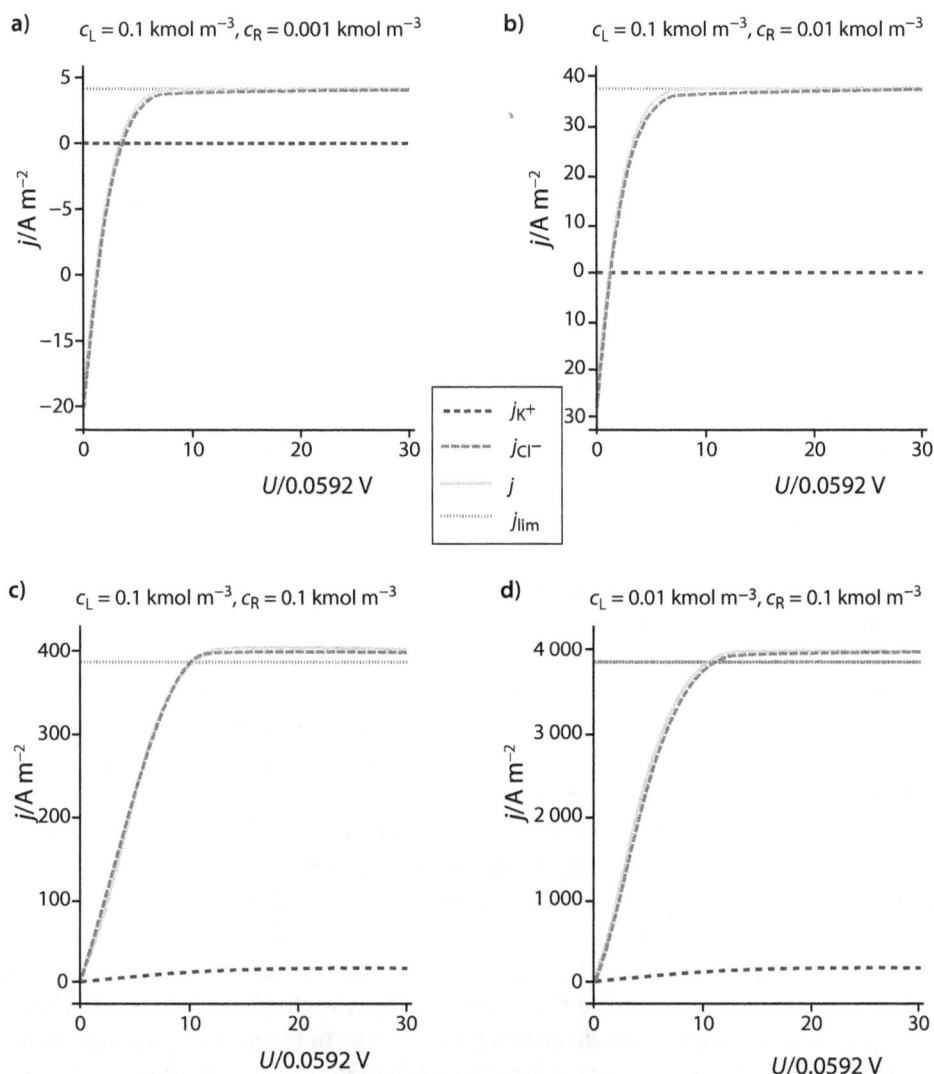

**a)** $c_L = 0.1$ kmol m$^{-3}$, $c_R = 0.001$ kmol m$^{-3}$

**b)** $c_L = 0.1$ kmol m$^{-3}$, $c_R = 0.01$ kmol m$^{-3}$

**c)** $c_L = 0.1$ kmol m$^{-3}$, $c_R = 0.1$ kmol m$^{-3}$

**d)** $c_L = 0.01$ kmol m$^{-3}$, $c_R = 0.1$ kmol m$^{-3}$

Legend:
- - - - - $j_{K^+}$
- - - - - $j_{Cl^-}$
.......... $j$
.......... $j_{lim}$

**Fig. 6.5: The dependence of the partial current densities and the limiting current on the applied voltage for anion-selective membrane.** (a) and (b) The concentration on the left is higher than that on the right. At the zero potential difference, anions flow through diffusion from the left to the right and a negative current is observed. (c) The concentration difference is zero, and at the zero voltage, the electric current is zero as well. (d) The concentrations on the left are lower than on the right. At the zero potential difference, anion flow through diffusion from the left to the right and a positive current is observed.

## 6.7 Three-dimensional model of fluid flow in the space between membranes

A key precondition of the reliable and efficient operation of electromembrane processes (including electrodialysis) is:
(a) uniformity of the flow hydrodynamics in the space between membranes,
(b) uniform distribution of process solutions by a hydraulic circuit between individual membrane chambers.

Issue (a) is discussed in this section, while case (b) will be explained in Chapter 7. Areas with a locally reduced velocity of electrolyte flow between membranes are often called hydraulic shadows (HS). Here, the solution becomes more desalinated due to a longer residence and concentration polarization becomes manifested as a result, leading to a decrease of the desalination efficiency. Besides, a local decrease of the ion conductivity and of the local current density appears. The current in the electrodialyser is then concentrated in more conductive parts of the unit where it can cause, in extreme cases, degradation of materials due to the locally increased current density; for details, see Chapter 7.

In an electromembrane device, individual membranes are separated by spacers (or distance inserts). The traditional spacer arrangement with direct flow used in ED is shown in Fig. 6.5. Its basic elements are distribution and collection channels on the sides, ensuring uniform distribution of the liquid along the inlet edge of the working space between the membranes. An integral part is also the turbulization net, filling the working space and consisting of two layers of fibres of different orientation. At the crossing point, these fibres can be interwoven or just overlaid. The net provides mechanical support for membranes and sets the distance between them. It also contributes to a more intensive mixing of the liquid, both in the direction parallel with the membrane surface (by forcing the fluid to flow parallel to the net fibres) and in the direction perpendicular to the membrane (because the fluid is forced to zigzag along the fibres, the so-called shear flow). This results in higher mass transfer coefficients from the solution volume towards the membrane surface and the concentration polarization is minimized. A typical net is also shown in Fig. 6.6.

Better uniformity of the liquid flow in the inter-membrane space (i.e. distribution of the liquid residence time) can be achieved by optimizing the geometry of spacers, that is spacer net and frame including distribution channels. This means that a conveniently designed spacer can significantly contribute to increasing the reliability and efficiency of the electromembrane process. The influence of the spacer geometry on the uniformity of the liquid flow can be studied using a convenient mathematical model, which is the subject of the following text.

First of all, it must be understood that the phenomena associated with the flow in the space between membranes, filled with the turbulization net, occur on two

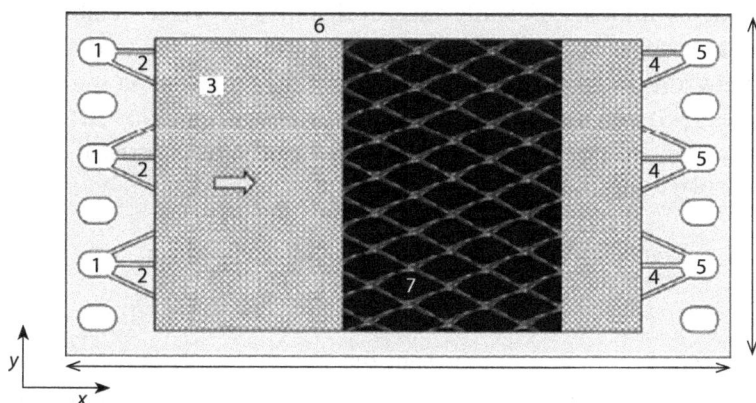

**Fig. 6.6: Schematic diagram of a spacer with direct liquid flow – top view.** (1) – slit in the spacer PE frame for liquid feed channels, (5) – slit in the spacer PE frame for liquid drain channels, (2) – distribution channels, (4) – collection channels, (3) – working space filled with a distance (turbulization) net, (6) – PE frame, (7) – photograph of a typical turbulization net used in electrodialysis, the arrow indicates the main direction flow of the liquid through the inter-membrane space.

spatial levels. From the point of view of optimizing the spacer geometry, the most important characteristics of the flow are in the order of ones or tens of centimetres. These phenomena are connected with the uniformity of the liquid distribution in the $x$–$y$ plane, that is on the membrane active surface (see Fig. 6.6), which is the focus of this text. Subtler characteristics of the flow occur on the sub-millimetre to millimetre levels, induced primarily by flowing around the turbulization net fibres. These phenomena are markedly reflected in the local values of the mass transfer coefficients between the liquid volume and membrane surface. However, a detailed mathematical description of the liquid flow on the sub-millimetre level in a system of the industrial device size is not feasible for calculation.

It is assumed that the local velocities of the liquid flow can be effectively averaged on the level of a characteristic net dimension, that is size of several meshes. Thus, the heterogeneous liquid–net system can be simplified to a homogeneous (continuous) system, while the hydraulic resistance of the net is proportionally higher than the resistance of empty space without the net. Besides, the net directs the liquid flow parallel with fibres, which is an important feature of the given model which may not be neglected. To solve this problem, we divide the model into two homogeneous hydraulic layers, each describing the flow in one layer of identically oriented parallel fibres. The liquid exchange between layers is enabled and solved in the form of a source term in the mass balance. This represents mixing the liquid in the radial direction due to the shear flow.

For the following steps, we assume ideally planar membranes (despite the fact they can be slightly deformed in the stack). In this ideal case, membranes form a rectangular channel with a significantly smaller distance between membranes (usually < 0.1 cm) in comparison with their active surface. The presented model describing the liquid flow in the space between membranes is based on the standard Navier–Stokes (NS) equation and together with the assumptions listed in the previous paragraph, NS equation can be modified to the final form of a dynamic equation for two hydraulic layers ($i = 1, 2$):

$$\rho(\bar{\boldsymbol{v}}_i \cdot \nabla \bar{\boldsymbol{v}}_i) = \underbrace{\eta \nabla^2 \bar{\boldsymbol{v}}_i}_{1} - \underbrace{\frac{12\eta}{h^2} \boldsymbol{A}_i^{-1} \bar{\boldsymbol{v}}_i}_{2} - \underbrace{\nabla \bar{p}_i}_{3} \quad \boldsymbol{A}_i = \frac{1}{r_t} \boldsymbol{t}_i \otimes \boldsymbol{t}_i + \frac{1}{r_n} \boldsymbol{n}_i \otimes \boldsymbol{n}_i \qquad (6.91)$$

Here, the quantities $\bar{\boldsymbol{v}}_i(x, y)$ and $\bar{p}_i(x, y)$ represent the velocity vector and pressure, dependent on the $x$ and $y$ positions (the bar above the symbol emphasizes that these are the mean quantities averaged in the radial direction), $h$ is the distance between membranes. The left side of the equation describes inertial forces. Under the viscous flow conditions ($Re < 100$), these forces are negligible and the left side of eq. (6.91) approaches zero. The term marked 1 takes into account the viscous forces and the term 3 pressure forces acting in the liquid. Contrary to the general NSt equation (6.127), there is the extra term 2 which takes into account the viscous forces between the layers of the liquid flowing in the direction perpendicular to membranes and the additional hydraulic resistance of the net against the flowing liquid. This resistance is significantly higher if the liquid flows perpendicularly to the net fibres than if it flows parallel with fibres. This fact is included in the transformation anisotropic matrix $\boldsymbol{A}_i$ different for both hydraulic layers due to the different orientation of the net fibres. The quantities $\boldsymbol{t}_i$ and $\boldsymbol{n}_i$ are unit vectors tangential and perpendicular to the net fibres in the $i$-th hydraulic layer, $r_t$ and $r_n$ represent the dimensionless hydraulic resistances of the net fibres in the tangential and perpendicular directions to the fibres. These resistances represent empirical parameters characteristic for a particular net type and must be determined experimentally.

The mass balance for this system is shown by eq. (6.92) for the first and second hydraulic layers. The right term of the mass balance described the liquid flow between hydraulic layers, which is proportional to the dimensionless hydraulic permeability $\kappa = 1$ and characteristic net size $o^2 = a^2 + b^2$, where $a$ and $b$ are the dimensions of one mesh of the net. The driving force of this interlayer flow is the pressure gradient between layers $(\bar{p}_1 - \bar{p}_2)/h$ (unit: Pa m$^{-1}$). The negative sign in the mass balance in layers 1 and 2 indicates the opposite direction of the interlayer flow of the liquid. For a more detailed description of the model, we refer to literature [7]

$$\nabla \cdot \bar{\boldsymbol{v}}_1 = -\nabla \cdot \bar{\boldsymbol{v}}_2 = \kappa \frac{o^2}{12\eta h} \frac{\bar{p}_1 - \bar{p}_2}{h} \qquad (6.92)$$

The basic result of this model is the velocity field of the liquid in the working space between membranes (see Fig. 6.7, corresponding to the linear velocity $v_{\text{lin}} = 5.3$ cm s$^{-1}$).

Attention is concentrated on the inlet side. Figure 6.7a shows the result of a model assuming negligible inertial forces, while Fig. 6.7b, c represents a more complex model considering inertial forces. Figure 6.7b shows the velocity field in the same hydraulic layer as Fig. 6.7a, while Fig. 6.7c in the second layer.

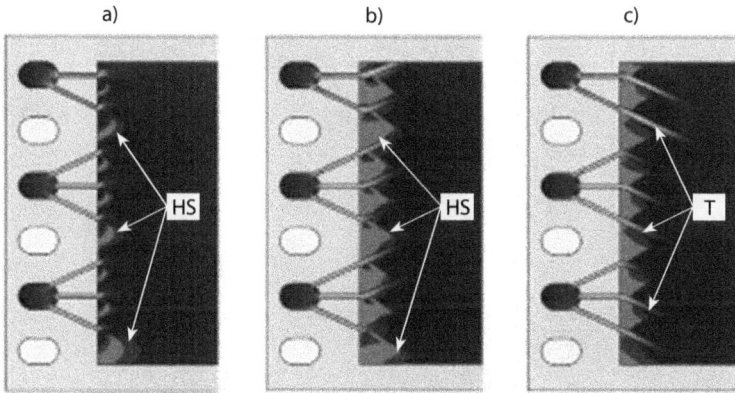

**Fig. 6.7: Field of the liquid flow velocity in the inter-membrane space of an electrodialysis device (see Fig. 6.1) on the inlet side.** (a) In hydraulic layer 1 inertial forces neglected, (b) and (c) in hydraulic layers 1 and 2 influence of inertial forces included; shown in shades of grey, white corresponds to the maximum velocity; hydraulic shadows (HS) located on the sides of distribution channels (some of them marked with arrows in (a) and (b)), darker areas HS correspond the liquid velocity less than $0.5v_{lin}$, lighter HS areas correspond the liquid flow velocity less than $0.25\ v_{lin}$; T are jets of the liquid penetrating deeper into the working space (some of them are marked with arrows in (c)); total flow rate is 50 dm$^3$ h$^{-1}$, linear velocity in the working space is $v_{lin} = 5.3$ cm s$^{-1}$, $h = 1.0$ mm.

We can see that the liquid flow in the working space is almost uniform, but significant differences of the flow velocity are observed near the inlet distribution channels. High velocities are observed especially at the point where distribution channels open into the working space. Conversely, on the sides of distribution channels are HS with a markedly longer liquid residence time. In this case, HS are defined as areas where the liquid flows at a velocity of less than $0.5v_{lin}$ (darker HS) or less than $0.25\ v_{lin}$ (lighter HS). The model with inertial forces considered predicts considerably larger areas of HS. This is caused by the fact that the liquid emerging from distribution channels penetrates deeper into the working space due to the inertia influence. The resulting jets are then directed according to the orientation of fibres in the hydraulic layer, that is they differ in the first and second hydraulic layers (see Fig. 6.6b, c). It shows that geometry of the distribution channels geometry can be optimized to minimize the HS area.

Another practical conclusion is that mathematical models with different complexity can provide quantitatively different results, although they correspond qualitatively.

## 6.8 Theory of similitude and dimensionless criteria

In the most cases, the membranes and electrodes in electromembrane processes are in planar arrangement. Under this assumption and for the purposes of this book, the membranes/electrodes are oriented in such a way that the x coordinate (axial, longitudinal) is parallel to the plane of membranes in the direction of the main liquid flow. The y coordinate is than parallel to the plane of membranes perpendicularly to the direction of the main liquid flow and the z coordinate (transverse) is oriented perpendicularly to the plane of membranes in the direction of the main flow of the electric current (see Fig. 6.8).

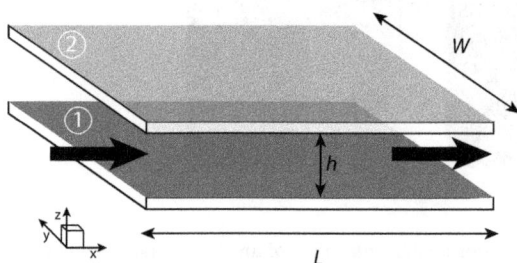

**Fig. 6.8: Orientation of membrane and main liquid flow in the Cartesian coordinate system, assuming a planar arrangement.** 1, 2 – membrane or electrode, L – length in the direction of the main liquid flow (indicated by white arrows), h – distance between membranes, W – width.

There are many criteria for selecting a model depending mainly on a particular application. In chemical engineering, characteristic times and dimensionless criteria of similarity are used. **Dimensionless criteria of similarity** allow to transfer information about another (similar) device to the studied system. **Characteristic time** (relaxation time) allows to estimate the time necessary to stabilize the process (chemical reaction, transport on various spatial levels) as a result of excitement from the equilibrium state by an external stimulus. By comparing different characteristic times, it is possible to evaluate the mutual rate of partial processes occurring simultaneously and to estimate the slowest of them – which determines the behaviour of the device as a whole. Based on this, the model type can be assessed (e.g. equilibrium or kinetic) for description of the studied device. To analyse electromembrane processes, the following characteristic times are important:

- **Residence time,**

$$\tau_R = L/v \qquad (6.93)$$

indicates the mean residence time of the liquid in the device in the axial direction of the x axis, for example in the space between membranes. Here, L is the characteristic trajectory length between the inlet and outlet points and v is the mean velocity of the liquid flow along this trajectory (along membranes). For normal electrodialysis, $L \approx 1$ m and $v = 0.05$ m s$^{-1}$ is $\tau_R \approx 1/0.05 = 20$ s.

- **Diffusion time in transverse (perpendicular) direction,**

$$\tau_{D,\updownarrow} = \frac{l_0^2}{D} = \left(\frac{h}{2}\right)^2 \frac{1}{D} \tag{6.94}$$

characterizes the transport time of diffusion components in the direction perpendicular to the membrane/electrode surface, where $l_0 = h/2$ is the characteristic length in the direction perpendicular to the membrane surface and to the prevailing flow direction (half of the distance between membranes) and $D$ is the effective diffusion coefficient of dissociated components. For electrodialysis of aqueous solutions without a net, typically $h = 0.001$ m and $D = 10^{-9}$ m$^2$ s$^{-1}$, then $\tau_{D,\updownarrow} = (0.001/2)^2/10^{-9} = 250$ s. Unfortunately, then $\tau_{D,\updownarrow} \gg \tau_R$, so transverse mixing is insufficient. **Therefore, transverse mixing is intensified, for example by inserting a net, which forces the fluid to "zigzag" between fibres;** see further, Sherwood number.

- **Diffusion time in longitudinal direction,**

$$\tau_{D,\leftrightarrow} = \frac{L^2}{D} \tag{6.95}$$

is defined similarly. For $L = 1$ m and $D = 10^{-9}$ m$^2$ s$^{-1}$ is $\tau_{D,\leftrightarrow} = 1^2/10^{-9} = 10^{-9}$ s $\gg \tau_R$, that is longitudinal mixing by molecular diffusion appears to be negligible. However, the effective diffusion coefficient in the longitudinal direction in the laminar flowing liquid can be significantly higher than the molecular diffusivity. This is caused by the so-called Aris–Taylor dispersion [3]:

$$D_{\text{eff}} = D\left(1 + \frac{\text{Pe}^2}{K}\right), \quad \text{Pe} = \frac{hv}{2D}, \quad \frac{\text{Pe}^2}{K} \approx 10^7 \tag{6.96}$$

Effective dispersion can be up to many orders higher than molecular diffusion. Then, the negligibility of longitudinal mixing need not be so obvious.

$$\tau_{D_{\text{eff}},\leftrightarrow} \approx \frac{L^2}{D_{\text{eff}}} \approx 100 \text{ s} \quad > \quad \tau_R \approx 20 \text{ s} \tag{6.97}$$

- **Relaxation (Debye) time,**

$$\tau_D = \frac{\lambda_D^2}{D} = \frac{1}{D}\frac{\varepsilon RT}{2cF^2} = \frac{\varepsilon}{\kappa} \tag{6.98}$$

describes the behaviour of the solution on the phase interface and is connected with the stabilization rate of the local electroneutrality. The characteristic dimension, by which we measure the proximity of the environment to electroneutrality, is the

Debye length $\lambda_D$. It depends on the environment permittivity $\varepsilon$, conductivity $\kappa$, molar concentration of dissociated components $c$ and temperature $T$. For $c = 0.1$ kmol m$^{-3}$ is $\lambda_D \approx 1$ nm and $\tau_D \approx 1$ ns.

– **Reynolds number,**

$$Re = hv\rho/\eta \tag{6.99}$$

characterizes the ratio between inertial and viscous forces acting in the liquid. It represents one of the most important criteria of fluid dynamics. Again, $v$ is the mean flow velocity, $\rho$ is the density and $\eta$ (unit: kg m$^{-1}$ s$^{-1}$) is the dynamics viscosity of the liquid. The Re value characterizes different flow modes: Re < 100 corresponds to viscous flow, 100 < Re < 1,500 to laminar flow, 1,500 < Re < 3,600 to transitional flow and Re > 3,600 to turbulent flow. (The numerical values are given for orientation only, as the modes are not sharply separated.) From the point of view of component transport in the radial direction, high values of Re are convenient but they lead to high pressure losses and consumption of electricity for pumping. The optimum Re values depend on the given application, especially for transitional flow. For electrolysis of aqueous solutions, these conditions are common: $h = 0.001$ m, $v = 0.05$ m s$^{-1}$, $\rho = 1,000$ kg m$^{-3}$ and $\eta = 0.001$ kg m$^{-1}$ s$^{-1}$. This corresponds to Re = 50. Because of the presence of the net in the inter-membrane spaces of the electrodialyser, the hydraulic losses can be high even for low Re.

– **Schmidt number,**

$$Sc_i = \eta/\rho D_i = v/D_i \tag{6.100}$$

expresses the ratio between the conduction of the momentum and mass in the radial direction, or in another words, it corresponds to the relationship between the thickness of the Prandtl hydrodynamic layer and the Nernst diffusion layer. It applies that the greater is the kinematic viscosity $v$ (unit: m$^2$ s$^{-1}$), the thicker is the hydrodynamic layer, and the greater $D$, the thinner is the diffusion layer. A typical Sc value for electrodialysis of aqueous solutions is Sc $= v/D = 10^{-6}/10^{-9} = 1,000$.

– **Sherwood number,**

$$Sh_i = k_i l_0/D_i \tag{6.101}$$

is the ratio between the actual mass transport intensity and the transport intensity in the hypothetical system without flow, where transport only occurs by diffusion. Here, $k_i$ (unit: m s$^{-1}$) is the coefficient of mass transfer of the component $i$, $l_0 = h/2$ is the characteristic length of the transport trajectory (e.g. in the case of a channel between two parallel desks, it is half of their distance) and $D_i$ is the diffusion coefficient of the component. The question arises whether we can use the concept of the Sherwood number, which was created based on the idea of mass transfer,

$$J_i = D_i \frac{dc_i}{dz}\bigg|_w = k_i(\tilde{c}_i - c_{i,w}), \qquad \tilde{D}_i \frac{d\tilde{c}_i}{d\tilde{z}}\bigg|_w = Sh_i(\tilde{c}_i - \tilde{c}_{i,w})$$

$$\tilde{D}_i = \frac{\tilde{D}_i}{D_0}, \ \tilde{c}_i = \frac{c_i}{c_0}, \ \tilde{z} = \frac{z}{l_0} \tag{6.102}$$

where the driving force is the concentration difference, also in the cases when another driving force, often predominant, is the electric potential difference. Here, $J_i$ is the surface density of the component flow, $d\tilde{c}_i/d\tilde{z}|_w$ is the outer normal derivative of the component concentration by the channel wall, $\tilde{c}_i$ is the mean value of the concentration in the flow channel and $c_w$ is the component concentration by the wall. For a symmetrical binary electrolyte $(z_+ = -z_-, c_+ = c_- = c, D_+ = D_- = D)$, local balance applies for cation

$$\frac{\partial c}{\partial \tau} = -\nabla \cdot \left( \mathbf{v}c - D\nabla c - Dcz \frac{F}{RT}\nabla\varphi \right) \tag{6.103}$$

and for anion

$$\frac{\partial c}{\partial \tau} = -\nabla \cdot \left( \mathbf{v}c - D\nabla c + Dcz \frac{F}{RT}\nabla\varphi \right) \tag{6.104}$$

The sum of these equations gives for a constant $D$ **Fick's second law** with a convective term

$$\frac{\partial c}{\partial \tau} = -\mathbf{v} \cdot \nabla c + D\nabla^2 c \tag{6.105}$$

It is obvious that migration contributions were eliminated. The development of the component concentration fields in the flowing electroneutral binary symmetrical electrolyte is controlled only by convection and diffusion, which fully corresponds to the standard concept of mass transfer. It need not apply for general electrolytes, and the analysis of possible deviations, that can be significant, is relatively demanding.

To describe the transfer, a model is often used of the ideally mixed core and stationary Nernst diffusion layer with the thickness $\delta$. The estimation of the Sherwood number value can be used to estimate the thickness of this layer:

$$k = \frac{D}{\delta}, \quad Sh = \frac{kl_0}{D} = \frac{l_0}{\delta} = \frac{h}{2\delta} \tag{6.106}$$

Note that we assume that the diffusion coefficient, mass transfer coefficient and Sherwood number do not have a component index. If the diffusion coefficients of components are different, we must replace them with some estimation of their mean effective value. The transport time in the transverse direction is then inversely proportional to the Sherwood number, which is expressed by the relation

$$\tau_{\updownarrow} = \frac{\tau_{D,\updownarrow}}{Sh} = \frac{1}{Sh}\left(\frac{h}{2}\right)^2\frac{1}{D} = \frac{2\delta}{h}\left(\frac{h}{2}\right)^2\frac{1}{D} = \frac{h\delta}{2D} \tag{6.107}$$

To estimate the Sherwood number, we usually use experimental data generalized in the form of equations

$$Sh = Sh(Re, Sc, \Gamma_1, \Gamma_2, \ldots) \tag{6.108}$$

where $\Gamma_1, \Gamma_2, \ldots$ are geometric parameters, describing, for example built-in, netting and so on.

– **Longitudinal Péclet number for mass,**

$$Pe_{i,\to\leftrightarrow} = \frac{\tau_{D_i,\leftrightarrow}}{\tau_R} = \frac{L^2}{D_i}\frac{v}{L} = \frac{Lv}{D_i} \tag{6.109}$$

characterizes the longitudinal (axial) dispersion of the component, that is the importance of diffusion and convective transport of components in the axial direction. The typical value of $Pe_{\to\leftrightarrow}$ in the mode of aqueous solutions electrodialysis is $Pe_{\to\leftrightarrow} = Lv/D_i = 1 \times 0.05/10^{-9} = 5 \times 10^7$. This value of Pe suggests that diffusion transport in the axial direction is negligible in the electrodialysis process. In general, we must bear in mind that transport in the transverse direction can be significantly increased by the influence of Taylor dispersion, see the diffusion time in the longitudinal direction.

– **Transverse Péclet number for mass,**

$$Pe_{\to\updownarrow} = \frac{\tau_{D,\updownarrow}}{\tau_R} = \frac{h^2}{4D}\frac{v}{L} \tag{6.110}$$

characterizes the ratio between the transverse (radial) dispersion (diffusion) and the convective axial transport of components. The typical value of $Pe_{\to\updownarrow}$ in the mode of aqueous solutions electrodialysis is $Pe_{\to\updownarrow} = h^2v/4DL = (0.001^2 \times 0.05)/(4 \times 10^{-9} \times 1) = 12.5$. This value suggests that diffusion transport in the transverse direction can be a limiting factor in the electrodialysis process. Transverse mixing is not perfect and must be increased with a convenient measure (netting in chambers), see the Sherwood number.

## 6.9 Local balance

The local balance of any extensive quantity $U$ can be expressed in the form

$$\underbrace{\frac{\partial \rho_Y}{\partial \tau}}_{\text{accumulation}} = \overbrace{-\nabla \cdot \Phi_Y}^{\text{input} - \text{output}} + \overset{\text{source}}{\sigma_Y} \tag{6.111}$$

where $\rho_Y$ is the (volume) density of the quantity $Y$, $\Phi_Y$ is the flux (surface density, intensity) of the quantity $Y$ and $\sigma_Y$ is the volume density of the source of the quantity $Y$. Accumulation is the increase of the quantity amount in the elementary volume per unit of time, divided by this volume. Input–output is the amount of the quantity transferred from the environment through the boundary into the elementary volume per unit of time, divided by this volume. Source is the quantity amount that is generated in the elementary volume per unit of time, divided by this volume.

### 6.9.1 Local balance of components

The balanced quantity is the amount of substance of the $i$-th component

$$\underbrace{\frac{\partial c_i}{\partial \tau}}_{\text{accumulation}} = \overbrace{-\nabla \cdot J_i}^{\text{input} - \text{output}} + \overbrace{\sum_j v_{i,j} r_j}^{\text{source}} \qquad (\text{unit: mol m}^{-3}\text{s}^{-1}) \tag{6.112}$$

where

$$J_i = \overbrace{vc_i}^{\text{convection}} \overbrace{-D_i \nabla c_i}^{\text{diffusion}} - c_i \overbrace{\frac{D_i}{RT} z_i F \nabla \varphi}^{\text{migration}} \quad (\text{unit: mol m}^{-2}\text{s}^{-1}) \tag{6.113}$$

is the molar flux described using Nernst–Planck equation, $v$ is the liquid flow velocity, $c_i$, $D_i$ and $z_i$ are the molar concentration, diffusion coefficient and charge number (valence) of the $i$-th component, $r_j$ (unit: mol m$^{-3}$ s$^{-1}$) is the reaction velocity of the $j$-th chemical reaction, $v_{i,j}$ is the stoichiometric coefficient of the $i$-th component in the $j$-th chemical reaction and $\varphi$ is the electric potential.

### 6.9.2 Local mass balance

$$\underbrace{\frac{\partial \rho}{\partial \tau}}_{\text{accumulation}} = -\nabla \cdot \overbrace{\left( \underset{\rho v}{\underbrace{\rho v}} \right)}^{\text{input} - \text{output}} \tag{6.114}$$

where $\rho$ is the density. The **mass source is always zero**. On the assumption of incompressible liquid ($\rho$ is constant), the equation is simplified into the **equation of continuity**

$$\underset{\substack{\text{velocity divergence}}}{\nabla \cdot \boldsymbol{v}} = 0 \tag{6.115}$$

This equation states that for any spatial area and at any time the volume flow into the area across the boundary from the surroundings equals to the volume flow out of the system across the boundary to the surroundings.

## 6.9.3 Local balance of electric charge

The charge balance is a linear combination of components balances

$$\frac{\partial \overbrace{\sum_i F z_i c_i}^{q}}{\partial \tau} = -\nabla \cdot \overbrace{\sum_i F z_i \boldsymbol{J}_i}^{\boldsymbol{j}} \left(\text{unit: C m}^{-3}\text{s}^{-1}\right) \tag{6.116}$$

therefore

$$\frac{\partial q}{\partial \tau} = -\nabla \cdot \boldsymbol{j} \tag{6.117}$$

where $q = F \sum_i z_i c_i$ is the charge density (unit: C m$^{-3}$) and

$$\boldsymbol{j} = F \sum_i z_i \boldsymbol{J}_i = \boldsymbol{j}_{\text{conv}} + \overbrace{\boldsymbol{j}_{\text{dif}} + \boldsymbol{j}_{\text{mig}}}^{\boldsymbol{j}_{\text{dis}}} = \overbrace{\boldsymbol{v} q}^{\text{convection}} \overbrace{- F \sum_i D_i z_i \nabla c_i}^{\text{diffusion}} - \overbrace{\sum_i c_i \frac{D_i}{RT} z_i^2 F^2 \nabla \varphi}^{\substack{\text{dissipation} \\ \text{migration} \\ \kappa = \text{conductivity}}} \tag{6.118}$$

is the surface current density. **The electric charge source is always zero.** The balance states that the charge accumulation in a volume unit is equal to the negative divergence of the vector field of the surface density (intensity) of the electric current. Divergence represents the discharge from the volume unit. In a steady state ($\partial q/\partial \tau = 0$) or under the **electroneutrality condition** ($q = 0$), we get

$$\nabla \cdot \boldsymbol{j} = 0 \tag{6.119}$$

It means that the electric current flowing into any volume element equals to the electric current flowing out. It is an analogy to the continuity equation for incompressible liquid

$$\nabla \cdot \boldsymbol{v} = 0 \tag{6.120}$$

where $\boldsymbol{v}$ is the flow velocity. If the diffusion term is negligible, we obtain

$$\nabla \cdot \boldsymbol{j} = \nabla \cdot (-\kappa \nabla \varphi) = 0 \qquad (6.121)$$

where

$$\kappa = \sum_i c_i \frac{D_i}{RT} z_i^2 F^2 \qquad (6.122)$$

is the (specific electric) conductivity (unit: S m$^{-1}$). If the conductivity is constant, we get **Laplace's equation**

$$\nabla^2 \varphi = 0 \qquad (6.123)$$

## 6.9.4 Local balance of internal energy

We record the balance in the form

$$\underbrace{\frac{\partial \rho u}{\partial \tau}}_{\text{accumulation}} = -\nabla \cdot \left( \overbrace{\underbrace{\boldsymbol{v} \rho u}_{\text{convection}} + \underbrace{\boldsymbol{q}}_{\text{conduction}}}^{\text{input}-\text{output}} \right) \overbrace{-\boldsymbol{j}_{\text{dissip}} \cdot \nabla \varphi}^{\text{source} > 0} \quad (\text{unit J m}^3\text{s}^{-1}) \qquad (6.124)$$

where $u$ is the specific internal energy (unit; J kg$^{-1}$) and

$$\boldsymbol{q} = -\lambda_Q \nabla T \qquad (6.125)$$

is the density of the heat flux intensity (unit: W m$^{-2}$), expressed from **Fourier's law**, where $\lambda_Q$ (unit: W m$^{-1}$ K$^{-1}$) is the heat conductivity and $T$ is the temperature. The source is caused by energy dissipation at the electric current passage (**Joule's heating effect**). The source of heat by dissipation of internal friction and adsorption of electromagnetic radiation is neglected, as well as the work in changes of volume (incompressible liquid).

## 6.9.5 Local balance of momentum

The balanced quantity is momentum $m\boldsymbol{v}$ (unit: kg m s$^{-1}$). The balance represents a crucial relation to describe the liquid dynamics and gets the form

$$\overset{(1)}{\frac{\partial(\rho \boldsymbol{v})}{\partial \tau}} + \overset{(2)}{\nabla \cdot (\rho \boldsymbol{v}\boldsymbol{v})} = -\overset{(3)}{\nabla \cdot \boldsymbol{\tau}} - \overset{(4)}{\nabla p} + \overset{(5)}{\rho \boldsymbol{f}} \qquad (6.126)$$

Its derivation is not fully trivial, so we refer to available literature written by Wilkes [5]. The product $\rho \boldsymbol{v}$ (unit: kg m$^{-2}$ s$^{-1}$) is the volume density of momentum, $\boldsymbol{\tau}$ is the tensor of the tangent tension and $p$ is the pressure. The meaning of individual terms is following: term (1) on the left side represents the momentum accumulation, terms

(2), (3) and (4) represent the momentum flux due to the liquid inertia (so-called inertia term), viscous friction in the tangent direction (so-called viscous term) and influence of the pressure force (so-called pressure term). The last term (5) can be explained as the source of momentum due to the influence of volume forces, most frequently gravity: $f = g$, where $g$ is the gravitational acceleration.

The solution of the momentum balance in its full form is very complicated and limited only to detailed studies within basic research. By adopting simplifying assumption, we can get much more soluble forms. For example, in Newtonian liquids, the tangent tension is directly proportional to the liquid dynamic viscosity $\eta$. Besides, many liquids (fluids) are incompressible. Then the equation transforms into the most important equation of liquid dynamics, **Navier–Stokes equation**

$$\frac{\partial v}{\partial \tau} + v \cdot \nabla v = \nu \nabla^2 v - \frac{\nabla p}{\rho} + f \qquad (6.127)$$

However, the solution of NS equation is still very complex and it can be further simplified depending on the prevailing flow mode. In the creeping flow mode, characterized by low values of the Reynolds number, the influence of inertia forces is negligible. Assuming the stationary flow and neglecting volume forces, we get **Stokes equation**

$$0 = \eta \nabla^2 v - \nabla p \qquad (6.128)$$

A special case is the flow in microporous and nanoporous membranes, where a simpler form from Darcy's law can replace the viscous term

$$0 = -\frac{\eta}{B} v - \nabla p \qquad (6.129)$$

where $B$ is the Darcy's coefficient, depending mainly on the pore size ($B \approx d^2$). In ion-selective membranes with a non-zero fixed charge $q_{fix}$, the electrostatic force inducing the electroosmotic flow often prevails (the charge of the electrolyte influenced by the Coulomb force $f_C$ is essentially equal to the fixed charge with the opposite sign)

$$f_C = -q\nabla\varphi = q_{fix}\nabla\varphi \qquad (6.130)$$

## 6.9.6 Balance of mechanical energy – Bernoulli's equation

In principle, Bernoulli's equation represents a consequence of NS equation describing the balance of mechanical energy in a steady state incompressible liquid and gets the form

$$\frac{1}{2}v_1^2 + \frac{p_1}{\rho} + gh_1 = \frac{1}{2}v_2^2 + \frac{p_2}{\rho} + gh_2 + e_{dissip} \qquad (6.131)$$

where $v$ is the flow velocity, $p$ is the pressure, $h$ is the geometric height, $g$ is the gravitational acceleration, $v^2/2$ is the specific kinetic energy, $p/\rho$ is the specific pressure energy, $gh$ is the specific potential (gravitational) energy and $e_{dis}$ is the specific loss (dissipation) of the mechanical energy by internal friction. Lower indexes 1 and 2 represent input and output. The equation of continuity (6.115), NS equation (6.127) and also Bernoulli's equation (6.131) are often referred to collectively as the **equations of motion**.

## 6.10 Electroneutrality condition

### 6.10.1 Substantiation of the electroneutrality condition

In the local balance of the electric charge, see eqs. (6.116) and (6.118), we can neglect in many cases (but not generally) some terms according to the equation

$$
\frac{\partial q}{\partial \tau} = \overbrace{-\nabla \cdot (vq)}^{} + F \sum_i D_i z_i \nabla^2 c_i + \left( \overbrace{\nabla \kappa \cdot \nabla \varphi}^{\text{to be neglected}} + \overbrace{\kappa \nabla^2 \varphi}^{\substack{\text{Poisson's equation} \\ -\kappa q/\varepsilon}} \right) \tag{6.132}
$$

By substituting for $\nabla^2 \varphi$ from **Poisson's equation**, we obtain an ordinary differential equation of the first order with a trivial analytic solution

$$
\frac{\partial q}{\partial \tau} = -\left(\frac{\kappa}{\varepsilon}\right) q = -\frac{q}{\tau_D}, \quad q(\tau) = q(0) \exp\left(-\frac{\tau}{\tau_D}\right) \tag{6.133}
$$

where $\tau_D = \varepsilon/\kappa = \lambda_D^2/D$ is the **relaxation time** – see eq. (6.98), $\kappa$ is the conductivity, $\varepsilon$ is the permittivity (dielectric constant) and $D$ is the characteristic value of the diffusion coefficient. The relaxation time also means the characteristic diffusion time related to the **Debye length** according to the relation (for uni-unipolar symmetrical electrolyte)

$$
\lambda_D = \sqrt{D\tau_D} = \sqrt{\frac{\varepsilon RT}{2cF^2}} \tag{6.134}
$$

The Debye length describes the thickness of the electric double layer. For concentrations in the order of 0.1 kmol m$^{-3}$, the values of the relaxation time about 1 ns and the Debye length values of the order of 1 nm. It means that the solution of eq. (6.133) approaches zero extremely fast, $q(\tau) \to 0$.

**Therefore, deviations from electroneutrality can only happen under "extreme" conditions, which are usually not possible in the electrolyte volume, but often occur in spatial discontinuities (singularities), especially at phase interfaces.**

## 6.10.2 Consequences of the electroneutrality condition

Now, let us deal with the analysis of the equations describing the surface density (intensity) of the electric current in electrolyte solutions **assuming local electroneutrality**. For simplicity, we will only deal with a **spatially 1D formulation**. The current density is expressed as a linear combination of the surface densities (intensities) of the molar fluxes described by Nernst–Planck equation, which leads to the relation

$$j = F\sum_{i=1}^{N} z_i J_i = F\sum_{i=1}^{N} z_i\left(-D_i\frac{dc_i}{dx} - c_i\frac{D_i}{RT}Fz_i\frac{d\varphi}{dx}\right) \tag{6.135}$$

In summations, we will write separately terms for the $N$-th component, expressed using the local electroneutrality condition

$$j = -F\sum_{i=1}^{N-1} D_i z_i\frac{dc_i}{dx} - FD_N z_N\frac{dc_N}{dx} - \left(\sum_{i=1}^{N} c_i\frac{D_i}{RT}z_i^2 F^2 + c_N\frac{D_N}{RT}z_N^2 F^2\right)\frac{d\varphi}{dx} \tag{6.136}$$

$$c_N = -\frac{1}{z_N}\sum_{i=1}^{N-1} z_i c_i \qquad \frac{dc_N}{dx} = -\frac{1}{z_N}\sum_{i=1}^{N-1} z_i\frac{dc_i}{dx} \tag{6.137}$$

By substituting in eq. (6.137) into (6.136), we get

$$j = -F\sum_{i=1}^{N-1}(D_i - D_N)z_i\frac{dc_i}{dx} - \sum_{i=1}^{N}(z_iD_i - z_ND_N)c_iz_i\frac{F^2}{RT}\frac{d\varphi}{dx} \tag{6.138}$$

where

$$\kappa = \sum_{i=1}^{N} c_i\frac{D_i}{RT}z_i^2 F^2 = \sum_{i=1}^{N-1}(z_iD_i - z_ND_N)c_iz_i\frac{F^2}{RT} \tag{6.139}$$

is the conductivity. Then, the electric potential derivative can be expressed explicitly in the form

$$\frac{d\varphi}{dx} = -\frac{j + Fz_i\sum_i D_i\frac{dc_i}{dx}}{\kappa} = -\frac{j + F\sum_{i=1}^{N-1}(D_i - D_N)z_i\frac{dc_i}{dx}}{\kappa} \tag{6.140}$$

The formula (6.140) can be substituted into Nernst–Planck equation (6.135), by which we get

$$J_i = -D_i\frac{dc_i}{dx} + c_i\frac{D_i}{RT}Fz_i\frac{j + F\sum_{j=1}^{N-1}(D_j - D_N)z_j\frac{dc_j}{dx}}{\kappa} \tag{6.141a}$$

$$J_i = -\sum_{j=1}^{N-1}\left(\overbrace{\delta_{i,j}D_i + D_i c_i z_i z_j \frac{F^2}{RT}\frac{(D_j - D_N)}{\kappa}}^{D_{i,j}}\right)\frac{dc_j}{dx} + D_i c_i z_i \frac{F}{RT}\frac{j}{\kappa} \tag{6.141b}$$

$$J_i = -\sum_{j=1}^{N-1}D_{i,j}\frac{dc_j}{dx} + D_i c_i z_i \frac{F}{RT}\frac{j}{\kappa} \tag{6.141c}$$

The local balance of non-reactive (inert) components has the form

$$\frac{dJ_i}{dx} = -\frac{d}{dx}\left(-\sum_{j=1}^{N-1}D_{i,j}\frac{dc_j}{dx} + D_i c_i z_i \frac{F}{RT}\frac{j}{\kappa}\right) = 0, \quad i = 1, \ldots, N-1 \tag{6.142}$$

It is a set of $N-1$ ordinary differential equations of the second order. Here, the current density $j$ is a parameter independent on $x$. The equations do not explicitly contain the electric potential. These equations can be practically used in modelling electromembrane processes and devices.

In a special case (all components have the same diffusion coefficient), we get the relation

$$D_i = D_N \quad \Rightarrow \quad J_i = -D_i\frac{\partial c_j}{\partial x} + D_i c_i z_i \frac{F}{RT}\frac{j}{\kappa} \tag{6.143}$$

In cases where diffusion is negligible compared to migration, it applies

$$\left|D_i\frac{\partial c_j}{\partial x}\right| \ll \left|c_i z_i \frac{F}{RT}\frac{j}{\kappa}\right| \quad \Rightarrow \quad J_i \cong c_i\frac{D_i}{RT}z_i F\frac{j}{\kappa} \tag{6.144}$$

# 6.11 Mathematical addendum

What follows is a brief introduction of the basic concepts of vector calculus and analysis, necessary for deeper understanding of the mathematical parts of the book. If the reader is not familiar with these concepts, we recommend reading an appropriate book [6] or textbook [8].

**Scalar field**

$$u(x, y, z, t) \equiv u(\mathbf{r}, t) \tag{6.145}$$

**Vector field**

$$\mathbf{v}(x, y, z, t) \equiv \mathbf{v}(\mathbf{r}, t) \equiv \left(v_x(\mathbf{r}, t), v_y(\mathbf{r}, t), v_z(\mathbf{r}, t)\right) \tag{6.146}$$

**Scalar product**

$$a = (a_x, a_y, a_z), \quad b = (b_x, b_y, b_z), \quad a \cdot b = a_x b_x + a_y b_y + a_z b_z \qquad (6.147)$$

**Operator $\nabla$**

$$\nabla \equiv \left( \frac{\partial}{\partial x}, \frac{\partial}{\partial y}, \frac{\partial}{\partial z} \right) \qquad (6.148)$$

**Gradient of a scalar field**

$$\nabla u(x, y, z, t) \equiv \nabla u(r, t) \equiv \left( \frac{\partial u}{\partial x}(r, t), \frac{\partial u}{\partial y}(r, t), \frac{\partial u}{\partial z}(r, t) \right) \qquad (6.149)$$

**Divergence of a scalar field**

$$\nabla \cdot v(x, y, z, t) \equiv \nabla \cdot v(r, t) \equiv \frac{\partial v_x}{\partial x}(r, t) + \frac{\partial v_y}{\partial y}(r, t) + \frac{\partial v_z}{\partial z}(r, t) \qquad (6.150)$$

**Local field outflow from a volume unit**

$$\nabla \cdot v(r, t) = \lim_{V \to 0} \frac{\int_A v \, dA}{V}, \quad r \in V \qquad (6.151)$$

**Line integral of a scalar field**, example: the mass of a wire as the integral of the mass of a length unit

$$\int_C u \, dc, \quad dc = |dC| \qquad (6.152)$$

**Line integral of a vector field**, example: the work of force $F$ at moving along the trajectory $s$

$$\int_s F \cdot ds \qquad (6.153)$$

Example: the line integral of the potential vector field does not depend on the trajectory (it only depends on the starting point $A$ and the end point $B$).

$$\int_A^B \nabla \varphi \cdot dC = \varphi_A - \varphi_B \qquad (6.154)$$

**Surface integral of a scalar field**, example: the mass of a metal sheet as the integral of the mass of an area unit $A$

$$\int_A u \, ds, \quad ds = |d\boldsymbol{A}| \tag{6.155}$$

**Surface integral of a vector field**, example: the volume flux as the surface integral of the velocity

$$\int_A \boldsymbol{v} \cdot d\boldsymbol{A} \tag{6.156}$$

**Volume integral of a scalar field**, example: the weight of a body as the integral of the density

$$\int_V \rho \, dV \tag{6.157}$$

**Volume integral of a vector field**, example: the momentum of a body as the integral of the momentum density

$$\int_V \rho \boldsymbol{v} \, dV \tag{6.158}$$

**Gauss's theorem** on the divergence of a vector field

$$\oint_A \boldsymbol{v} \cdot d\boldsymbol{A} = \int_V (\nabla \cdot \boldsymbol{v}) \, dV \tag{6.159}$$

# References

[1]  J. Newman, Electrochemical Systems. 1st Edition. Prentice-Hall, New York, 1973.
[2]  I. Roušar, K. Micka, and A. Kimla, Technická Elektrochemie II (in Czech). Academia Prague, 1981.
[3]  R. B. Bird, W. E. Stewart, and E. N. Lightfoot, Transport Phenomena. 2nd Edition. Wiley, New York, 2002. ISBN 0-471-07392-X.
[4]  R. Taylor and R. Krishna, Multicomponent Mass Transfer. John Wiley, New York, 1993. ISBN 0471574171.
[5]  J. O. Wilkes, Fluid Mechanics for Chemical Engineers. Prentice-Hall, New York, 1999. ISBN 13 9780134712826.
[6]  E. Kreyszik, Advanced Engineering Mathematics. 9th Edition. John Wiley, New York, 2006. ISBN 0470646136.

[7]   R. Kodým, F. Vlasák, D. Šnita, A. Černín, and K. Bouzek, Spatially two-dimensional
       mathematical model of the flow hydrodynamics in a channel filled with a net-like spacer.
       J. Membr. Sci. 368: 171–183, 2010.
[8]   Š. Porubský, Fundamental Mathematics for Engineers, Vol. I. VŠCHT Praha, 2001.
       ISBN 80-7080-418-1.

Part II: **Electromembrane separation and synthesis processes**

**Electromembrane separation processes** utilize the selectivity of ion-selective membranes for cations and anions to separate the electrolyte from one liquid phase and transfer it to the second liquid phase. Electrodialysis (ED) is the most important among these processes and, therefore, the entire Chapter 7 is devoted to it. The chapter also describes the mathematical modelling of the processes. Electrodeionization (EDI), capacitive deionization (CDI), electrodialysis with reversal of the polarity of electrodes, also called electrodialysis reversal (EDR), and electrodialysis to concentrate electrolyte solutions (EDC) are discussed in Chapters 8 and 9. This group of processes serves to demineralize solutions, to concentrate the electrolytes in the solution or to separate the electrolytes from the non-electrolytes. Next, Chapter 10 deals with diffusion dialysis, although, strictly speaking, it is not an electromembrane process.

**Electromembrane synthesis processes** combine the principles of electromembrane separation process with chemical or electrochemical reactions, including water splitting in bipolar membranes or ion-exchange, for the production of certain chemical substances as products. These processes include electrodialysis for ion substitution, also called electrodialysis metathesis (EDM), electrodialysis with bipolar membranes (EDBM), electrophoresis (EFC) and membrane electrolysis (ME), which is applied, for example, in the production of chlorine and sodium hydroxide or in the production of hydrogen. The electrolysis process is discussed in Chapter 11.

Chapter 12 describes the main industrial applications of electromembrane separation and synthesis processes.

https://doi.org/10.1515/9783110739466-008

David Tvrzník, Aleš Černín, Luboš Novák

# 7 Electrodialysis

## 7.1 Basic principles

**Electrodialysis** (**ED**) is the most important electromembrane separation process, which is used to partially demineralize electrolyte solutions or to concentrate electrolytes in such solutions. The equipment for technical realization of the ED process is called **electrodialysis stack** or **electrodialyser**.

The principle of the process is illustrated in Fig. 7.1. **The membrane stack** of the electrodialyser is formed by regularly alternating the **cation-selective membranes** (CM) and the **anion-selective membranes** (AM), with a space between them to allow the flow of liquid. We call these spaces **flow compartments** (flow chambers). In practice, these compartments are established by **spacers**. The membrane stack is placed between the **end plates** fitted on the inner side with **electrodes**. The flow compartments, delimited by AM on the anode side and by CM on the cathode side, are **diluting chambers**. The liquid flowing through the diluate chambers is referred to as the **diluate**. The flow compartments, delimited by CM on the anode side and by AM on the cathode side, are **concentrating chambers**. The liquid flowing through the concentrate chambers is referred to as the **concentrate**. The driving force of the electrodialysis process is the electric potential gradient, which, when DC voltage is applied to electrodes, causes direct electric current to pass through the electrodialysis stack, thereby removing electrolytes from the diluate and transferring them through the membranes to the concentrate.

The capability of the ED to demineralize and, at the same time, concentrate the electrolyte solutions is given by the selectivity of CM for cations and AM for anions, when using direct electric current, in combination with the capability of ion-selective membranes to hydraulically separate two solutions of different compositions or concentrations and sequences of CM and AM in the membrane stack.

From Fig. 7.1 it is also evident that in the membrane stack of the electrodialyser, the structural element is regularly repeated sequentially from anode to cathode. The structural element consists of CM, concentrate chamber (concentrate spacer), AM and diluate chamber (diluate spacer). This basic structural element is called a **cell pair** (Fig. 7.2). By increasing the number of cell pairs, the electrodialysis stack can achieve a higher process capacity. The maximum number of cell pairs in the electrodialysis stack is limited by manufacturing capabilities, technical characteristics of individual components, electrodialysis stack design or parameters of DC power supply. The number of cell pairs is also limited by the uneven distribution of liquid flow among the flow chambers, by the shape of chambers or by the risk of thermal damage to the membrane stack. Currently, the largest industrial electrodialysis stacks consist of up to 2,000 cell pairs.

https://doi.org/10.1515/9783110739466-009

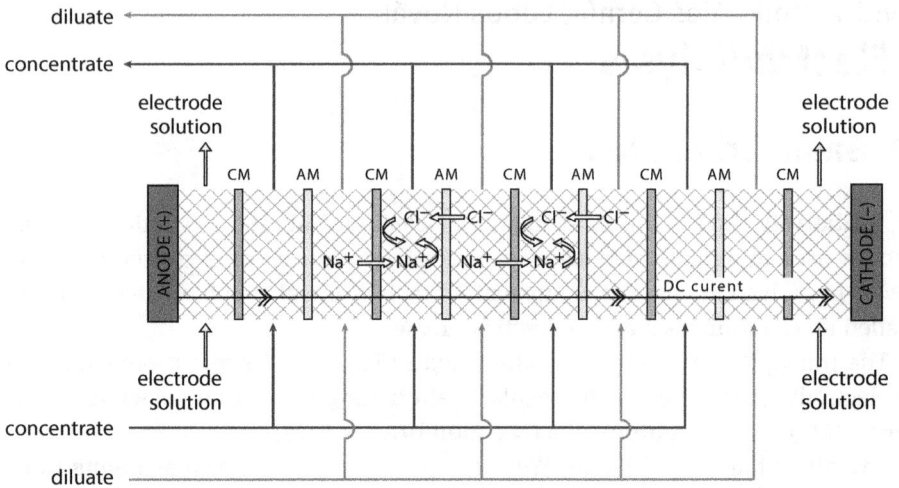

**Fig. 7.1: Principle of electrodialysis.** CM – cation-selective membrane, AM – anion-selective membrane; diluate chambers are bordered by AM on the anode side and by CM on the cathode side; concentrate chambers are bordered by CM on the anode side and by AM on the cathode side; electrode chambers are adjacent to electrodes.

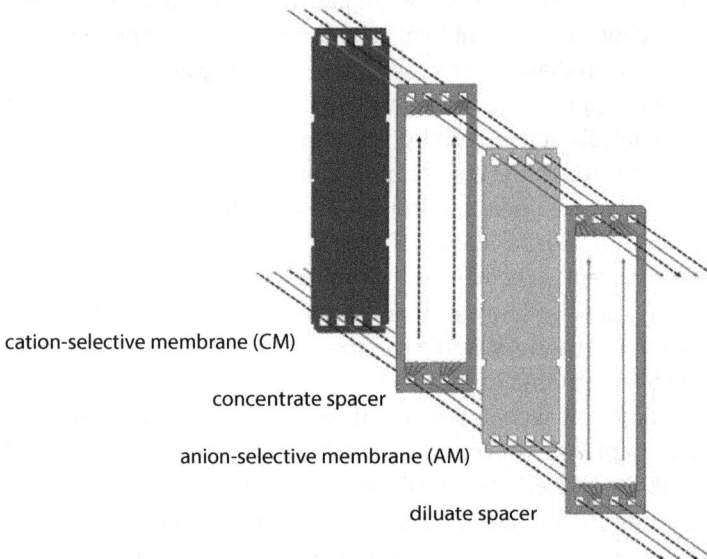

cation-selective membrane (CM)

concentrate spacer

anion-selective membrane (AM)

diluate spacer

**Fig. 7.2:** Cell pair.

**Electrode chambers**, through which an **electrode solution** is passed, are separated from both main streams, especially because of the necessity to remove the gases generated at the electrodes during electrode reactions (particularly $O_2$ and $H_2$)

or due to associated pH changes. An aqueous solution of a suitable electrochemically inert electrolyte (such as $Na_2SO_4$, $NaNO_3$) is used as the electrode solution. Under the current load, the following electrode reaction happens on the anode

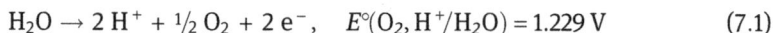

$$H_2O \rightarrow 2\,H^+ + \frac{1}{2}\,O_2 + 2\,e^-, \quad E^\circ(O_2, H^+/H_2O) = 1.229\,V \qquad (7.1)$$

and on the cathode

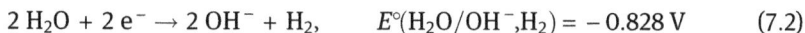

$$2\,H_2O + 2\,e^- \rightarrow 2\,OH^- + H_2, \quad E^\circ(H_2O/OH^-, H_2) = -0.828\,V \qquad (7.2)$$

or

$$2\,H^+ + 2\,e^- \rightarrow H_2, \quad E^\circ(H^+/H_2) = 0.0\,V \qquad (7.3)$$

depending on the pH. The pH value of an electrode solution is usually artificially maintained in an acidic region, reducing or eliminating the risk of so-called **scaling** (precipitation of insoluble organic substances) on the cathode.

Solutions processed by the ED process usually have a TDS in the order of $10^{-1}$–$10^1$ g $dm^{-3}$. By means of demineralization by electrodialysis, the TDS of these solutions can be reduced to the order of $10^{-1}$ g $dm^{-3}$, sometimes even to the order of $10^{-2}$ g $dm^{-3}$. By means of concentrating by electrodialysis, the TDS of these solutions can be increased to the order of $10^1$ g $dm^{-3}$ and, in extreme cases, even to 200–300 g $dm^{-3}$. The ability to work with higher electrolyte concentrations in the concentrate as well as tolerances to a higher level of supersaturation by low-soluble substances are considered as certain benefits of electrodialysis compared to reverse osmosis (RO) in some applications.

## 7.2 Electrodialysis process variants

### 7.2.1 Ion exchange electrodialysis

The function of standard electrodialysis stacks according to the diagram in Fig. 7.1 is restricted only to desalination and concentration of electrolyte solutions. However, a relatively wide range of electrodialysis stacks do not adhere to the standard structure of the membrane stack or hydraulic streams while using the same or similar components and at the same time also relying on the same mass transfer mechanisms. Such devices do not necessarily maintain the number or type of the main hydraulic streams (except for the electrode solution) and are not only capable of desalinating or concentrating the solutions, but are also capable of changing their chemical nature. That is why we consider electromembrane synthesis processes as opening up a number of new application options.

There are, for example, installations in which the membrane stack consists of only one type of ion-selective membrane (cation- or anion-selective membrane). In this particular case, it is no longer possible to talk about diluate and concentrate because desalination or concentration of the solution does not occur here at all. The system acts as a continuous ion exchanger, which can be used, for example when softening water, see Fig. 7.3 [1]. However, the efficiency of water softening by electrodialysis is strongly linked to the properties of the ion-selective membranes used.

Another example is the stabilization of pH (removal of citric acid from fruit juices). See Fig. 7.4 [1]. This is, on the other hand, a very interesting area of ED application from an economic point of view, as usually only a small area of ion-selective membranes is needed for processing of large volumes of juices over a given time.

In other types of equipment, cation- or anion-selective membranes can always be placed in pairs in the membrane stack. In addition to the diluate and concentrate, there is also a main hydraulic stream, in which only ion exchange occurs. This principle can be used for a number of reactions, for example in the pharmaceutical industry. The example of the use of a three-circuit electrodialysis stack to prepare 3-(2,2,2-trimethylhydrazinium) propionate dihydrate from the compound $Br^-(CH_3)_3N^+NHCH_2CH_2COOR$ (where R is the lower alkyl, e.g. methyl) is demonstrated in Fig. 7.5. From the diluate (NaOH solution), $OH^-$ ions are transferred through the anion-selective membrane into the hydraulic circuit with the solution $Br^-(CH_3)_3N^+NHCH_2CH_2COOR$, which leads to hydrolysis of the compound at formation of the product $(CH_3)_3N^+NHCH_2CH_2COO^-$, while the released $Br^-$ ions are transferred through the second anion-selective membrane into the concentrate, where NaBr solution is formed by simultaneous transfer of $Na^+$ ions from the diluate. The process described is an adaptation of a three-chamber membrane electrolysis process presented in [2, 3].

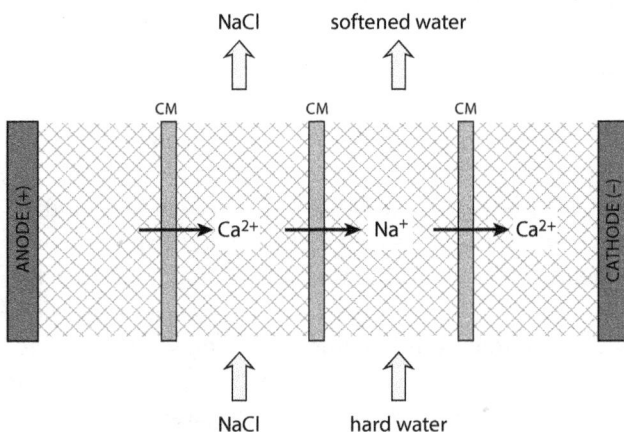

**Fig. 7.3:** Water softening by electrodialysis.

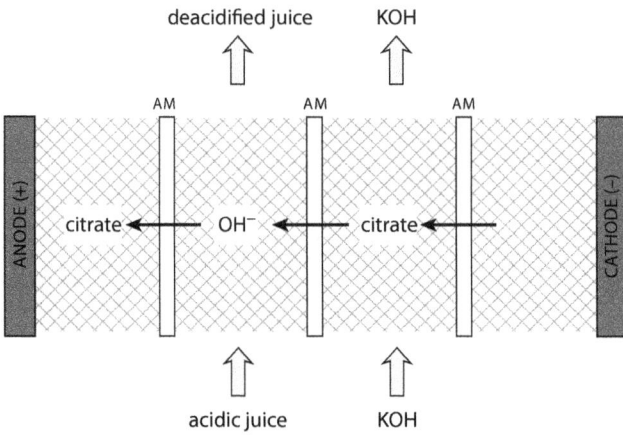

**Fig. 7.4:** Deacidification of fruit juices by electrodialysis.

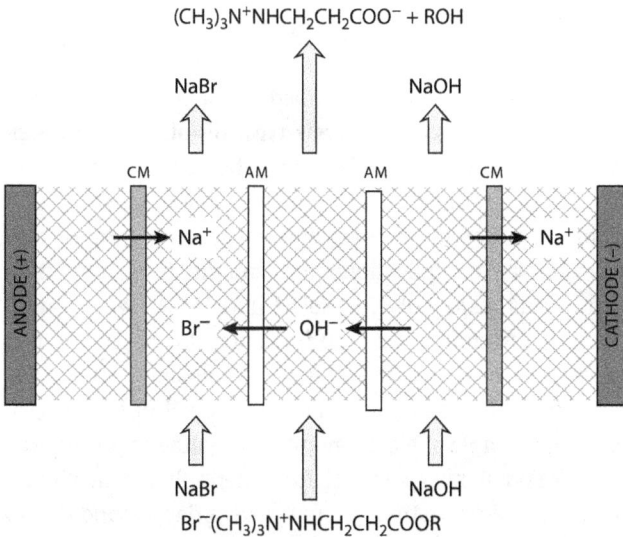

**Fig. 7.5:** Example of a three-circuit electrodialysis stack.

Another type of device retains the standard structure of the membrane stack, but not the number of main hydraulic streams, as it works with two diluates and two concentrates. In such an electrodialysis stack, a double displacement reaction is possible. When electrolyte AB enters the device in hydraulic stream diluate 1 and electrolyte XY in hydraulic stream diluate 2, electrolyte XB will leave the device in hydraulic stream concentrate 1 and electrolyte AY in hydraulic stream concentrate 2. This process is sometimes referred to as **electrodialysis metathesis (EDM)**. An example of the EDM process is shown in Fig. 7.6.

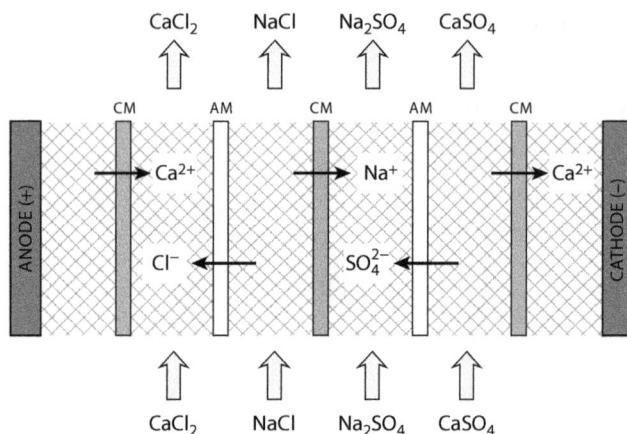

**Fig. 7.6:** Example of four-circuit electrodialysis stack.

## 7.2.2 Electrodialysis with bipolar membranes

When **bipolar membranes** (**BM**, see their properties described in Chapter 4) are used in the ED process, the process is referred to as **electrodialysis with bipolar membranes** (EDBM, or **bipolar membrane electrodialysis**, BMED). Here, the function of bipolar membrane is not based on the selective transport of cations and anions as in case of cation- or anion-selective membranes, but on the water splitting reaction to form $H^+$ and $OH^-$ ions. This can be used in a number of applications, for example, for salt splitting into the relevant hydroxides and acids, especially for NaOH and $H_2SO_4$ regeneration from $Na_2SO_3/Na_2SO_4$ solution resulting from flue gas desulfurization, for NaOH and HCl regeneration from spent regenerant waste from ion exchange technologies and in the production of weak organic acids from their salts, for adjusting the pH of fruit juices, ciders and wines, etc. This is another example of electromembrane synthesis processes.

The minimum equilibrium thermodynamic potential for water splitting in bipolar membrane is 0.83 V at 25 °C. In order to achieve the required current density and due to the real kinetics of water splitting, a voltage of 0.9–2.0 V is, however, common in practice.

The concept of using the EDBM process for splitting of salts to the relevant acids and hydroxides was developed by Allied Chemical Corporation, which – based on the basic engineering analysis – showed the economic potential of this process, especially due to its low energy consumption [4]. In laboratory tests of $Na_2SO_4$ splitting by EDBM process at a temperature of 50 °C, solutions with a concentration of 12–16 % (m/m) NaOH and 5–10 % (m/m) $H_2SO_4$ were produced, with power consumption of only 1,450–1,700 kW h t$^{-1}$ NaOH [5].

The main disadvantages of the EDBM technology include high price and short service lifetime of bipolar membranes as well as the limitation of NaOH concentration by the chemical stability of anion-selective membranes and anion-selective part of

BM. Another disadvantage is the rather low selectivity of ion-selective membranes in the environment of strong acids and hydroxides. In particular, the easy transport of $H^+$ ion through the anion-selective membrane reduces the rate and current efficiency, increases energy consumption and reduces the achievable acid and hydroxide concentration in the EDBM process [6].

The EDBM process is used in the classic three-circuit arrangement, see Fig. 7.7. However, due to the problem of selectivity of the ion-selective membranes or depending on the products required, it is sometimes preferable to choose only the two-circuit arrangement [5], see Fig. 7.8.

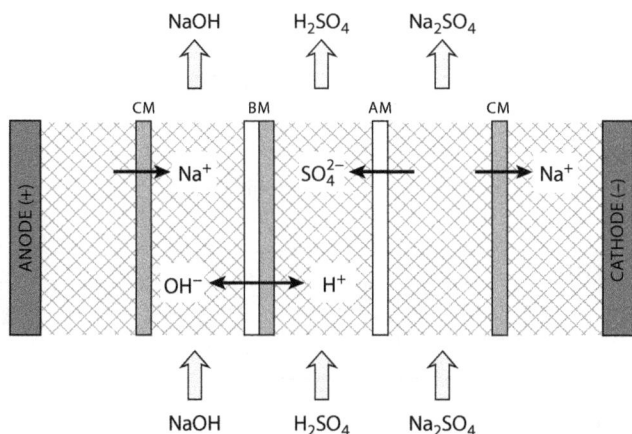

Fig. 7.7: Three-circuit electrodialysis stack with bipolar membranes.

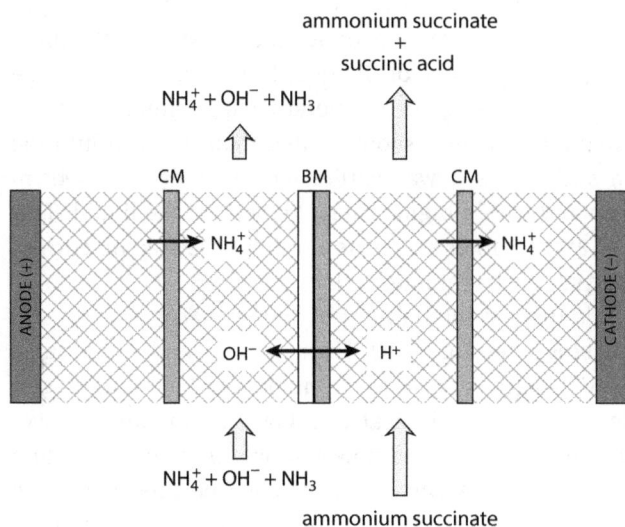

Fig. 7.8: Two-circuit electrodialysis stack with bipolar membranes.

## 7.2.3 Basic technology arrangement

The general scheme of the simple ED technology is shown in Fig. 7.9. The scheme does not include the electrode circuit. The electrode circuit participates in the mass transfer, causing problems in many cases, but the mass transfer between the diluate or concentrate and the electrode circuit is negligible compared to that between the diluate and the concentrate. Therefore, it is not necessary to include this exchange in the basic schemes. Of course, other hydraulic streams can be introduced into the scheme, such as dosing of chemicals into the concentrate.

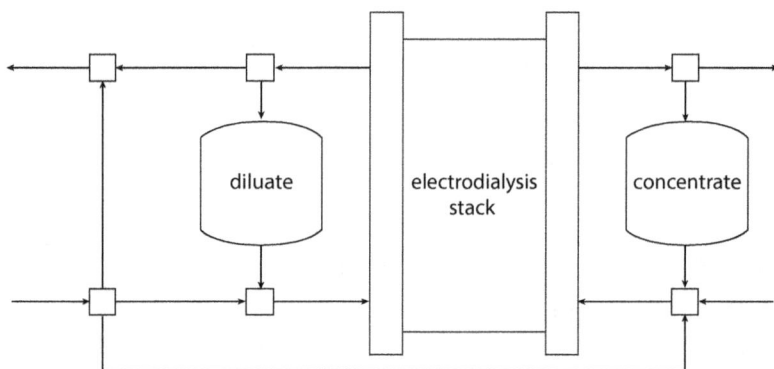

**Fig. 7.9:** General scheme of simple ED technology.

Each electrodialysis stack contains one or more **hydraulic stages**, see Fig. 7.10a. By increasing the number of hydraulic stages, the flow path length is extended and – at the given linear velocity of the liquid – higher level of demineralization of the processed solution is achieved by a single passage of the liquid through the stack. The number of cell pairs in each hydraulic stage can generally vary. Diluate entering the second and every other hydraulic stage is more diluted than the feed into the previous stage. Therefore, in such cases it was earlier recommended to work in these stages with less cell pairs and with a higher linear velocity of the liquid. This increases the electrical current and so other stages do not limit the electric current passing through the electrodialysis stack that much [7].

If the electrodialysis stack is divided into several hydraulic stages, it may also contain more than one **electric stage**, see Fig. 7.10b. The maximum and, at the same time, optimum number of electric stages is equal to the number of hydraulic stages. The method of dividing to electric stages, as indicated in Fig. 7.10b, requires power supply to both electrical stages from the same DC power supply. A block separating the two electric stages can be fitted with a pair of independent electrodes and the function of both electric stages controlled by two independent DC power supplies.

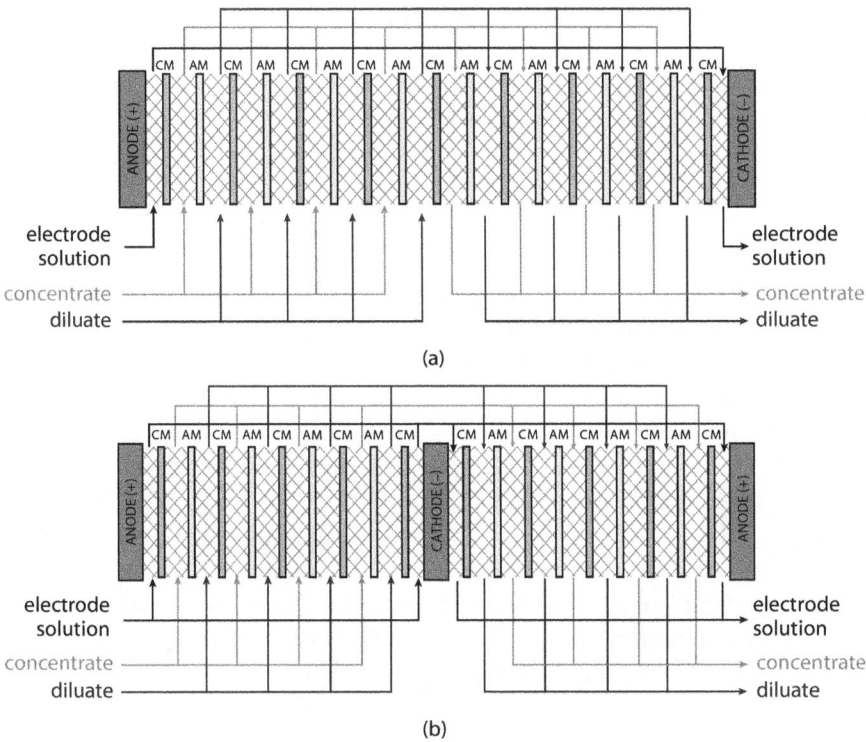

**Fig. 7.10:** Hydraulic scheme of electrodialysis stack; a) with two hydraulic and one electric stages, b) with two hydraulic and two electric stages.

Currently, the separation to hydraulic and electric stages is most often achieved by connecting several electrodialysis stacks in series.

Electrodialysis stacks can be operated independently or, depending on the capacity and quality requirements of particular application, multiple electrodialysis stacks may be grouped into **lines hydraulically connected in parallel**; each line containing several electrodialysis stacks hydraulically connected in series and electrically connected in parallel (hydraulic and electric stages).

In the treatment of water with electrolyte concentration at the level of brackish and river water (0.5–3 kg m$^{-3}$) by ED process in a three-stage or a four-stage arrangement, excessive current leakage may cause thermal damage or burning of the membrane stack in the third and fourth stages. The reason is the low level of electrolyte concentration in the diluate and the relatively high concentration of electrolyte in the concentrate. To reduce this risk, it is appropriate to split the concentrate circuit into two separate circuits. The make-up feed water is first injected into the second concentrate circuit and the resulting blowdown is then directed as a make-up into the first concentrate circuit.

Based on the general scheme of simple ED technology presented in Fig. 7.9, the following basic operating modes can be performed:
-   batch,
-   feed and bleed,
-   continuous single-pass or once-through (with bypass).

Each main hydraulic stream (diluate, concentrate) is operated in one of these modes.

In some cases, operating modes can also be changed during the operation, if this results in the facilitation of mass transfer in the ED process. As the quality – for example, in the batch operation – reaches the required level at the outlet from the ED line, the diluate outlet may be redirected to another tank, which changes the operating mode to a continuous single-pass operation until the original tank is completely empty. In many cases, this will improve the ED technology performance.

For each individual application, the optimal configuration and operating mode of ED technology can be determined based on the data from the pilot device or the mathematical modelling of the mass transfer. Here, the ED technology configuration means the number of parallel lines, number of hydraulic and electric stages, type of electrodialysis stack and number of cell pairs.

# 7.3 Approaches to mathematical modelling

## 7.3.1 Mass transfer in electrodialysis

Mass transfer in electromembrane devices is a complex process as individual components gradually pass through the flow chambers and ion-selective membranes. The transport of components in electrolyte solutions and ion-selective membranes is discussed in Chapters 3 and 5.

Except for extreme cases of excessive current leakage, see Section 7.3.4.2, herein, we will assume a quasi-homogeneous electric field in the electrodialysis stack, with the vector of intensity of the electric field (gradient of electric potential) oriented perpendicular to the surface of ion-selective membranes (parallel to the normal line of membrane surface, in the direction of $x$ axis). The problem of electric potential distribution in the electrodialysis stack is thus reduced to a one-dimensional (1D) case, which greatly simplifies all other calculations. The validity of this assumption can be proved by experimental data or by results of 2D calculations for most practical operating conditions.

The key phenomena in the ED process, like in many other chemical systems, take place at phase interfaces – in this case, at the interface between the electrolyte solution and the ion-selective membrane – where electrolyte is separated from the diluate and transferred into the concentrate. The liquid flows through the flow chambers of the electrodialysis stack, parallel to the surface of ion-selective membranes (perpendicular to the normal line of membranes, in the direction of $y$ axis), while the transfer of electrolyte from the diluate to the concentrate takes place in the direction parallel to the orientation of the vector of electric field intensity.

If, in accordance with the assumptions mentioned above, the electric potential gradient in the direction of the liquid flow is zero and the diffusion rate in the same direction can usually be neglected – due to the commonly used linear velocities of the liquid in the order of $10^0$–$10^1$ cm s$^{-1}$ – according to the Nernst–Planck equation (3.47), the absolutely dominant mechanism of mass transfer in the working chambers in the direction of the liquid flow is convection. In the direction perpendicular to the surface of ion-selective membranes, the mass transfer in the liquid phase is exclusively done by diffusion and migration – according to eq. (3.47) and due to the absence of convection.

The mass transfer in the direction perpendicular to the surface of ion-selective membranes is thus, obviously, influenced by the nature of fluid flow in the working chambers.

### 7.3.1.1 Effect of hydrodynamics in flow chambers

In the simplest case (i.e. ideal, which means practically unfeasible), it would be a piston-type flow. This means that the transverse mixing would be perfect and in the transverse direction (perpendicular to the membrane plane), the composition would be uniform (constant in the given cross-section of the chamber). If we do not want to or cannot solve the equations of flow for a given geometric arrangement, including the influence of the inserted netting, we must use some simplified model. For example, we can introduce a function or functions of position in a transverse direction, which describes the deviation from the ideal behaviour, for example,

- functions describing by polynomials of different degrees, transition between the piston-type flow and laminar flow with parabolic profile of the velocity (Fig. 7.11a),
- functions that are in the vicinity of the membranes (in diffusion, laminar, boundary layers or sublayers) linear and in the core of chambers constant (Fig. 7.11b),
- functions that are in the vicinity of the membranes zero and in the core of chambers constant (Fig. 7.11c).

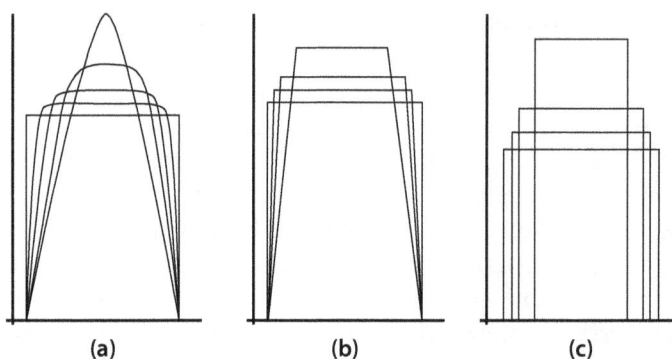

**Fig. 7.11:** Examples of functions describing uneven transverse distribution of quantities in flow chambers.

Such functions may be used to describe the velocity profile and the transverse mixing by means of an additional dispersion coefficient, which is added to the diffusion coefficient. However, the amended coefficient can only be used in diffusion terms and not in migration, as in simple terms, electric field cannot be intermixed. These functions should be constructed in such a way so that their mean value equals 1.

For the purpose of mathematical modelling of mass transfer in ED, it is necessary to adopt a suitable simple model of hydrodynamics of the fluid flow in flow chambers. Some authors [8–10] work with the model of laminar fluid flow in an empty rectangular chamber with a characteristic parabolic profile of fluid linear velocity across the chamber. The corresponding mass transfer model is referred to as a **convection-diffusion** model. The paper [11] suggested that this concept is not capable of explaining the real intensity of the mass transfer in ED with net-like spacers when processing dilute electrolyte solutions.

Based on practical experience with net-like spacers, the hydrodynamics model as shown in Fig. 7.11c appears to be the most appropriate. In this model, each flow chamber consists of a total of three liquid phases. Assume that the coordinates of the left edge of the flow chamber is $x = 0$ and of the right edge $x = d$, respectively. The middle phase ($\delta \leq x \leq d - \delta$) is a flow phase with constant linear velocity of the liquid in the direction of axis $y$ and perfect transverse (in the direction of axis $x$) intermixing of the fluid. The two edge phases are stationary (diffusion layers) with thickness $\delta$, that is, $v_y(x) = 0$ in the intervals ($0 \leq x < \delta$) and ($d - \delta < x \leq d$). It is a **model of stagnant diffusion layers and a perfectly intermixed core**. This model satisfactorily explains not only the observed mass transfer in electrodialysis stack but also the phenomenon of concentration polarization (limiting current density, limiting current).

The link between mass transfer and convection in the flow phase is expressed in this hydrodynamics model of the liquid flow by the Sherwood number,

$$Sh = a \, Re^b \, Sc^c \tag{7.4}$$

where Re is the Reynolds number, $Sc$ is the Schmidt number and $a$, $b$, $c$ are constants. For net-like spacers, the value of coefficient $b$ was determined as $b = \frac{2}{3}$ [12]. Considering the definition of the Sherwood number,

$$\text{Sh} = kd/D = d/\delta \tag{7.5}$$

where $k$ is the mass transfer coefficient, $d$ is the characteristic length (channel thickness) and $D$ is the diffusion coefficient, the following applies to the dependence of $\delta$ on the linear velocity of the fluid:

$$\delta \sim v^{-2/3} \tag{7.6}$$

The adoption of this fluid flow model in the flow chambers greatly simplifies the mass transfer modelling. The local ion flux through ion-selective membranes is – at the given gradient of electric potential, flow mode in the flow chambers and temperature – only the function of local concentrations of ions in the core (flow phase) of the diluate and the concentrate.

## 7.3.1.2 Local balance

The balance scheme of the volume elements of diluate and concentrate chamber is shown in Fig. 7.12.

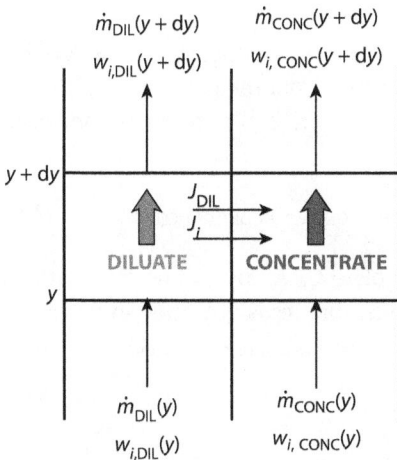

Fig. 7.12: Local mass balance in the diluate and concentrate.

The local mass balance for the diluate and individual component $i$ shall be

$$\frac{d\dot{m}_{\text{DIL}}}{dy} = -nWJ_{\text{DIL}} \tag{7.7}$$

$$\frac{d}{dy}(\dot{m}_{DIL}w_{i,DIL}) = \frac{d\dot{m}_{DIL}}{dy}w_{i,DIL} + \dot{m}_{DIL}\frac{dw_{i,DIL}}{dy} = -nWJ_i \quad \Rightarrow \frac{dw_{i,DIL}}{dy} = -\frac{nW(J_i - w_{i,DIL}J_{DIL})}{\dot{m}_{DIL}}$$

$$(7.8)$$

where $\dot{m}_{DIL}$ is the local mass flow of the diluate through the respective working chambers (unit: kg s$^{-1}$) and $w_{i,DIL}$ is the local mass fraction of the $i$-th component in the flow phase of the diluate; $J_{DIL}$ and $J_i$ (unit: kg m$^{-2}$ s$^{-1}$) are the local fluxes of the diluate and its $i$-th component, respectively, through the ion-selective membranes in direction to the concentrate, $n$ is the number of cell pairs and $W$ is the width of flow channel (flow chamber). The value of $J_{DIL}$ is the sum of fluxes of all components that are transported through ion-selective membranes and, thus, they include the fluxes of all ions, solvent and some neutral components present in the diluate.

### 7.3.1.3 Total (integral) balance of electrodialysis stack

The differences between the input and output mass flow and the composition of the diluate and concentrate relate to the mass transfer between the two main hydraulic streams through the ion-selective membranes. For the total molar flow of the $i$-th component from the diluate to the concentrate $\dot{N}_i$ (mol s$^{-1}$), the following can be derived from the input and output balance

$$\dot{N}_i = \frac{1}{M_i}(\dot{m}_{DIL,0}w_{i,DIL,0} - \dot{m}_{DIL,1}w_{i,DIL,1}) = \frac{1}{M_i}(\dot{m}_{CONC,1}w_{i,CONC,1} - \dot{m}_{CONC,0}w_{i,CONC,0}) \quad (7.9)$$

where $M_i$ is the molar mass of the $i$th component and subscripts 0 and 1 symbolize the input into the electrodialysis stack and the output from the electrodialysis stack, respectively. With the use of volumetric flow rates $\dot{V}$ (m$^3$ s$^{-1}$) and molar concentrations $c$, this equation has the following form:

$$\dot{N}_i = \dot{V}_{DIL,0}c_{i,DIL,0} - \dot{V}_{DIL,1}c_{i,DIL,1} = \dot{V}_{CONC,1}c_{i,CONC,1} - \dot{V}_{CONC,0}c_{i,CONC,0} \quad (7.10)$$

Let's consider the processing of the binary electrolyte $C_{v_+}A_{v_-}$ by means of electrodialysis. If the effect of diffusion in the ion-selective membranes is neglected in this stage, the following shall apply to the local flux of cation through the cation-selective membrane $J_{C,CM}$ (unit: mol m$^{-2}$ s$^{-1}$)

$$J_{C,CM} = \frac{t_{C,CM}\,j_x}{z_C\,F} \quad (7.11)$$

and analogously to the flux of cation through the anion-selective membrane $J_{C,AM}$

$$J_{C,AM} = \frac{t_{C,AM}\,j_x}{z_C\,F} \quad (7.12)$$

where $t_{C,CM}$ and $t_{C,AM}$ means the transport number of cation in the cation-selective membrane and anion in the anion-selective membrane, respectively, $z_C$ is the charge of cation, $j_x$ is local **current density** and $F$ is the Faraday constant. The local flux of the electrolyte through ion-selective membranes from the diluate to the concentrate $J_S$ can be expressed as

$$J_S = \frac{(t_{C,CM} - t_{C,AM})}{v_C z_C} \frac{j_x}{F} \tag{7.13}$$

where $v_C$ is a stoichiometric coefficient of the cation.

The electric current passing through the active area of the ion-selective membrane $I_E$ is expressed by the integral

$$I_E = \int_0^{A_{eff}} j_x \, dA \tag{7.14}$$

where $A_{eff}$ is the active area of the ion-selective membrane.

The electric current measured on the mains of electrode terminals $I$ is always higher than the electric current passing through the active area of ion-selective membranes $I_E$ by the **leakage electric current (shunt current)** $I_L$ and by the **ground fault current** $I_G$. Shunt current passes through the liquid flow distributors, ion-selective membranes in the area of liquid flow distributors and flow distribution manifolds. Ground fault current $I_G$ passes through the grounding elements in piping to the ground. The total electric current is therefore the following sum:

$$I = \bar{I}_E + \bar{I}_L + I_G \tag{7.15}$$

In this equation, average values were used for $I_E$ and $I_L$, as in Section 7.3.4, we showed that both values are changing across the electrodialysis stack but their sum is constant.

By integrating eq. (7.13) based on the active area of the ion-selective membrane, we get the following expression for the total molar flow of electrolyte from the diluate to the concentrate $\dot{N}_S$ (unit: mol s$^{-1}$)

$$\dot{N}_S = \frac{n \eta I}{v_C z_C F} \tag{7.16}$$

where $n$ is the number of cell pairs and $\eta$ is the **current efficiency** which we used to replace – due to practical reasons – the difference of cation transport numbers in both types of ion-selective membranes. The current efficiency is always less than 1, which is given not only by the non-ideal selectivity of ion-selective membranes, see eq. (7.13), but also by the effect of the back diffusion (see Section 7.3.3.5) and the leakages of the electric current, as discussed above.

## 7.3.2 Voltage loss in electrodialysis stack

The loss of voltage on a membrane stack of electrodialyser, as a function of the local current density, is expressed by the equation

$$\Delta\varphi(j_x) = n\left[\Delta\varphi_{AM}(j_x) + \Delta\varphi_{DIL}(j_x) + \Delta\varphi_{CM}(j_x) + \Delta\varphi_{CONC}(j_x)\right] \tag{7.17}$$

where $\Delta\varphi$ terms indicate the voltage losses on particular components of the membrane stack, that is, on the anion-selective membrane (AM), in diluate chamber (DIL), on the cation-selective membrane (CM) and in the concentrate chamber (CONC). For these terms, the following expression applies

$$\Delta\varphi_{AM} = \Delta\varphi_{Don,CONC-AM}(j_x) + \Delta\varphi_{Don,AM-DIL}(j_x) - \int_0^{d_{AM}}\left[\frac{j_x}{\kappa_{AM}(x)} + \frac{RT}{F}\sum_{k=1}^{n}\frac{t_{k,AM}}{z_k}\frac{d\ln c_{k,AM}}{dx}\right]dx \tag{7.18}$$

$$\Delta\varphi_{DIL} = -\int_0^{d_{DIL}}\left[\frac{\alpha j_x}{\kappa_{DIL}(x)} + \frac{RT}{F}\sum_{k=1}^{n}\frac{t_{k,DIL}}{z_k}\frac{d\ln c_{k,DIL}}{dx}\right]dx \tag{7.19}$$

$$\Delta\varphi_{CM} = \Delta\varphi_{Don,DIL-CM}(j_x) + \Delta\varphi_{Don,CM-CONC}(j_x) - \int_0^{d_{CM}}\left[\frac{j_x}{\kappa_{CM}(x)} + \frac{RT}{F}\sum_{k=1}^{n}\frac{t_{k,CM}}{z_k}\frac{d\ln c_{k,CM}}{dx}\right]dx \tag{7.20}$$

$$\Delta\varphi_{CONC} = -\int_0^{d_{CONC}}\left[\frac{\alpha j_x}{\kappa_{CONC}(x)} + \frac{RT}{F}\sum_{k=1}^{n}\frac{t_{k,CONC}}{z_k}\frac{d\ln c_{k,CONC}}{dx}\right]dx \tag{7.21}$$

whereas the following applies to Donnan potentials

$$\Delta\varphi_{Don,CONC-AM} = -\frac{RT}{z_iF}\ln\frac{c_{i,AM}(0)}{c_{i,CONC}(d_{CONC})} \tag{7.22}$$

$$\Delta\varphi_{Don,AM-DIL} = -\frac{RT}{z_iF}\ln\frac{c_{i,DIL}(0)}{c_{i,AM}(d_{AM})} \tag{7.23}$$

$$\Delta\varphi_{Don,DIL-CM} = -\frac{RT}{z_iF}\ln\frac{c_{i,CM}(0)}{c_{i,DIL}(d_{DIL})} \tag{7.24}$$

$$\Delta\varphi_{Don,CM-CONC} = -\frac{RT}{z_iF}\ln\frac{c_{i,CONC}(0)}{c_{i,CM}(d_{CM})} \tag{7.25}$$

In a special case of a binary electrolyte solution and in the absence of diffusion in the ion-selective membranes, the following shall apply

$$\Delta\varphi_{AM} = -R_{A,AM}j_x - \frac{RT}{F}\left[\frac{t_{C,AM}}{z_C} + \frac{(1-t_{C,AM})}{z_A}\right] \ln\frac{c_{S,DIL}(0)}{c_{S,CONC}(d_{CONC})} \tag{7.26}$$

$$\Delta\varphi_{DIL} = -\alpha j_x \int_0^{d_{DIL}} \frac{dx}{\kappa_{DIL}(x)} - \frac{RT}{F}\left[\frac{t_{C,S}}{z_C} + \frac{(1-t_{C,S})}{z_A}\right] \ln\frac{c_{S,DIL}(d_{DIL})}{c_{S,DIL}(0)} \tag{7.27}$$

$$\Delta\varphi_{CM} = -R_{A,CM}j_x - \frac{RT}{F}\left[\frac{t_{C,CM}}{z_C} + \frac{(1-t_{C,CM})}{z_A}\right] \ln\frac{c_{S,CONC}(0)}{c_{S,DIL}(d_{DIL})} \tag{7.28}$$

$$\Delta\varphi_{CONC} = -\alpha j_x \int_0^{d_{CONC}} \frac{dx}{\kappa_{CONC}(x)} - \frac{RT}{F}\left[\frac{t_{C,S}}{z_C} + \frac{(1-t_{C,S})}{z_A}\right] \ln\frac{c_{S,CONC}(d_{CONC})}{c_{S,CONC}(0)} \tag{7.29}$$

In eqs. (7.26)–(7.29), $R_{A,AM}$ and $R_{A,CM}$ symbols indicate area resistances of anion-selective and cation-selective membranes, respectively; $\kappa_{DIL}$ and $\kappa_{CONC}$ indicate local conductivity of the diluate and of the concentrate, respectively; and $\alpha$ is the coefficient taking into account the effect of the spacer net on the ohmic resistance of the membrane stack. For example, for the net-like spacers by MEGA a.s., the value of $\alpha = 1.5$ was determined based on experimental data.

If we express individual terms of the equation for the voltage loss in the electrodialysis stack as the function of the current density, then the equation

$$U + \Delta\varphi(j_x) = 0 \tag{7.30}$$

in which $U$ is total voltage on the membrane stack of the electrodialyser can be used to calculate the local current density.

The total voltage loss in the electrodialysis stack (voltage measured on the electrode terminals) is by few volts higher than the voltage measured on the membrane stack, as it includes the decomposition voltage on the electrodes and the voltage loss in the electrode chambers.

## 7.3.3 Limiting factors of the electrodialysis process

### 7.3.3.1 Limiting current density, limiting current

In Section 7.3.1.3 we could see the link between the electrolyte flux through the ion-selective membranes and the applied electric current. We also assumed that the processes at the phase interface between the liquid phase and the ion-selective membrane do not restrict the mass transfer in any way. Let us now investigate these phenomena, provided that the hydrodynamics of the liquid flow as described in Section 7.3.1.1, apply here as well. We still neglect the influence of diffusion in ion-selective membranes and simplify the problem to the case of binary electrolyte $C_{v_+}A_{v_-}$.

For the flux of cation C in the ion-selective membrane and in the diffusion layer of the liquid phase at the interface with the ion-selective membrane, the following equations are applied:

$$J_{C,M} = \frac{t_{C,M}}{z_C} \frac{j_x}{\mathbf{F}} \tag{7.31}$$

$$J_{C,S} = -v_C D_S \frac{dc_S}{dx} + \frac{t_{C,S}}{z_C} \frac{j_x}{\mathbf{F}} \tag{7.32}$$

In these equations, $t_{C,M}$ and $t_{C,S}$ indicate the cation transport numbers in the ion-selective membranes, respectively in the liquid phase, $v_C$ is a stoichiometric coefficient and $z_C$ is the cation charge, $j_x$ is the local current density, $\mathbf{F}$ – Faraday constant, $D_S$ – diffusion coefficient of binary electrolyte in the liquid phase and $c_S$ is the binary electrolyte concentration in the liquid phase.

In a steady state, the cation balance at the phase interface between the flow chamber (liquid phase) and the ion-selective membrane is expressed as follows:

$$J_{C,S} = J_{C,M} \tag{7.33}$$

so with substitution from the transport eqs. (7.31) and (7.32) we get

$$-v_C D_S \frac{dc_S}{dx} + \frac{t_{C,S}}{z_C} \frac{j_x}{\mathbf{F}} = \frac{t_{C,M}}{z_C} \frac{j_x}{\mathbf{F}} \tag{7.34}$$

and from here, the following term can be deduced for the change of the electrolyte concentration in the diffusion layer

$$\frac{dc_S}{dx} = -\frac{(t_{C,M} - t_{C,S})j_x}{v_C z_C D_S \mathbf{F}} \tag{7.35}$$

and for the current density

$$j_x = -\frac{v_C z_C \mathbf{F} D_S}{(t_{C,M} - t_{C,S})} \frac{dc_S}{dx} \tag{7.36}$$

Equation (7.35) is important for the calculation of the electrolyte concentration profile across the flow chamber, which must be known when calculating the voltage loss in the flow chamber, see Section 7.3.2, as the solution conductivity is the function of the electrolyte concentration. You can see that according to eq. (7.35), with the increasing value of $j_x$, the electrolyte concentration in the diffusion layer of the diluate decreases in the direction to the ion-selective membrane. This means there is a certain maximum value of $j_x$, at which $c_{S,DIL}(0)$ or $c_{S,DIL}(d_{DIL})$ drops to zero. This state corresponds to reaching the **limiting current density** $j_{x,lim}$, defined as

$$j_{x,\text{lim}} = \frac{v_C z_C F D_S \, c_{S,\text{DIL}}}{\delta |t_{C,M} - t_{C,S}|} \tag{7.37}$$

where $t_{C,M}$ stands for the cation transference number in the ion-selective membrane (anion-selective or cation-selective), for which $j_{x,\text{lim}}$ is lower. The ratio between the local current density $j_x$ and its limit value $j_{x,\text{lim}}$ is often referred to as the **concentration polarization level**

$$P = j_x / j_{x,\text{lim}} \tag{7.38}$$

The relevant integral quantity to $j_{x,\text{lim}}$ is referred to as **limiting electric current** $I_{\text{lim}}$, which is – according to eq. (7.14) and using eq. (7.37) – defined as follows:

$$I_{\text{lim}} = \int_0^{A_{\text{eff}}} j_{x,\text{lim}} \, dA = \frac{v_C z_C F D_S}{\delta |t_{C,M} - t_{C,S}|} \int_0^{A_{\text{eff}}} c_{S,\text{DIL}} \, dA \tag{7.39}$$

The dependence of $c_{S,\text{DIL}}$ on the active surface of the ion-selective membrane can be obtained from the balance of the diluate chamber volume element. If the transport of the solvent through the ion-selective membranes and the effect of the electrolyte concentration on the solution density are neglected, the balance is expressed as follows:

$$\dot{V}_{\text{DIL}} \, c_{S,\text{DIL}}(A) = \dot{V}_{\text{DIL}} \, c_{S,\text{DIL}}(A + dA) + J_S \, dA \tag{7.40}$$

where $\dot{V}_{\text{DIL}}$ is the volume flow of the diluate through the diluate chamber. With adjustment and substitution from eqs. (7.13) and (7.37), we get the following expression for the change of the electrolyte concentration in the diluate across the active surface of the ion-selective membrane:

$$\frac{dc_{S,\text{DIL}}}{dA} = -\frac{J_S}{\dot{V}_{\text{DIL}}} = -\frac{D_S}{\delta} \frac{(t_{C,CM} - t_{C,AM})}{|t_{C,M} - t_{C,S}|} \frac{c_{S,\text{DIL}}}{\dot{V}_{\text{DIL}}} \tag{7.41}$$

After the necessary adjustment and integration, we finally get the following equation:

$$c_{S,\text{DIL}} = c_{S,\text{DIL},0} \, \exp\left[ -\frac{D_S}{\delta} \frac{(t_{C,CM} - t_{C,AM})}{|t_{C,M} - t_{C,S}|} \frac{A}{\dot{V}_{\text{DIL}}} \right] \tag{7.42}$$

By substituting this equation into eq. (7.39), and after the necessary adjustment and integration, we finally get the following expression for $I_{\text{lim}}$:

$$I_{\text{lim}} = \frac{v_C z_C F \dot{V}_{\text{DIL}} \, c_{S,\text{DIL},0}}{(t_{C,CM} - t_{C,AM})} \left\{ 1 - \exp\left[ -\frac{D_S}{\delta} \frac{(t_{C,CM} - t_{C,AM})}{|t_{C,M} - t_{C,S}|} \frac{A_{\text{eff}}}{\dot{V}_{\text{DIL}}} \right] \right\} \tag{7.43}$$

The example of the volt-ampere characteristic of the electrodialysis stack, indicating the existence of the limiting current, is shown in Fig. 7.13. The graph further shows the level of concentration polarization at the inlet to the diluate chamber and at the outlet from it. The concentration polarization level grows from the inlet towards the outlet. The limiting current is achieved when the level of concentration polarization at the inlet to the electrodialysis stack (i.e. also across the whole flow path length) reaches one (100 %).

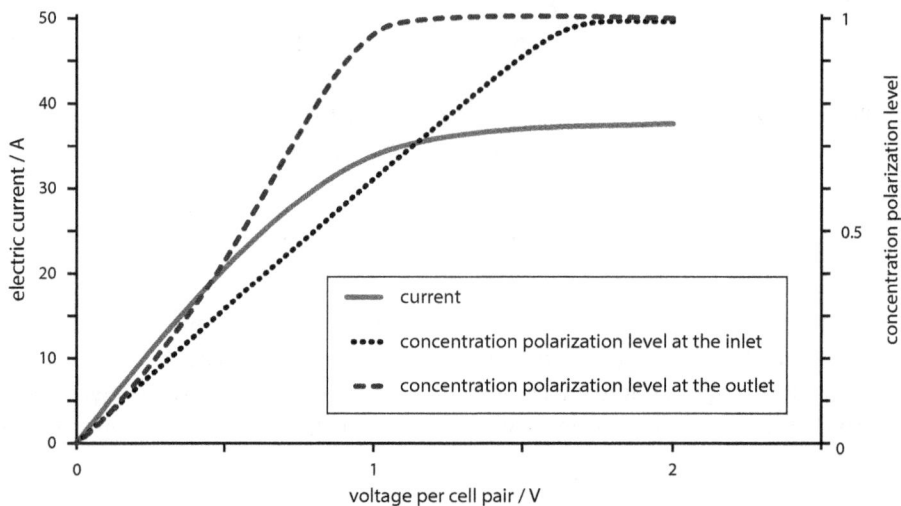

**Fig. 7.13:** Example of volt-ampere characteristic of the electrodialysis stack.

For completeness, it is necessary to indicate that the curve of the electric current in Fig. 7.13 may be at higher voltages, increasing again due to the water splitting on the surface of the ion-selective membrane under conditions of a strong electrical field and the catalytic effect of the ion exchange groups in the membrane or the components of the solution and the transport of water splitting products, $H^+$ and $OH^-$ ions. We call such electric current, **overlimiting**. The operation of the electrodialysis stack in the overlimiting region is extremely undesirable due to the inefficient use of electric current, increased consumption of electricity, shift of the diluate and concentrate pH value, increased risk of scaling (in some cases) or burning of the ion-selective membrane in the active area.

### 7.3.3.2 Transport of solvent

The transport of solvent through the ion-selective membranes can be described, for instance, by the Schlögl equation, the qualitative analysis of which indicates that the solvent is transported through the ion-selective membrane, in general, by electroosmotic and osmotic mechanism. Along with the solvent, some other neutral components

present in the solution may be transported through the ion-selective membranes. The transport of solvent at electrodialysis limits the concentration of electrolyte:
- in the solution that can be further desalted by the process,
- in the concentrate that can be achieved by the process.

The electroosmotic transport of water through the ion-selective membrane is significant. As for electrolytes of type NaCl, $Na_2SO_4$, $NH_4NO_3$, $Ca(NO_3)_2$, in electrodialysis with heterogeneous ion-selective membranes RALEX® CM-PES and AM-PES available from MEGA a.s., the water transport from diluate to concentrate through the ion-selective membranes of 3–4 kg per 1 kg of electrolyte was experimentally found.

If the transport of the solvent occurs exclusively by the mechanism of electroosmosis during electrodialysis, in practice – with small differences in the concentration of electrolyte between the concentrate and the diluate – the following equation applies to the change of the diluate or concentrate mass, $m$:

$$\frac{m}{m_0} = \frac{1-(a+1)w_{S,0}}{1-(a+1)w_S} \tag{7.44}$$

where $w_S$ is the mass fraction of electrolyte in the given hydraulic circuit and the coefficient $a$ expresses the ratio between the mass flux of the solvent and electrolyte through the ion-selective membranes (subscript 0 symbolizes the initial state). The analogous equation applies to the change of mass flow in the given hydraulic circuit.

Figure 7.14 shows the example of the effect of electrolyte concentration in the diluate at the inlet to the electrodialysis stack on the concentration of electrolyte in the concentrate when the concentrate was formed solely by a simultaneous transfer of water and electrolyte, double salt $5Ca(NO_3)_2 \cdot NH_4NO_3 \cdot 10H_2O$. The tests were performed on the pilot ED-Y electrodialysis stack by MemBrain s.r.o. at constant linear liquid velocity of 6.9 cm s$^{-1}$ and a voltage of 1.0 V per cell pair. The graph clearly demonstrates the existence of two interesting areas. In the lowest concentration range of nitrate nitrogen ($NO_3$–N), in the order of $10^{-1}$ kg m$^{-3}$, a step change in the electrolyte concentration in the concentrate occurs. This can be explained by the fact that in this area, the electric current (and, therefore, the electroosmotic transport of water) considerably depends on the electrolyte concentration in the diluate at the inlet to the electrodialysis stack. At low electric currents, the contribution of the electroosmotic transport of water is small and the concentration of electrolyte in the concentrate is significantly affected by osmosis. By increasing the electrolyte concentration in the diluate at the inlet to the electrodialysis stack, the maximum electric current is gradually reached and thus the contribution of the electroosmotic transport of solvent is constant. At the same time, the difference in osmotic pressures between the diluate and concentrate decreases, which reduces the contribution of osmosis, and the electrolyte concentration in the concentrate increases.

For completeness, it is necessary to add that the solvent does not have to be transported from the diluate to the concentrate or vice versa by ion-selective membranes only, but also due to internal leakages in the electrodialysis stack. This issue is discussed in more detail in Section 7.5 as one of the limiting factors of electrolyte concentration in the concentrate.

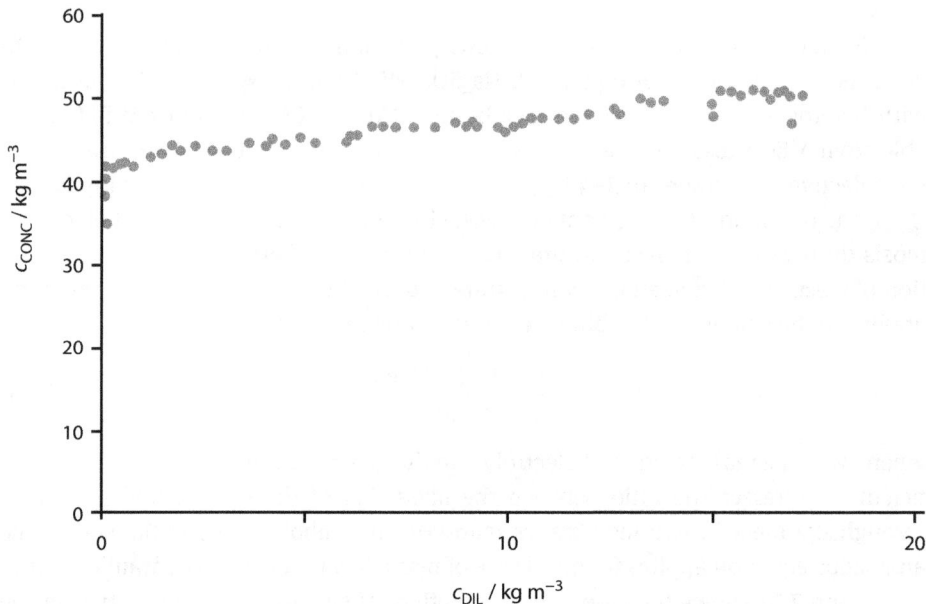

**Fig. 7.14:** Effect of nitrate nitrogen (NO$_3$-N) in the diluate (c$_{DIL}$) at the inlet to electrodialysis stack on the concentration of NO$_3$-N in the concentrate (c$_{CONC}$) at desalination of 5Ca(NO$_3$)$_2$·NH$_4$NO$_3$·10H$_2$O solution.

### 7.3.3.3 Selectivity of ion-selective membranes for counter-ions and co-ions

In Section 3.5.3, the balance between the ion-selective membrane and the electrolyte solution was discussed. By combining the conditions of electroneutrality in the liquid and solid phases with the Donnan equilibrium, we can estimate the equilibrium concentrations of counter-ions and co-ions in the solid phase, depending on the electrolyte concentration in the liquid phase.

An example of solution is shown in Fig. 7.15a, where $a_1 \approx c_1$ was applied for simplicity. From the graph it is evident that in very diluted solutions, the moving ions are formed almost exclusively by counter-ions; however, with the increasing concentration of external electrolyte, the concentration of both counter-ions and co-ions in the solid phase increases.

From the definition of the transport number of ion, see eq. (3.52), an important qualitative conclusion can be made: the ion-selective membranes are almost selective for counter-ions only in very diluted solutions of electrolytes, but their selectivity decreases with the increasing concentration of electrolyte in the solution – see Fig. 7.15b. In practice, this means a reduction of ED process rate as well as an increase in power consumption.

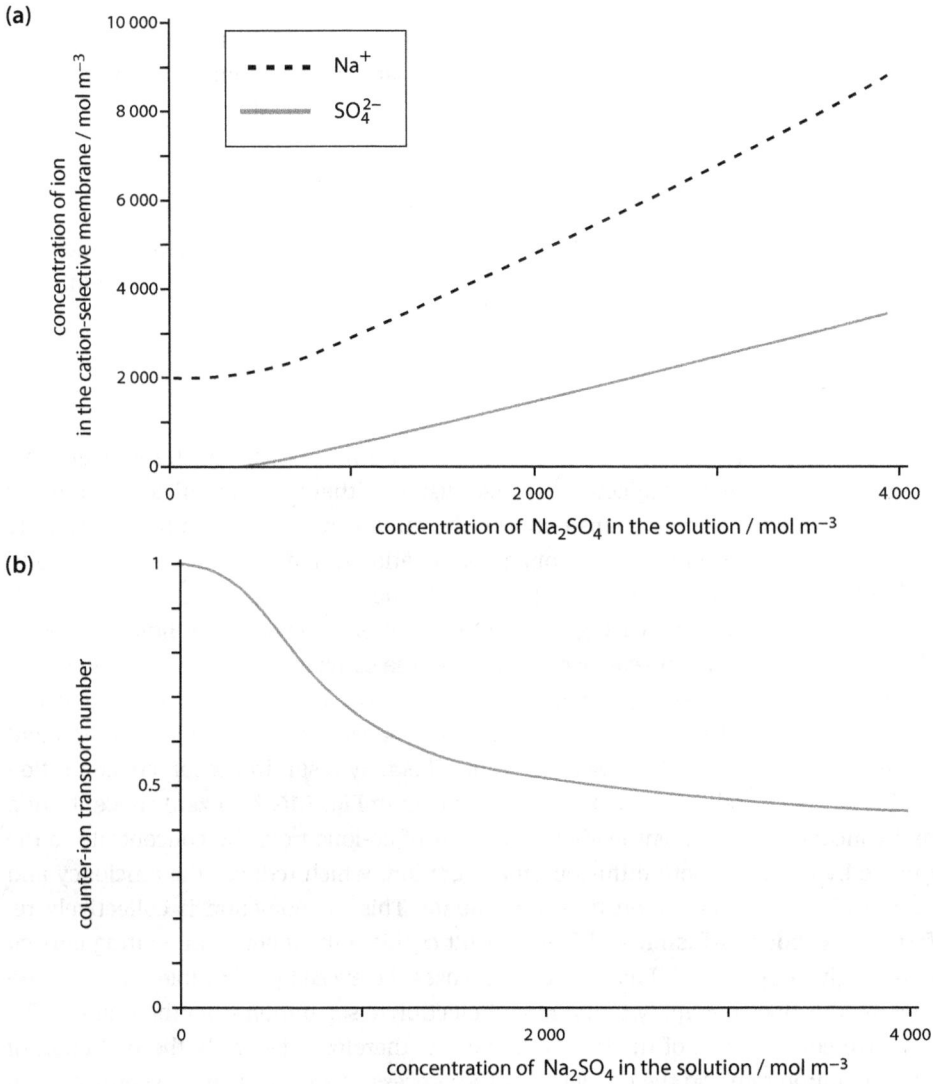

**Fig. 7.15:** (a) Concentration of $Na^+$ and $SO_4^{2-}$ ions in cation-selective membrane with $c_M = 2,000$ mol m$^{-3}$ and $z_M = -1$; (b) transport number of counter-ion vs. $Na_2SO_4$ concentration in the solution.

### 7.3.3.4 Selectivity of ion-selective membranes for two counter-ions

Let's have two ions, $A^{z_A}$ and $B^{z_B}$, for which $z_A z_B > 0$ applies. Due to the different selectivities of a membrane for both ions as well as their different charges, mobilities and concentrations in the solution and in the membrane, these two ions are transferred through the membrane at a different rate. This phenomenon is termed the **permselectivity** (see Chapter 5) between ions having the same sign of charge. Permselectivity coefficient of ion B against ion A, $\psi_{B,A}$ is defined as follows [13]:

$$\psi_{B,A} = \frac{(J_B/c_{B,S})}{(J_A/c_{A,S})} = \frac{z_B u_{B,M}}{z_A u_{A,M}} \frac{(c_{B,M}/c_{A,M})}{(c_{B,S}/c_{A,S})} \tag{7.45}$$

where $J$ denotes molar flux, $c$ denotes molar concentration and $u$ denotes mobility of ions A and B in the solution (subscript S) and in the membrane (subscript M).

### 7.3.3.5 Back diffusion

In practice, the transport of ions by the diffusion mechanism in the ion-selective membranes cannot be neglected because, usually, there is a significant difference in the electrolyte concentrations between the diluate and the concentrate, naturally resulting from the ED principle. From a combination of the conditions of electroneutrality in solution and the ion-selective membrane with the Donnan equilibrium, it is clear that with the increasing concentration of electrolyte in the liquid phase at the interface with the ion-selective membrane, the concentration of counter-ions and co-ions in the ion-selective membrane increases, see Fig. 7.15. In the ion-selective membrane located between the diluate and the concentrate, a concentration gradient of both types of ions will necessarily occur. This may result in the ion concentration profile in the ion-selective membranes, as shown in Fig. 7.16. Non-zero concentration and concentration gradient leads to transport of co-ions from the concentrate to the diluate by means of both diffusion and migration, which reduces the efficiency and rate of electrolyte separation from the diluate. This phenomenon is collectively referred to as **back diffusion** and – as a result of this – the theoretical limiting current density given by eq. (7.37) may, in certain cases, be exceed by multiple times. Nevertheless, this does not apply to the rate of electrolyte separation from the diluate. The negative consequence of the back diffusion is, therefore, generally the reduction of the separation rate and the efficiency of the process, the increase in the consumption of electricity and the limitation of the diluate quality.

In the case of binary electrolyte $C_{\nu_+}A_{\nu_-}$, the cation flux in the ion-selective membrane is given by the equation

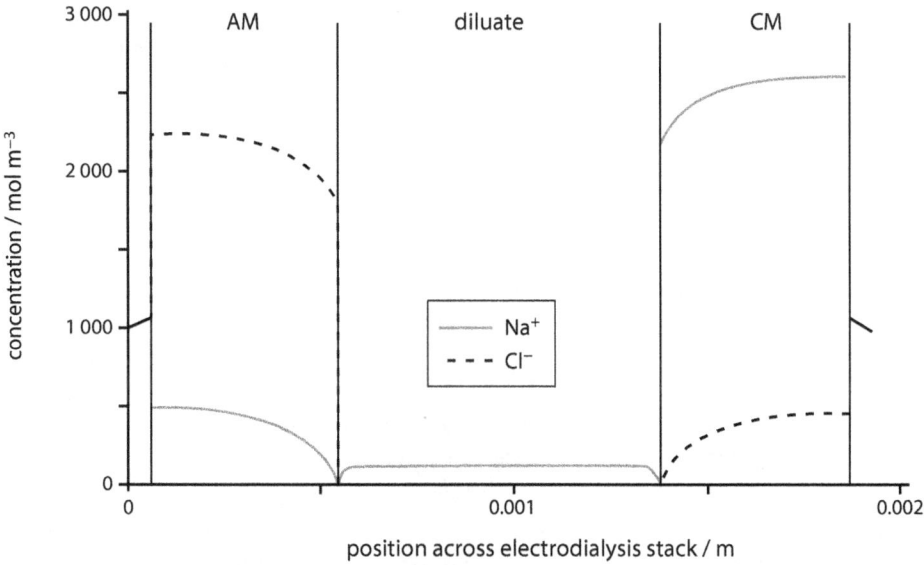

**Fig. 7.16:** Example of the ion concentrations profile in the cell pair. Electrolyte concentration is 100 mol m$^{-3}$ in the diluate and 1,000 mol m$^{-3}$ in the concentrate.

$$J_C = -D_{C,M} \frac{dc_{C,M}}{dx} + \frac{t_{C,M}}{z_C} \left( \frac{j_x}{F} + z_C D_{C,M} \frac{dc_{C,M}}{dx} + z_A D_{A,M} \frac{dc_{A,M}}{dx} \right) \qquad (7.46)$$

and in the case of anion flux in the ion-selective membrane, the following equation is applied:

$$J_A = \frac{\frac{j_x}{F} - z_C J_C}{z_A} \qquad (7.47)$$

By substituting the electroneutrality condition in the ion-selective membrane into eq. (7.46), we get the following term for the change of the cation concentration across the ion-selective membrane:

$$\frac{dc_{C,M}}{dx} = \frac{J_C - \frac{t_{C,M}}{z_C} \frac{j_x}{F}}{t_{C,M} (D_{C,M} - D_{A,M}) - D_{C,M}} \qquad (7.48)$$

where $t_{C,M}$ must be calculated in each integration step in accordance with eq. (3.52) because of the variable concentration profile across the ion-selective membrane.

Equations (7.32) and (7.48) express the change in the cation concentration in the diffusion layer and in the ion-selective membrane according to $x$ coordinate as a function of local composition of the phase, $J_C$ and $j_x$. The equilibrium between the liquid phase and the ion-selective membrane is described by the Donnan equilibrium. At known concentrations of electrolyte in the flow phase of diluate and the concentrate and for known $j_x$, the values of $J_C$ and $J_A$, and the corresponding profile of cation and anion concentrations between both flow-through phases can be subsequently calculated numerically. It is actually a solution of the differential equation with boundary condition.

### 7.3.3.6 Electrolyte solubility

Solubility is one of the factors limiting the possibility of electrolyte concentration in the concentrate. This may result in a limited water recovery or higher consumption of make-up water for the concentrate dilution. Well-soluble electrolytes limit the degree of concentration only exceptionally, essentially only in the case of extreme concentration of solutions by the ED process, see Section 12.3. As an example, we can mention $Na_2SO_4$. Solubility of this salt in water at temperatures up to 20 °C is lower than the concentration of the solution generated by electroosmotic transport of water, together with the electrolyte (holds for RALEX® membranes). In general, however, the possibilities of concentration are limited by the solubility of weakly soluble substances, characterized by the **solubility product**, $K_{SP}$, for which, in the case of binary electrolytes of $C_{v_+}A_{v_-}$ type, the following equation is applied:

$$K_{SP} = a_C^{v_C} \, a_A^{v_A} \qquad (7.49)$$

where $a$ represents activities of ions in a saturated solution. Exceeding the solubility product results in scaling in concentrate chambers and, possibly, also in the concentrate pipeline or on the cathode.

In practice, the risk of scaling is reduced by real kinetics of precipitation. In many cases, a significant supersaturation of the solution $S$ can be tolerated in the concentrate:

$$S = \frac{a_C^{v_C} \, a_A^{v_A}}{K_{SP}} \qquad (7.50)$$

As for insoluble carbonates, the risk of scaling may be reduced or completely eliminated by acidification of the concentrate, which shifts the dissociation equilibrium from $CO_3^{2-}$ towards $HCO_3^-$ and $CO_2$. In this case, we are talking about the **Langelier saturation index** (LSI), defined by the equation

$$LSI = pH - pH_S \qquad (7.51)$$

where $pH_S$ is the pH value at which the solution becomes saturated with $CaCO_3$.

The tolerance for the supersaturation of a solution with weakly soluble substances may be increased by dosing **antiscalants** into the concentrate. The most commonly used antiscalants include sodium hexametaphosphate, phosphonates and various polymeric antiscalants such as polyacrylate, etc. The use of antiscalants may be limited in the case of partial demineralization of water for potable or irrigation water and in the food industry. The overview of the most important weakly soluble substances, the scaling of whose is experienced during demineralization of surface water by ED process, is shown in Tab. 7.1. In other areas of application, we witness scaling of other substances, for example, $CaHPO_4 \cdot 2H_2O$ at demineralization of whey.

**Tab. 7.1:** Overview of weakly soluble substances and supersaturation tolerances (according to [14]).

| Substance | Solubility product* | Saturation limit | |
|---|---|---|---|
| | | Without antiscalant | With antiscalant |
| $CaCO_3$ | $8.7 \times 10^{-9}$ | LSI < 2.2 | LSI < 2.7 to 3 |
| $CaSO_4$ | $4.93 \times 10^{-5}$ | < 2.25 to 4 $K_{SP}$ | < 12.5 $K_{SP}$ |
| $BaSO_4$ | $1.08 \times 10^{-10}$ | < 100 $K_{SP}$ | < 150 to 225 $K_{SP}$ |
| $SrSO_4$ | $2.81 \times 10^{-7}$ | < 4 to 8 $K_{SP}$ | – |
| $CaF_2$ | $3.95 \times 10^{-11}$ | < 500 $K_{SP}$ | – |

*For the concentration units of mol $dm^{-3}$.

In one of the applications by MEGA a.s., $CaF_2$ scaling was experienced in the concentrate chambers at the level of $CaF_2$ supersaturation in the order of $10^3$ $K_{SP}$ despite the reversal of electrode polarity at the frequency of once per 30–60 min. At even higher levels of $CaF_2$ supersaturation (in the order of $10^4$ $K_{SP}$) and electrolyte concentration (in the concentrate range of 50–100 kg $m^{-3}$), this insoluble substance precipitated directly in the phase of cation-selective membrane RALEX® CM-PES at the interface with the concentrate, which resulted in its irreversible damage. In accordance with the Donnan equilibrium and due to the concentration conditions in the concentrate, $F^-$ ions penetrate into the phase of the cation-selective membrane and $CaF_2$ scaling was experienced. However, the same situation was not observed in the case of the anion-selective membrane, where the penetration into the solid phase is prevented – in accordance with the same principle – by $Ca^{2+}$ ion charge.

The scaling is manifested by a decrease in the electric current, that is, deterioration of the demineralization capability of the electrodialysis stack and by an increase in pressure loss. In the case of insoluble carbonates, hydroxides, hydrogen phosphates, phosphates and fluorides, scaling can be eliminated by chemical cleaning using an acid (in the case of $CaF_2$, it is necessary to use $HNO_3$). The extreme solution would be a rebuild of the electrodialysis stack and mechanical cleaning of spacers and ion-selective membranes. Operational possibilities to prevent scaling include a perfect pre-treatment of the feed water (e.g. by softening), limiting the water recovery or more intensive dilution of the concentrate by water, using antiscalants, acid dosing into the concentrate, and technological discipline and finally the reversal of the electrode polarity.

### 7.3.3.7 Pressure loss and distribution of liquid flow between individual flow chambers

Due to the limited external tightness, the operating range of each industrial electrodialysis stack is restricted by certain pressure, to which a leak-free operation (or operation with minimum leaks) is guaranteed. In the case of horizontal industrial electrodialysis stack EDR-III-0.68 available from MEGA a.s., which is similar to Fig. 7.26, the declared maximum operating pressure is 250 kPa. The hydraulic characteristics of this electrodialysis stack are shown in Fig. 7.17a. On the horizontal axis you can see an average linear velocity of the water in flow chambers across the entire electrodialysis stack. In many ED industrial applications, electrodialysis stacks are connected in series into lines with multiple (in the case of an electrodialysis stack of EDR-III type with up to four) hydraulic stages, see Section 7.2.3, where pressure losses in individual hydraulic stages are summed up.

The logical consequence of limitation by maximum pressure loss can, therefore, be a limitation of the volumetric flow rate of the liquid, at which the electrodialysis line can be operated. Under these conditions, the suitability of the given line configuration may be lost due to the negative impact of low liquid flow rate on the mass transfer kinetics in the ED process. At the same time, the volumetric flow rate of the liquid must not fall below the minimum value, see discussion in Section 7.5. Due to the limitation by the maximum pressure and minimum volume flow of the liquid, the maximum number of hydraulic stages for the given electrodialysis stack is also limited.

From Fig. 7.17a, the influence of the number of cell pairs on the hydraulic characteristics of the electrodialysis stack is further evident. Electrodialysis stacks with higher number of cell pairs are logically operated at a proportionally higher volumetric flow rate of the liquid. However, by increasing the number of cell pairs and flow rate, the pressure loss in the manifold for the distribution of the liquid flow across the electrodialysis stack increases due to the higher flow rate and the extension of the manifold length. This may result in an uneven distribution of the liquid flow between the individual flow chambers, as indicated in Fig. 7.17b. It was shown experimentally that under certain operating conditions, this non-uniformity of the fluid flow distribution has a negative impact on the demineralization performance of the electrodialysis stack.

**(a)**

**(b)**

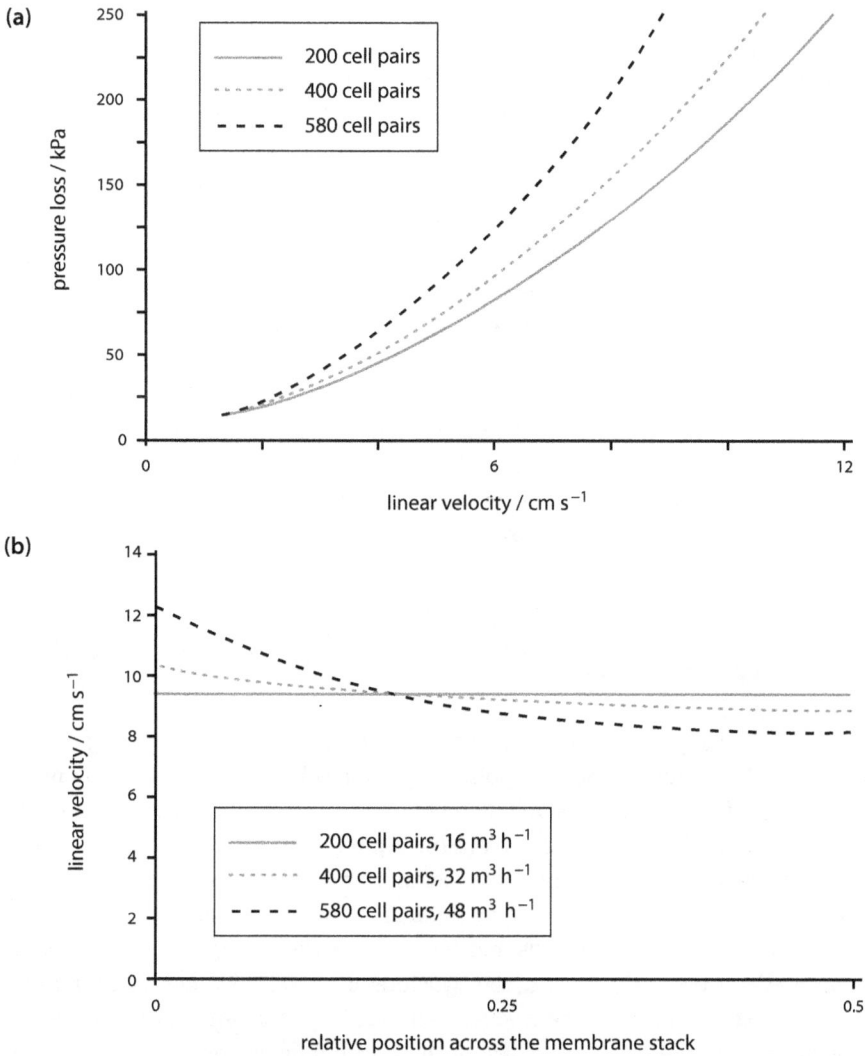

**Fig. 7.17:** Example of hydraulic characteristics of the electrodialysis stack (a) and distribution of linear velocity of the water across the electrodialysis stack (b).

### 7.3.3.8 Transport of neutral components

The mechanism of the ED process is to separate the electrically charged components (ions). Along with electrolytes, the solvent is also transported through the ion-selective membranes by means of electroosmosis mechanism, see Section 7.3.3.2. However, a number of other neutral components may be present in the treated solution, such as dissolved gases ($CO_2$), silica ($SiO_2$), sacharides, etc. According to the Donnan equilibrium, these substances may freely enter the ion-selective membrane and be subjected to transport through the ion-selective membrane by the mechanism of convection (along with the solvent), or possibly diffusion. In practice, of course, it also depends on the size, structure and polarity of molecules of these neutral components.

The amount of transported neutral components is, therefore, apparently related to the amount of electrolyte transferred through the ion-selective membranes. If a solution containing both, higher concentration of electrolytes and the valuable neutral components, is demineralized by the ED process, the losses of these precious components during the transfer from the diluate to the concentrate may be carefully monitored. This is, among others, the case of demineralization of whey by electrodialysis, where significant losses of lactose and proteins are experienced.

The fact that electrodialysis does not concentrate these substances is essential. This is sometimes used as a competitive advantage of the ED process compared to reverse osmosis (RO). Surface waters in volcanically active areas such as the Canary Islands contain, for example, high concentrations of $SiO_2$. The reverse osmosis process – in contrast to the ED process – rejects this substance in the retentate, which may result in $SiO_2$ scaling.

### 7.3.3.9 Feed water composition

Like any membrane separation process, electrodialysis is sensitive to plugging of flow chambers, to scaling, fouling and also poisoning of ion-selective membranes by membrane poisons. **Scaling** refers to precipitation of weakly soluble inorganic substances in the concentrate circuit or on the cathode. **Fouling** refers to the blocking of the ion-selective membrane surface by organic substances or colonies of microorganisms. Ion-selective membranes have a limited chemical stability, for example, against aggressive chemicals (concentrated acids and hydroxides). **Poisoning** refers to the effect of substances that irreversibly damage the structure of ion-selective membranes or deteriorate their electrochemical properties, for example, by permanent binding to the ion exchange groups, by breaking them down or by contributing to breakdown of the ion-selective membrane matrix. Examples include oxidizing agents (chlorine, ozone), heavy metal cations (Fe, Mn), (hydrogen) sulphides, polyelectrolytes, ionic surfactants, etc.

Therefore, strict limits are defined for feed water composition for each application area. Also, the substances which cause damage to DSA/ATA/MMO-anodes cannot be ignored, for example, fluorides or bromides. An example of such limits is shown in Tab. 7.2.

Failure to comply with these limits can, in extreme cases, result in the loss of the ED technology supplier guarantee towards the end user, reduced service lifetime of the system, higher frequency of chemical cleaning cycles or rebuild, that is, generally in the increase of operating costs (chemicals, operators, spare parts), not to mention the failure to meet the requirements for the technology performance.

**Tab. 7.2:** Limits of feed water composition for EDR-III electrodialysis stack by MEGA a.s.

| Parameter | Limit | |
|---|---|---|
| | Common | Critical |
| TDS (mg dm$^{-3}$) | 0.5–2.5 | 8.0 |
| Conductivity (mS cm$^{-1}$) | 0.5–5.0 | 10 |
| pH (operation) | 6–8 | > 10 |
| pH (chemical cleaning) | 1–14 | No limits |
| Temperature (°C) | 15–25 | 5–35 |
| Turbidity (NTU) | 0.5 | 5.0 |
| SDI$_5$ | 5.0 | 10 |
| TSS (mg dm$^{-3}$) | 1.0 | 10 |
| COD$_{Cr}$ (mg dm$^{-3}$ O$_2$) | 50 | 100 |
| TOC (mg dm$^{-3}$ O$_2$) | 15 | 50 |
| Free chlorine (mg dm$^{-3}$) | 1.0 | 5.0 |
| Fe$^{2(3)+}$ (mg dm$^{-3}$) | 0.3 | 3.0 |
| Mn$^{2+}$ (mg dm$^{-3}$) | 0.1 | 1.0 |
| Al$^{3+}$ (mg dm$^{-3}$) | 0.1 | 1.0 |
| Hydrogen sulphides (mg dm$^{-3}$) | 0.1 | 1.0 |
| F$^-$ (mg dm$^{-3}$) | 1.0 | 2.5 |
| As (mg dm$^{-3}$) | 0.1 | 1.0 |
| Oils (mg dm$^{-3}$) | 1.0 | 2.5 |

## 7.3.4 Distribution of electric current in electrodialysis stack

This section discusses the issue of possible uneven distribution of electric current. In the "ideal" case, the electric current only passes through the active area of membranes perpendicularly to their plane, with a uniform distribution of current density. In such a hypothetical case, the increase in the total active area of membranes would be sufficient enough to increase the performance (production). However, this situation cannot be achieved, in particular because of the following reasons:
- current density decreases with decreasing concentration of the diluate
- the distribution of process solutions between individual membrane chambers is uneven
- during passage through the membrane, electric current passes through flow distributors and flow distribution manifolds (leakage current, shunt or parasitic current)

Liquid flow distributors, ion-selective membranes in the area of liquid flow distributors and flow distribution manifolds form a conductive bridge for the parallel passage of electric current outside the active area of the ion-selective membranes. This phenomenon is undesirable, as it increases the power consumption without any contribution to the mass transfer and, at the same time, threatens to damage the membrane stack by so-called **burning**, since there are large differences in electrical potential between the active area and the flow distribution manifolds, often up to few tens of volts. The leakage electric current not only passes through the liquid in the flow distributors, but also through the adjacent ion-selective membranes. The conductivity of the ion-selective membrane is determined by its swelling and compression, which affects the amount of swelling water or other solvent. The ion-selective membrane may also be – to some extent – swollen in the sealing area around the liquid flow distributors. In this area, the risk of burning is particularly high because heat is generated due to dissipation of electric energy by passing leakage current as it cannot be carried away in the liquid flow.

However, these conclusions presented in connection with the electrodialysis process can be generalized to most electromembrane processes. Studying this phenomenon experimentally and at an industrial scale would be considerably costly and time-consuming. The appropriate mathematical model is, therefore, a very useful tool in this regard. To study the current distribution in the device, two types of mathematical models were used:
- 2D macrohomogeneous model based on local balances in the form of partial differential equations,
- model based on the equivalent electric diagram.

The first type of model provides more information than the other (local distribution of quantities and flows), but it is rather difficult. The second model is computationally easier. However, it provides good qualitative and semiquantitative information on the possibility of leakage in current generation.

### 7.3.4.1 Two-dimensional macrohomogeneous model

The mathematical description of the flow of mass and electrical charge in the industrial electrodialysis stack is not simple. This is mainly due to the size of the device which is in the order of meters, while the key phenomena take place in a submillimetre scale in the space between the membranes and, especially, at the membrane interface. For this purpose, a fundamental but realistic simplification has been adopted to address this issue. This simplification is based on the fact that with the increasing number of cell pairs in the stack, the width of one pair is minor compared to the width of the entire membrane stack. In a more distant view of the entire membrane stack, human eye cannot clearly recognize the membrane interface, that is, the membrane and solution phases blend together. The system appears to be homogeneous and such model approach is referred to as macrohomogeneous. The stack consists of three continuous phases:
- phase of desalted solution (diluate),
- phase of concentrated solution (concentrate),
- phase of membranes.

However, this is an anisotropic environment, that is, efficient transport properties of the environment depend on the position and direction. This is because the membranes are hydraulically impermeable, that is, they prevent from intermixing of diluate and concentrate solutions, and these solutions flow only in parallel with the plane of membranes. In contrast, the electric current passes spontaneously in all directions, depending on the local conductivity of the environment. This also relates to anisotropic ion transport. Ions can only be transported in an axial direction by convection, while in the radial direction, they are only transported by diffusion and migration. This anisotropy must be considered in the model.

For simplicity, it is sufficient to deal with the distribution of mass and charge in a two-dimensional coordinate system in a cross-section given by two main directions, that is, the coordinate $z$, which is perpendicular to the membranes and indicates the main direction of the flow of electric current and the distribution of the liquid into individual membrane chambers, and the coordinate $x$, which corresponds to the main flow of process solutions in the space between the membranes. The model is stationary and isothermal. Some other simplifications were adopted too, in order to facilitate the calculation without a significant impact on the result:

- The real process solution containing various anions and cations was replaced by ideal model NaCl solution of identical electrical properties.
- Membranes are considered as ideally permselective and the membrane-specific conductivity is independent of the salinity of surrounding solutions. The permeability of membranes for water (solvent) is neglected.
- The concentration gradients in the inter-membrane space in radial directions are neglected. This corresponds to the assumption of an ideal radial dispersion.

The schematic drawing of the model geometry is shown in Fig. 7.18 and it represents the cross-section of the electrodialysis stack in the $x$-$z$ plane. The membranes are oriented horizontally, with the anode positioned at the top and the cathode below. The flow of the diluate and the concentrate is identical.

The main model equations represent the balance of the charge given by the eq. (7.52), where $j(x,z)$ is the vector of the local current density, $G$ is the tensor of specific conductivity of the environment, $\varphi(x,z)$ is the electrical potential, $\varphi_M(x,z)$ is the Donnan potential, reducing the driving force and therefore the flow of the charge, $h_P$ is the thickness of a single cell pair and $c_{n,DIL}(x,z)$ and $c_{n,CONC}(x,z)$ are molar concentrations of NaCl in the diluate and concentrate, respectively, $R$ is molar gas constant, $T$ is temperature and $F$ is the Faraday constant. As for the calculation of $G$, we refer to the literature [15]. Naturally, the conductivity of the solution changes with the local NaCl concentration.

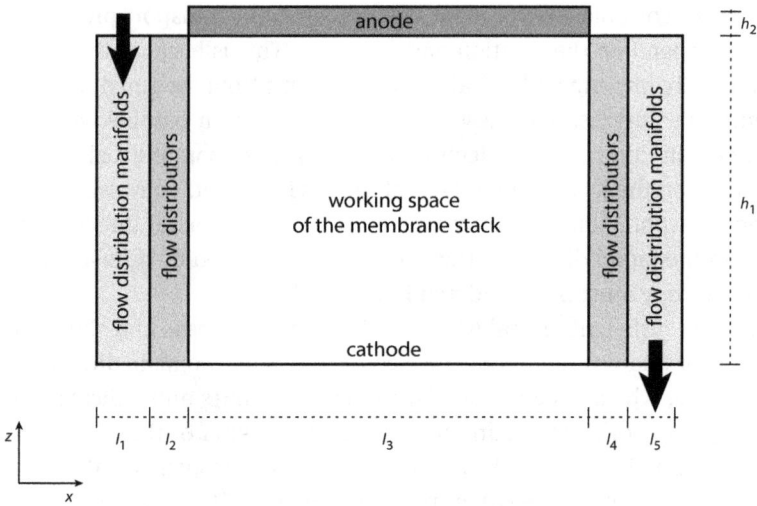

**Fig. 7.18: Schematic two-dimensional drawing of the model geometry representing the cross-section of the electrodialysis stack in *x-z* plane.** The membranes are oriented horizontally; black arrows indicate inflow and discharge of process solutions. Dimensions: $l_1 = l_5 = 8.7$ cm, $l_2 = l_4 = 6.8$ cm, $l_3 = 65.0$ cm, $h_1 = 54.8$ cm, $h_2 = 2.0$ cm.

$$\nabla \cdot \boldsymbol{j} = 0 \qquad -\boldsymbol{G} \left( \begin{array}{c} \dfrac{\partial \varphi}{\partial x} \\[2mm] \dfrac{\partial \varphi}{\partial z} - \dfrac{\varphi_{\mathrm{M}}}{h_{\mathrm{p}}} \end{array} \right) = \boldsymbol{j} \qquad \varphi_{\mathrm{M}} = -\frac{2RT}{F} \ln \left( \frac{c_{\mathrm{n,CONC}}}{c_{\mathrm{n,DIL}}} \right) \qquad (7.52)$$

The material balance of NaCl in the concentrate and diluate circuit is represented by eqs. (7.53) and (7.54), where $\boldsymbol{J}_{n,\mathrm{CONC}}$ and $\boldsymbol{J}_{n,\mathrm{DIL}}$ are molar fluxes of NaCl in the concentrate and diluate by diffusion and convection, respectively:

$$\nabla \cdot \boldsymbol{J}_{n,\mathrm{CONC}} = +N \qquad \boldsymbol{J}_{n,\mathrm{CONC}} = -D \, \nabla c_{\mathrm{n,CONC}} + c_{\mathrm{n,CONC}} \boldsymbol{v}_{\mathrm{CONC}} \qquad (7.53)$$

$$\nabla \cdot \boldsymbol{J}_{n,\mathrm{DIL}} = -N \qquad \boldsymbol{J}_{\mathrm{n,DIL}} = -D \, \nabla c_{\mathrm{n,DIL}} + c_{\mathrm{n,DIL}} \boldsymbol{v}_{\mathrm{DIL}} \qquad (7.54)$$

In these equations, the value of $N$ corresponds to the generation/consumption of NaCl due to the transport of the salt from the diluate circuit to the concentrate circuit, see the eq. (7.55). The first term of the right side describes the migration flux of ions through membranes, while the second term describes the diffusion flux. It is determined by the component of the current density vector in $z$ direction, $j_z$, the current efficiency of desalination $\eta$ and permeability of salt through the membrane due to the diffusion $\mu$:

$$N = \frac{\eta}{Fh_p} j_z - \mu(c_{n,CONC} - c_{n,DIL})$$  (7.55)

For the convective flow of solutions, the equation of continuity applies in the form (7.56) for the concentrate and in the form (7.57) for the diluate:

$$\nabla \cdot v_{CONC} = 0 \qquad v_{CONC} = -P \, \nabla p_{CONC}$$  (7.56)

$$\nabla \cdot v_{DIL} = 0 \qquad v_{DIL} = -P \, \nabla p_{DIL}$$  (7.57)

In these equations, $P$ (unit: $m^2 \, s^{-1} \, Pa^{-1}$) is the tensor of hydraulic permeability, $v$ is the vector of velocity of the fluid flow and $p$ is hydraulic pressure. For a more detailed description of the model, we refer to the literature [15]. Figure 7.19 shows the calculated behaviour of the electrodialysis stack in two distinctly different situations:
- in the case of ideally uniform distribution,
- in the case of significantly uneven distribution of the liquid flow among individual flow chambers.

The nature and uniformity of the flow are shown by the streamlines in Fig. 7.19a1, a2 for the first and the second case. It is evident that in the second case, Fig. 7.19a2, significantly less solution (diluate and concentrate) flows through the centre of the membrane stack, which results in the increased desalination of the diluate solution in this part of the stack. This is also evident from Fig. 7.19b1, b2 showing the concentration fields in the electrodialysis stack.

However, if the distribution of the solutions is uniform, the concentration decreases evenly in all flow chambers, away from the inlet to the working space in the direction to the outlet. In the case of uneven distribution of flow, NaCl concentration in the diluate solution decreases in the central part of the stack to almost zero. Here, the solution is almost non-conductive and no electric current passes through this area. It is then forced to pass through at a considerably different trajectory, preferably concentrating in the area of entry to the diluate and concentrate chambers, where NaCl concentration is high – see Fig. 7.19c2. For comparison, Fig. 7.19c1 shows the flow of the current in the ideal case. The combination of locally increased intensity of electric current and low flow rate of the solution may cause an enormous increase in the local temperature and the subsequent degradation of materials. Such a situation is extremely dangerous in terms of the reliability of the electrodialysis stack operation. Another consequence of uneven hydrodynamics is the increase in the parasitic current in the flow distribution manifolds in the part of the membrane stack, isolated due to the decrease in the concentration of dissolved salts. This leads to a significant decrease in the efficiency of the process.

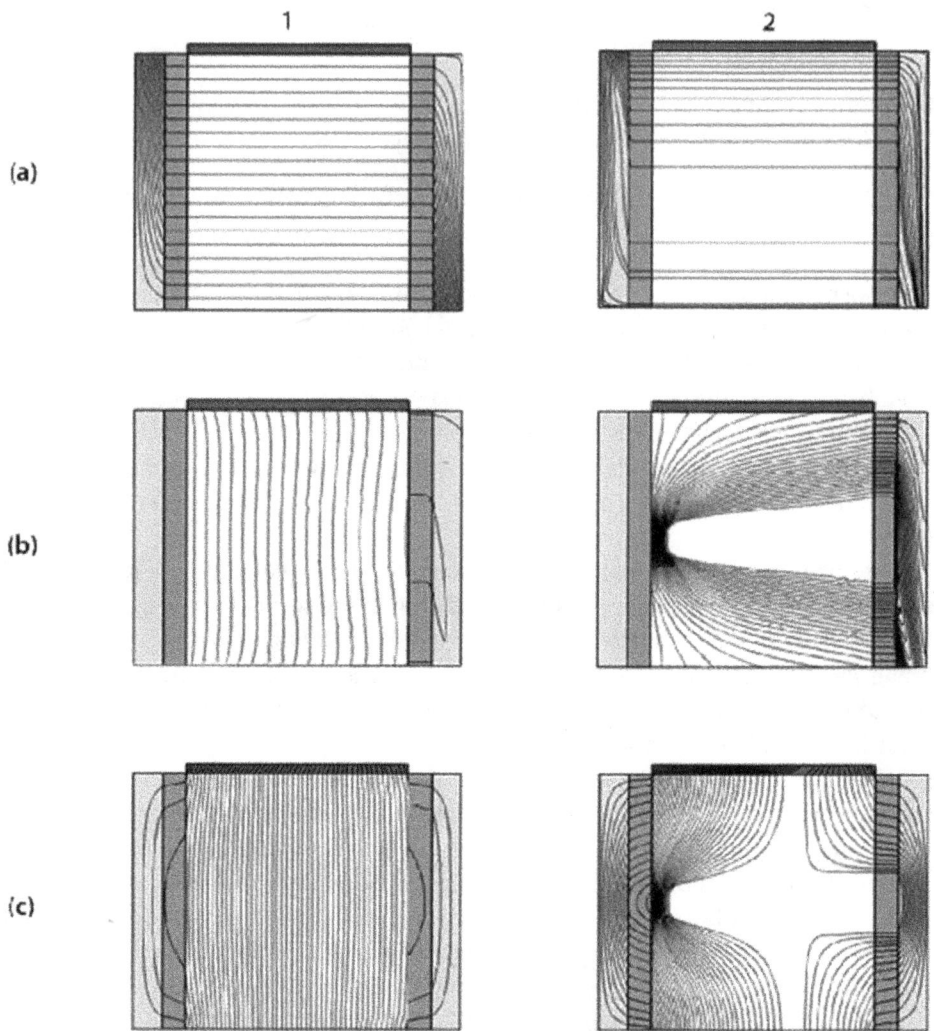

Fig. 7.19: (a) Streamline diagram showing the flow of process solutions of diluate and concentrate (b) Field of molar concentration of NaCl in diluate (c) Trajectory of electric current flow through the electrodialysis stack Series (1) – uniform distribution of process solutions, level of desalination 64.5 %, voltage 653 V, voltage per cell pair 3.3 V; series (2) – extremely uneven distribution of process solutions, level of desalination 49 %, voltage 1,710 V, voltage per cell pair 8.6 V; common operating parameters: current load 200 A, diluate/concentrate flow rate 8 m³ h⁻¹, NaCl concentration in diluate and concentrate 0.257 mol dm⁻³, pressure loss in ED unit is 40 kPa.

### 7.3.4.2 Leakage electric current – equivalent electric diagram

In the following text we assume a standard two-circuit electrodialysis stack. Let's divide this electrodialysis stack into three domains: the working space (subscript E), the area of liquid flow distributors (subscript D) and the area of flow distribution manifolds (subscript L). In addition, we assume that both the liquid flow distributors and flow distribution manifolds have the same geometry for the diluate and the concentrate. To calculate the leakage electric currents, the equivalent electric diagram can be used, see Fig. 7.20, which is presented in differential form here. The basic assumption of this approach is the neglect of the effect of leakage currents on the homogeneity of electric field in the working space. The ion-selective membranes around the liquid flow distributors are considered as electric resistors. Ignoring the real mechanism of the electric current flow from the liquid to the solid phase, which has its limits, see Section 7.3.3.1, which is another simplification of this approach. The total electric current passing through the electrodialysis stack is given by the sum of the electric currents passing through the working space ($I_E$) and the leakage electric current ($I_L$):

$$I = I_E + I_L \tag{7.58}$$

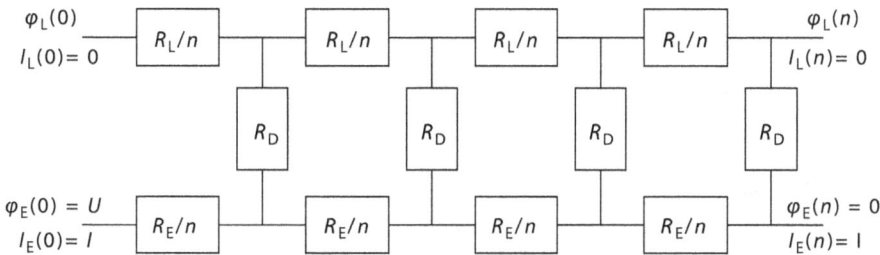

**Fig. 7.20:** Equivalent electric diagram for calculation of leakage electric currents.

To change the electrical potential in the active area $\varphi_E$ across the membrane stack, the following is based on Ohm's law

$$\frac{d\varphi_E}{dx} = -\frac{R_E I_E}{L_{MS}} \tag{7.59}$$

where $R_E$ is the electrical resistance of the working space of the membrane stack and $L_{MS}$ is the thickness of the membrane stack. Analogously, for electrical potential change in the area of flow distribution manifold, $\varphi_L$, it applies

$$\frac{d\varphi_L}{dx} = -\frac{R_L I_L}{L_{MS}} \tag{7.60}$$

where $R_L$ is the electrical resistance of the flow distribution manifolds – it implies that the resulting electrical resistance of the four resistors connected in parallel correspond to the flow distribution manifolds of the diluate and concentrate at the inlet and outlet, that is,

$$
\begin{aligned}
R_L &= \left( \frac{1}{R_{L,DIL,in}} + \frac{1}{R_{L,CONC,in}} + \frac{1}{R_{L,DIL,out}} + \frac{1}{R_{L,CONC,out}} \right)^{-1} \\
&= \frac{L_{MS}}{A_L \left( \kappa_{DIL,in} + \kappa_{CONC,in} + \kappa_{DIL,out} + \kappa_{CONC,out} \right)}
\end{aligned}
\tag{7.61}
$$

where in and out subscripts denote the diluate resistance (DIL) and concentrate resistance (CONC) at the inlet to and at the outlet from the membrane stack of the electrodialyser, $A_L$ is the area of the cross-section of the flow distribution manifold and $\kappa$ is the conductivity of diluate (DIL) and concentrate (CONC).

For the change of electric current passing through the flow distribution manifolds, the following applies

$$
\frac{dI_E}{dx} = - \frac{dI_L}{dx} = \frac{\varphi_E - \varphi_L}{R_D L_{MS}} n
\tag{7.62}
$$

where $R_D$ is electrical resistance of flow distributors and adjacent parts of ion-selective membranes and $n$ is the number of cell pairs. The value of $R_D$ generally has the meaning of twelve resistors connected in parallel:

$$
R_D = \left( \sum_{i=1}^{12} \frac{1}{R_{D,i}} \right)^{-1} = \left( \sum_{i=1}^{12} \frac{\kappa_i A_i}{L_i} \right)^{-1}
\tag{7.63}
$$

These components of electrical resistance in the area of flow distributors are the distributors of the diluate and concentrate flow at the inlet and outlet themselves (four contributions), and the cation-selective and the anion-selective membranes adjacent to the distributors of the diluate and concentrate flow at the inlet and outlet (eight contributions). By solving the eq. (7.62) and by combining it with the eqs. (7.59) and (7.60), we get the second derivative $I_L$ across the membrane stack

$$
\frac{d^2 I_L}{dx^2} = \frac{n}{R_D L_{MS}^2} \left[ (R_E + R_L) I_L - R_E I \right]
\tag{7.64}
$$

Equation (7.64) is a second order linear nonhomogeneous differential equation. Its particular solution can be found by taking into account the following conditions: $I_L(0) = 0$ and $dI_L/dx(L_{MS}/2) = 0$. For $I_L$, we get,

$$
I_L = I(1 - b) \left[ 1 - \frac{e^{\lambda x} + e^{\lambda(L_{MS} - x)}}{1 + e^{\lambda L_{MS}}} \right]
\tag{7.65}
$$

where

$$b = -\frac{n}{R_D L_{MS}^2} \quad \text{and} \quad \lambda = \sqrt{\frac{n}{R_D L_{MS}^2}(R_E + R_L)}$$

Coefficients $b$ and $\lambda$ are independent on the number of cell pairs. Corresponding $I_E$ can be obtained from eq. (7.58). An alternative analytical approach to this problem is presented in [13].

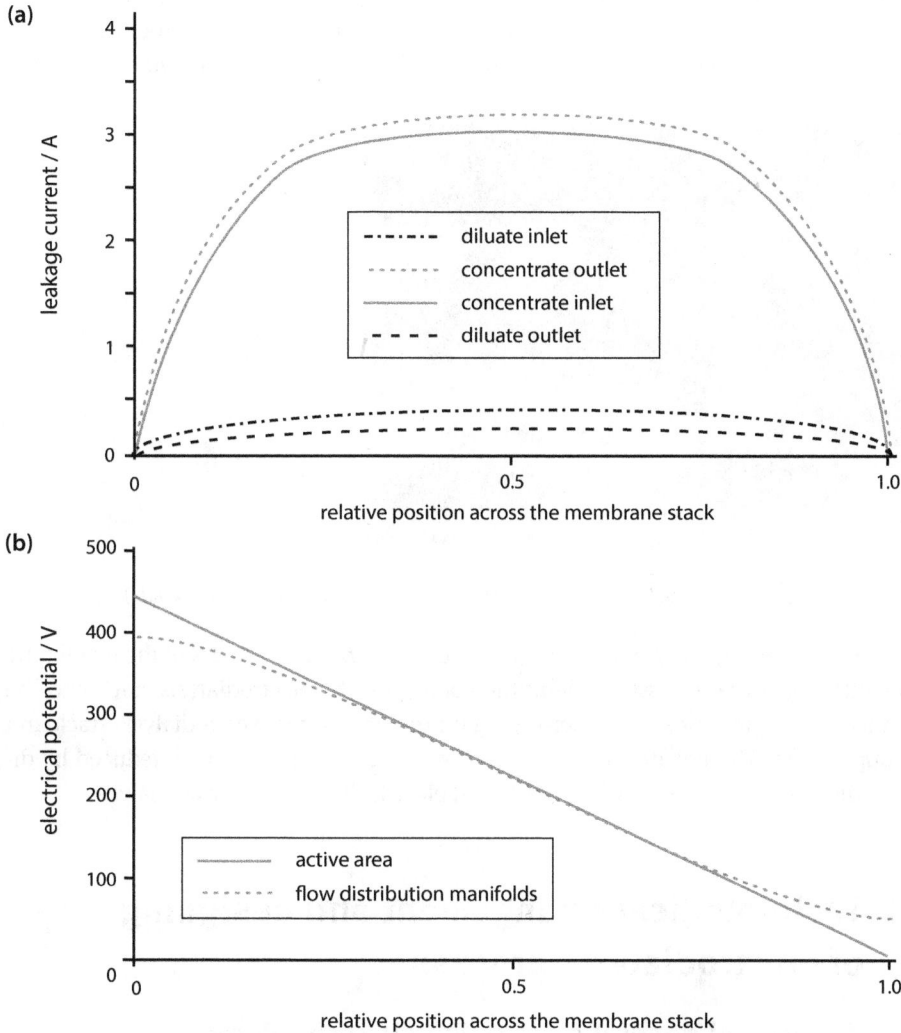

**Fig. 7.21:** Profile of leakage electric currents in flow distribution manifolds (a) and profile of electrical potential in the working space and flow distribution manifolds (b) across the membrane stack of the electrodialyser.

The method of calculation of leakage electric current presented above, among others, enables to reach the qualitative conclusion that the highest risk of the electrodialysis stack burning is at the edge of the membrane stack and the leakage electric currents increase with increasing voltage on the electrodialysis stack and the number of cell pairs, which are all in agreement with the experiments.

The example of leakage electric currents distribution in flow distribution manifolds is shown in Fig. 7.21a. Figure 7.21b shows the corresponding profile of electrical potential in the working space and flow distribution manifolds. It is clear from the graphs why burning of electrodialysis stacks usually occurs at the edge of the membrane stack. The example of damage to the peripheral part of the electrodialysis stack due to burning, in association with the leakage electric current, is shown in Fig. 7.22.

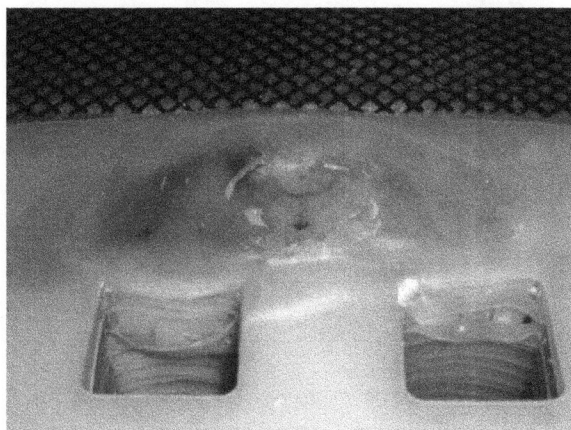

**Fig. 7.22:** Example of damage to electrodialysis stack due to burning (photo MEGA a.s.).

The practical consequence of the leakage electric currents and the risk of the membrane stack burning is the necessity to limit the voltage on the electrodialysis stack and the conductivity of the concentrate, depending on the type of the electrodialysis stack and the application. The negative effect of leakage electric currents may be reduced by dividing the membrane stack hydraulically and electrically into two equal halves.

# 7.4 Technological arrangement and designing of electrodialysis process

## 7.4.1 Balance of general electrodialysis technology

The general scheme of simple ED technology was shown in Fig. 7.9. Here, it is repeated with detailed description (Fig. 7.23) to facilitate understanding. The scheme does not include the electrode circuit, which does not participate in the mass transfer between the hydraulic circuits. In accordance with the general principles of modelling, see Chapter 6, the following total mass balances apply to this scheme:

$$\dot{m}_{DIL,5} + \dot{m}_{CONC,1} = \dot{m}_{DIL,8} + \dot{m}_{CONC,2} + \frac{dm_{DIL}}{d\tau} + \frac{dm_{CONC}}{d\tau} \tag{7.66}$$

$$\dot{m}_{DIL,5} = \dot{m}_{DIL,1} + \dot{m}_{DIL,6} + \dot{m}_{DIL,7} \tag{7.67}$$

$$\dot{m}_{DIL,1} + \dot{m}_{DIL,4} = \dot{m}_{DIL,in} \tag{7.68}$$

$$\dot{m}_{DIL,in} + \dot{m}_{CONC,in} = \dot{m}_{DIL,out} + \dot{m}_{CONC,out} \tag{7.69}$$

$$\dot{m}_{DIL,out} = \dot{m}_{DIL,2} + \dot{m}_{DIL,3} \tag{7.70}$$

$$\dot{m}_{DIL,2} + \dot{m}_{DIL,6} = \dot{m}_{DIL,8} \tag{7.71}$$

$$\dot{m}_{DIL,3} = \dot{m}_{DIL,4} + \frac{dm_{DIL}}{d\tau} \tag{7.72}$$

$$\dot{m}_{CONC,1} + \dot{m}_{DIL,7} + \dot{m}_{CONC,4} = \dot{m}_{CONC,in} \tag{7.73}$$

$$\dot{m}_{CONC,out} = \dot{m}_{CONC,2} + \dot{m}_{CONC,3} \tag{7.74}$$

$$\dot{m}_{CONC,3} = \dot{m}_{CONC,4} + \frac{dm_{CONC}}{d\tau} \tag{7.75}$$

and these component mass balances:

$$\dot{m}_{DIL,5}w_{i,DIL,5} + \dot{m}_{CONC,1}w_{i,CONC,1} = \dot{m}_{DIL,8}w_{i,DIL,8} + \dot{m}_{CONC,2}w_{i,CONC,2} +$$
$$+ w_{i,DIL}\frac{dm_{DIL}}{d\tau} + m_{DIL}\frac{dw_{i,DIL}}{d\tau} + w_{i,CONC}\frac{dm_{CONC}}{dt} + m_{CONC}\frac{dw_{i,CONC}}{d\tau} \tag{7.76}$$

$$\dot{m}_{DIL,1}w_{i,DIL,1} + \dot{m}_{DIL,4}w_{i,DIL,4} = \dot{m}_{DIL,in}w_{i,DIL,in} \tag{7.77}$$

$$\dot{m}_{DIL,in}w_{i,DIL,in} + \dot{m}_{CONC,in}w_{i,CONC,in} = \dot{m}_{DIL,out}w_{i,DIL,out} + \dot{m}_{CONC,out}w_{i,CONC,out} \tag{7.78}$$

$$\dot{m}_{DIL,3}w_{i,DIL,3} = \dot{m}_{DIL,4}w_{i,DIL,4} + w_{i,DIL}\frac{dm_{DIL}}{d\tau} + m_{DIL}\frac{dw_{i,DIL}}{d\tau} \tag{7.79}$$

$$\dot{m}_{DIL,2}w_{i,DIL,2} + \dot{m}_{DIL,6}w_{i,DIL,6} = \dot{m}_{DIL,8}w_{i,DIL,8} \tag{7.80}$$

$$\dot{m}_{CONC,1}w_{i,CONC,1} + \dot{m}_{DIL,7}w_{i,DIL,7} + \dot{m}_{CONC,4}w_{i,CONC,4} = \dot{m}_{CONC,in}w_{i,CONC,in} \tag{7.81}$$

$$\dot{m}_{CONC,3}w_{i,CONC,3} = \dot{m}_{CONC,4}w_{i,CONC,4} + w_{i,CONC}\frac{dm_{CONC}}{dt} + m_{CONC}\frac{dw_{i,CONC}}{d\tau} \tag{7.82}$$

$$w_{i,DIL,6} = w_{i,DIL,5} \tag{7.83}$$

$$w_{i,DIL,1} = w_{i,DIL,5} \tag{7.84}$$

$$w_{i,\text{DIL},7} = w_{i,\text{DIL},5} \tag{7.85}$$

$$w_{i,\text{DIL},4} = w_{i,\text{DIL}} \tag{7.86}$$

$$w_{i,\text{DIL},3} = w_{i,\text{DIL,out}} \tag{7.87}$$

$$w_{i,\text{DIL},2} = w_{i,\text{DIL,out}} \tag{7.88}$$

$$w_{i,\text{CONC},4} = w_{i,\text{CONC}} \tag{7.89}$$

$$w_{i,\text{CONC},3} = w_{i,\text{CONC,out}} \tag{7.90}$$

$$w_{i,\text{CONC},2} = w_{i,\text{CONC,out}} \tag{7.91}$$

In eqs. (7.66) to (7.91), $m$ denotes mass, $\dot{m}$ is the mass flow rate and $w_i$ is the mass fraction of the $i$th component and there are a total of 40 unknowns. In the mass balances, the ideal mixing of the content of the diluate and the concentrate tanks is assumed.

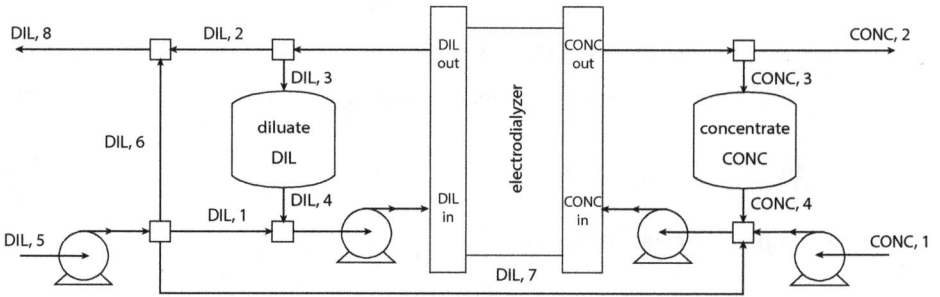

**Fig. 7.23:** Scheme of electrodialysis for the mass balance calculation.

Furthermore, the mass flow rates and compositions of output streams from the line of electrodialysis stacks through the kinetics of the ED process are the functions of mass flow rates and compositions of input streams into the line of electrodialysis stacks, that is,

$$\dot{m}_{\text{DIL,out}} = f_1(\dot{m}_{\text{DIL,in}}, \dot{m}_{\text{CONC,in}}, w_{i,\text{DIL,in}}, w_{i,\text{CONC,in}}) \tag{7.92}$$

$$w_{i,\text{DIL,out}} = f_2(\dot{m}_{\text{DIL,in}}, \dot{m}_{\text{CONC,in}}, w_{i,\text{DIL,in}}, w_{i,\text{CONC,in}}) \tag{7.93}$$

while $\dot{m}_{\text{CONC,in}}$ and $\dot{m}_{\text{DIL,in}}$ are – because of the equal pressure between the diluate and the concentrate at the inlet to the line of electrodialysis stacks – bound by the following relation

$$p_{\text{CONC,in}}(\dot{m}_{\text{CONC,in}}) = p_{\text{DIL,in}}(\dot{m}_{\text{DIL,in}}) \tag{7.94}$$

where $p$ denotes pressure.

## 7.4.2 Designing electrodialysis technologies

Design of a simple ED technology generally involves the following steps:
- gathering input data (assignment)
- assessing the feasibility of the assignment
- laboratory verification or piloting of the technology
- custom design of technology

The first step is to gather the input data. There must always be a complete specification of the composition of all feed media and products, which is related to the requirements for quality of feed media, see Section 7.3.3.9, and process guarantees provided by the technology supplier, which are necessary for the prediction of ED process kinetics. In the case of products (diluate or concentrate), the specification of key parameters in the form of a suitable summary property is usually sufficient, for example, total dissolved solids or conductivity; in the case of diluate, the concentrations of the monitored components, if relevant, are also required, for example, ash in whey, $NO_3^-$ or $F^-$ ions in potable water, ratios between concentrations of $Na^+$, $Ca^{2+}$ and $Mg^{2+}$ ions in water for irrigation, etc. Another mandatory input is the technology throughput specification. The bond between the mass or volume flow rate of the product and the processed solution, in particular in the applications of partial demineralization of water (river, brackish, waste), is often given in the form of so-called **water recovery** $R_w$, defined as follows:

$$R_w = \frac{\dot{m}_{DIL}}{\dot{m}_{in}} \cong \frac{\dot{V}_{DIL}}{\dot{V}_{in}} \qquad (7.95)$$

where subscripts in and DIL symbolize the feed water and the demineralized product (diluate), respectively. Incomplete assignment needs to be completed, while the remaining unknown parameters in balances are usually determined by the technologist, based on the specific design of the ED technology.

The second step in designing the ED technology is to assess the feasibility of the assignment: to verify whether the given medium can be processed by electrodialysis and the required parameters of products can be achieved. Every supplier of the ED technology must be provided with the specifications of limits of composition of the solution to be processed, which usually includes the limits of concentrations of total dissolved solids or conductivity, pH, temperature and the content of heavy metals (Fe, Mn), oxidizing agents (chlorine, chloramine, ozone), hydrogen sulphides, suspended and colloidal matter, fatty and oily substances, organic substances, turbidity, viscosity, etc. An example of such a specification is shown in Tab. 7.2. In this step, it is further verified whether the removal of components that are not at all affected by ED process or only partially (neutral components, ions with low mobility) affected, is required, and whether the required concentrations of selected components in the produced diluate or concentrate are realistically achievable (potential problems include too low residual concentrations of components in the diluate, too high concentrations of components in the concentrate, supersaturation of concentrate with weakly soluble components, etc.).

The third step in designing the ED technology is the laboratory verification or piloting of ED technology. This time and cost-intensive step is, nowadays, often replaced by mathematical modelling of mass transfer in ED, especially in demineralization of surface or even ground water. In laboratory or pilot tests, it is necessary to proceed as follows:
- operate the diluate always in the batch mode
- operate the concentrate in the same mode and observe the same concentration conditions as planned in the proposed ED technology
- ensure perfect mixing of the diluate in the tank
- the test must cover the entire concentration range of the diluate, which may occur in the proposed industrial ED technology
- at regular intervals or even continuously, monitor the voltage, electric current, batch volume/weight, pH, conductivity and composition (concentration of monitored components) in the diluate

We will deal with the processing of experimental data from laboratory and pilot tests and their use in the design of ED technology, in more details hereinafter. Alternative methods of ED technology designing are discussed in the paper [16].

The fourth step may comprise determination of the configuration and operating mode of the ED system, calculation of material balances, power consumption, consumption of chemicals such as acids for pH adjustment of the concentrate or antiscalants, creating PFD and P&ID diagrams, specification of the system and its parts, calculation of operating expenses (OPEX) including consumption of spare parts, etc.

## 7.4.3 Transferring operating data from one electrodialysis stack to another

The industrial ED technology design is based on the available experimental data, obtained in the laboratory or pilot scale. First of all, it is necessary to find a suitable method to transfer operating data from the relevant laboratory or pilot equipment to industrial electrodialysis stacks.

The composition of the concentrate affects the electrical resistance of the electrodialysis stack, back diffusion of co-ions and transfer of solvent through the ion-selective membranes by the mechanism of osmosis. Therefore, we will assume in our reasoning that the ED process kinetics is controlled either by local conditions in the diluate stream or any local composition of the concentrate corresponds to specific local composition of the diluate. However, the second condition is generally not met in the transfer of operating data between two different electrodialysis stacks.

The transfer of operating data is possible only if the working spacers and ion-selective membranes used in the laboratory or pilot equipment are of the same type as the ones in electrodialysis stacks to which the operating data are transferred. In the case of working spacers, this means the same thickness of the spacer frame as also the same type of spacer net.

In addition, it is necessary to note that together with the operating data, operating conditions such as voltage per cell pair, linear velocity of liquid and temperature are transferred as well.

## 7.4.4 Method of diluate chambers division

The basis of all other calculations is the expression of the dependence of mass fraction of the selected component in the diluate at the outlet from the pilot electrodialysis stack, $w_{i,\text{DIL,out}}$, on the mass fraction of this component in the diluate at the inlet to the pilot electrodialysis stack $w_{i,\text{DIL,in}}$. This dependence will be further referred to as $w_{i,\text{DIL,out}}(w_{i,\text{DIL,in}})$ and can be usually easily approximated by the polynomial,

$$w_{i,\text{DIL,out}} = \sum_{k=0}^{m} a_k\, w_{i,\text{DIL,in}}^{k} \tag{7.96}$$

where $a_k$ are coefficients and $m$ is a suitably selected degree of the polynomial or other appropriately selected function. The suitability of the use of polynomials arises from the study of the real ED process kinetics.

Next, let's consider another electrodialysis stack with the flow path length $L_2$ same as the line of $n$ pilot electrodialysis stacks hydraulically connected in series and electrically connected in parallel with the flow path length $L_1$, see Fig. 7.24. The following applies:

$$n = L_2/L_1 \tag{7.97}$$

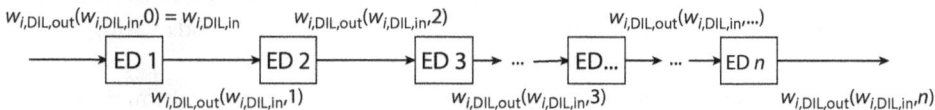

Fig. 7.24: Scheme for illustration of calculation algorithm of the value of function $w_{i,\text{DIL,out}}(w_{i,\text{DIL,in}})$.

In this case, the mass fraction of the $i$th component in the diluate at the outlet from one stage of the line of pilot electrodialysis stacks represents the inlet value for another stage of the same line of pilot electrodialysis stacks. As we know the dependence (7.96) for the pilot electrodialysis stack, we can calculate $w_{i,\text{DIL,out}}$ after the passage of the liquid through any number of hydraulic stages $n$ of the line of pilot electrodialysis stacks, that is, after passage along any flow path length.

We get the function $w_{i,\text{DIL,out}}(w_{i,\text{DIL,in}}, n)$ or $w_{i,\text{DIL,out}}(w_{i,\text{DIL,in}}, L)$, respectively. The calculation of the value of this function is given by the recursive algorithm,

$$w_{i,\text{DIL,out}}\left(w_{i,\text{DIL,in}}, n\right) = \sum_{k=0}^{m} a_k \, w_{i,\text{DIL,out}}^{k}\left(w_{i,\text{DIL,in}}, n-1\right) \tag{7.98}$$

whereas

$$w_{i,\text{DIL,out}}\left(w_{i,\text{DIL,in}}, 0\right) = w_{i,\text{DIL,in}} \tag{7.99}$$

According to the algorithm (7.98), when transferring data from one electrodialysis stack to another, $n$ is usually not an integral number. It is, therefore, necessary to make a substitute polynomial dependence:

$$w_{i,\text{DIL,out}}\left(w_{i,\text{DIL,in}}, n\right) = \sum_{k=0}^{l} b_k n^k \tag{7.100}$$

where $l$ is the degree of the polynomial within the interval $< 2, m >$ and coefficient $b_k$ of this polynomial can be calculated based on the quantification of $l + 1$ values of the function $w_{i,\text{DIL,out}}(w_{i,\text{DIL,in}}, n)$ for integral numbers $n$ close to the real $n$ by solving the system of linear equations,

$$\begin{bmatrix} 1 & n_0^1 & . & n_0^{l-1} & n_0^l \\ 1 & n_1^1 & . & n_1^{l-1} & n_1^l \\ . & . & . & . & . \\ 1 & n_{l-1}^1 & . & n_{l-1}^{l-1} & n_{l-1}^l \\ 1 & n_l^1 & . & n_l^{l-1} & n_l^l \end{bmatrix} \cdot \begin{bmatrix} b_0 \\ b_1 \\ . \\ b_{l-1} \\ b_l \end{bmatrix} = \begin{bmatrix} w_{i,\text{DIL,out}}\left(w_{i,\text{DIL,in}}, n_0\right) \\ w_{i,\text{DIL,out}}\left(w_{i,\text{DIL,in}}, n_1\right) \\ . \\ w_{i,\text{DIL,out}}\left(w_{i,\text{DIL,in}}, n_{l-1}\right) \\ w_{i,\text{DIL,out}}\left(w_{i,\text{DIL,in}}, n_l\right) \end{bmatrix} \tag{7.101}$$

whereas we select

$$n_{i+1} - n_i = 1 \tag{7.102}$$

Based on this procedure, it is possible to transfer data from one pilot electrodialysis stack not only to the devices with longer flow path lengths, but also with shorter flow path lengths.

The situation is complicated by a partial transfer of the solvent through the ion-selective membranes, together with ions, which results in a gradual prolongation of the residence time of the liquid in diluate chambers in each additional hydraulic stage of the line of pilot electrodialysis stacks, and, thus, in the increase of the deviation of the calculated value $w_{i,\text{DIL,out}}(w_{i,\text{DIL,in}}, n)$ from reality. This deviation increases with the increasing difference $w_{i,\text{DIL,in}} - w_{i,\text{DIL,out}}(w_{i,\text{DIL,in}}, n)$, that is, with the increasing amounts of electrolyte transferred through the ion-selective membranes at the passage of the fluid through the diluate chamber of the electrodialysis stack at a flow path length $L_2$ (line consisting of $n$ pilot electrodialysis stacks with flow path length $L_1$).

The key benefit of using this method is that instead of mass fraction of the relevant component, the concentration of the component can be used or an appropriate summary characteristic of the solution such as conductivity, total dissolved solids, etc.

## 7.4.5 Method based on local balance of diluate

The mass balance of the diluate chamber volume element was described in Section 7.3.1.2, from where differential equations are derived. These express the change in the mass flow rate of the diluate and the mass fraction of the $i$th component in the diluate along the flow path length. The values $J_{DIL}$ and $J_i$ in the equations, hereinafter, are the functions of local conditions in the electrodialysis stack, that is, local composition of diluate and concentrate, current density, linear fluid velocity and temperature. In accordance with the presumptions for the transmission of operating data between two electrodialysis stacks (see above), we consider $J_{DIL}$ and $J_i$ only as functions of $w_{i,DIL}$.

The situation is simplest if there are, during the passage of diluate through the electrodialysis stack, only negligible changes in the mass flow rate of the diluate and the mass fraction of the $i$-th component of the diluate along the flow path length, in practice at very short flow path length of the fluid (at very short residence time of the diluate in the electrodialysis stack). The recorded operating data can be used – based on the balance of the diluate tank, see eqs. (7.107) and (7.108) and compare them with eq. (7.9) – to calculate the relevant flows of the diluate $\dot{N}_{DIL}$ and of the $i$-th component $\dot{N}_i$ through the ion-selective membranes from the diluate to the concentrate in the electrodialysis stack, as the function of the mass fraction of the $i$-th component in the diluate feed $w_{i,DIL,in}$ (in the diluate at the inlet to electrodialysis stack) and from them – by dividing them with the active area of the ion-selective membrane – the corresponding dependencies $J_{DIL}(w_{i,DIL})$ and $J_i(w_{i,DIL})$. Based on the study of the ED process kinetics in such cases, it seems to be suitable to approximate $J_{DIL}$ by a directly proportional function:

$$J_{DIL} = a J_i \tag{7.103}$$

and $J_i$ by a polynomial:

$$J_i = \sum_{j=0}^{m} b_j w_{i,DIL}^{j} \tag{7.104}$$

In other cases, the method of solving is more complex. The evaluation of coefficient $a$ and vector $\boldsymbol{b}$ in functions (7.103) and (7.104) is possible if functions $\dot{m}_{DIL,out}$ $(\dot{m}_{DIL,in})$ and $w_{i,DIL,out}$ $(w_{i,DIL,in})$ are known. These functions are obtained again from the record of operating data in the laboratory or pilot test, using the diluate tank balance, see eqs. (7.107) and (7.109). The system of differential eqs. (7.7) and (7.8) can be symbolically written as the function returning a two-dimensional vector:

$$\begin{bmatrix} \dot{m}_{\text{DIL,out}} \\ w_{i,\text{DIL,out}} \end{bmatrix} = \mathbf{f}(w_{i,\text{DIL,in}}, a, \mathbf{b}) \tag{7.105}$$

Coefficient $a$ and vector $\mathbf{b}$ can be then calculated using the non-linear regression analysis, in which the minimizing functions are defined as follows:

$$\sum_{k=1}^{n} \left[ \dot{m}_{\text{DIL,out, exp},k} - f_1(w_{i,\text{DIL,in},k}, a, \mathbf{b}) \right]^2 = \min$$

$$\sum_{k=1}^{n} \left[ w_{i,\text{DIL,out, exp},k} - f_2(w_{i,\text{DIL,in},k}, a, \mathbf{b}) \right]^2 = \min \tag{7.106}$$

As soon as we know the value of coefficient $a$ and vector $\mathbf{b}$, we can calculate the local mass flow rate of the diluate $\dot{m}_{\text{DIL}}$ and the mass fraction of the $i$-th component of the diluate $w_{i,\text{DIL}}$ for any flow path length $L$ by solving the system of differential eqs. (7.8) and (7.9). Therefore, the notation of function (7.105) can be symbolically extended by the parameter $L$.

## 7.4.6 Mathematical solutions for individual operating modes

### 7.4.6.1 Batch mode

In the batch operation, the diluate/concentrate batch recirculates between the tank and the electrodialysis stack until its quality meets the requirements (e.g. until the diluate conductivity falls below a certain value). As soon as the operation is completed, it is necessary to discharge the batch and fill the tank with a new batch or switch to another tank. The second option allows for easy continuation of the operation.

Let's assume that the diluate batch is perfectly mixed and the dead volume of the diluate is negligible compared with the batch volume. In this case, it is possible to identify the mass fraction of the $i$th component in the diluate batch and at the inlet to the line of electrodialysis stacks $w_{i,\text{DIL,in}}$.

By adjusting the equation for the total mass balance, we get the following term for the change in the diluate batch weight:

$$\frac{\mathrm{d}m_{\text{DIL}}}{\mathrm{d}\tau} = \dot{m}_{\text{DIL,out}} - \dot{m}_{\text{DIL,in}} \tag{7.107}$$

In ED, instead of mass flow rate, volume flow rate is often controlled at the inlet to the line. If the dependence of the liquid density on the concentration of its key element is known from pilot tests, the weight or the mass flow rate can be calculated from the volume and volume flow rate of the liquid, respectively.

The mass balance of the $i$-th component in the diluate batch is expressed as follows:

$$\frac{d(m_{DIL}w_{i,DIL,in})}{d\tau} = \dot{m}_{DIL,out}w_{i,DIL,out} - \dot{m}_{DIL,in}w_{i,DIL,in} \qquad (7.108)$$

which can be used to define the following term for the time-change of the mass fraction of the $i$-th component in the batch:

$$\frac{dw_{i,DIL,in}}{d\tau} = \frac{\dot{m}_{DIL,out}w_{i,DIL,out} - \dot{m}_{DIL,in}w_{i,DIL,in} - w_{i,DIL,in}\frac{dm_{DIL}}{d\tau}}{m_{DIL}} = \frac{\dot{m}_{DIL,out}(w_{i,DIL,out} - w_{i,DIL,in})}{m_{DIL}}$$
$$(7.109)$$

In eqs. (7.107) to (7.109), $m_{DIL}$ represents the instantaneous weight of the diluate batch.

The mathematical task for batch operation is defined as follows:

$$\text{for } \tau_1: \ w_{i,DIL,in}(\tau_1) \leq w_{i,P} \qquad (7.110)$$

where $\tau_1$ is the batch processing time and $w_{i,P}$ is the required mass fraction of the $i$-th component in the demineralized diluate (product).

Batch operation is usually terminated as soon as the condition (7.110) is met. However, in many cases, it is more efficient to redirect the diluate at the outlet from the line into another tank at the moment when the following condition is met:

$$\text{for } \tau_1: \ w_{i,DIL,out}(\tau_1) \leq w_{i,P} \qquad (7.111)$$

and then continue in a single-pass mode until the first tank is completely empty. This step requires additional time:

$$\tau_2 = \frac{m_{DIL}(\tau_1)}{\dot{m}_{DIL,in}(\tau_1)} \qquad (7.112)$$

However, the total batch processing time

$$\tau = \tau_1 + \tau_2 \qquad (7.113)$$

is usually shorter than in a simple batch mode operation. Hereinafter, we will refer to this modification of the batch mode as the "modified batch mode".

The performance of the technology operated in a simple batch mode in the form of the mass flow rate of demineralized diluate (product) $\dot{m}_P$ is:

$$\dot{m}_P = \frac{m_{DIL}(\tau_1)}{\tau_1} \qquad (7.114)$$

and in the modified batch mode:

$$\dot{m}_P = \frac{\dot{m}_{DIL,out}(\tau_1)\,\tau_2}{\tau} \tag{7.115}$$

A special case of the batch mode is the variant in which a piston-type flow of the fluid is simulated in the diluate tank (there is no mixing of the batch content; diluate is fed to ED line from the bottom part and then returned back to the upper part of the tank). Such a case can be implemented, for example, by using cylindrical tanks with a high ratio between the tank height and its diameter. As for the principle, this case is equivalent to a repeated ED. The mass flow rate of the product corresponds (in ideal case) to a continuous single-pass mode of operation and is independent of the selected flow path length.

### 7.4.6.2 Feed and bleed mode

In feed and bleed operation, a part of the diluate/concentrate at the outlet from the line of electrodialysis stacks is returned back to the inlet to this line. This circuit is called a **diluate/concentrate recycle loop**. The part of the diluate/concentrate which leaves the loop as the product/waste, so-called **blowdown** or **overflow**, is supplemented by the feed of the processed solution at the inlet to the line of electrodialysis stacks, so-called **make-up**. This operating mode is used in the case of demand for the continuous operation, where one pass of the processed liquid through the line of electrodialysis stacks would not ensure the achievement of the required product quality.

If we realize the dependence of the mass fraction of the $i$th component in the diluate at the outlet from the line of electrodialysis stacks $w_{i,DIL,out}$ on the mass fraction of the same component at the inlet to the line $w_{i,DIL,in}$, the task consists in the finding of such $w_{i,DIL,in}$, for which the following applies:

$$w_{i,DIL,out}\left(w_{i,DIL,in}\right) - w_{i,P} = 0 \tag{7.116}$$

where $w_{i,P}$ is the required mass fraction of the $i$th component in the product (diluate at the outlet from the line). For the diluate make-up mass flow rate, we get $\dot{m}_{DIL,MU}$ from the balance:

$$\dot{m}_{DIL,MU} = \dot{m}_{DIL,in}\,\frac{w_{i,DIL,in} - w_{i,P}}{w_{i,DIL,MU} - w_{i,P}} \tag{7.117}$$

The effect of the mass fraction of the monitored component $i$ in the diluate on the flux of the diluate through the ion-selective membranes is known from the pilot test. The diluate blowdown mass flow rate from the relevant loop is:

$$\dot{m}_P = \dot{m}_{DIL,BD} = \dot{m}_{DIL,MU} - \dot{m}_{DIL,in} + \dot{m}_{DIL,out} \tag{7.118}$$

### 7.4.6.3 Continuous single-pass mode

In the continuous single-pass operation, the required quality of the diluate is achieved as early as after one pass through the line of electrodialysis stacks so that the diluate exits the line directly as a product. Because of the dependence of the mass fraction of the $i$th component in the diluate $w_{i,\mathrm{DIL}}$ on the diluate flow path length $L$, see eq. (7.8), the mathematical task for this operating mode consists in finding such $L$, for which the following applies:

$$w_{i,\mathrm{DIL}}(L) - w_{i,\mathrm{P}} = 0 \tag{7.119}$$

In some cases, such a configuration of the line of electrodialysis stacks is selected, where

$$w_{i,\mathrm{DIL}}(L) < w_{i,\mathrm{P}} \tag{7.120}$$

In this case, it is possible to partially **bypass** the line of electrodialysis stacks, whereas by mixing the bypass with the diluate at the outlet from the line of electrodialysis stacks, a product of the required quality is produced. For the mass flow rate of the diluate bypass $\dot{m}_{DIL,\mathrm{BD}}$, the following term can be deduced from the balance:

$$\dot{m}_{\mathrm{DIL,BP}} = \dot{m}_{\mathrm{DIL}}(L)\,\frac{w_{i,\mathrm{P}} - w_{i,\mathrm{DIL}}(L)}{w_{i,\mathrm{DIL,in}} - w_{i,\mathrm{P}}} \tag{7.121}$$

Thus, the product mass flow rate is expressed as follows

$$\dot{m}_{\mathrm{P}} = \dot{m}_{\mathrm{DIL}}(L) + \dot{m}_{\mathrm{DIL,BP}} = \dot{m}_{\mathrm{DIL}}(L)\left[1 + \frac{w_{i,\mathrm{P}} - w_{i,\mathrm{DIL}}(L)}{w_{i,\mathrm{DIL,in}} - w_{i,\mathrm{P}}}\right] = \dot{m}_{DIL}(L)\left[\frac{w_{i,\mathrm{DIL,in}} - w_{i,\mathrm{DIL}}(L)}{w_{i,\mathrm{DIL,in}} - w_{i,\mathrm{P}}}\right] \tag{7.122}$$

## 7.4.7 Optimization of flow path length and operating mode in a specific application

The discussion in the previous chapters shows that it is possible to calculate the mass fraction of the monitored component $i$ for any liquid flow path length $L$, that is, for any ED line configuration, the mass flow rate of the product can be calculated for any operating mode. To optimize the configuration and operating mode of the ED line, appropriate optimization criteria must be defined. For these purposes, we introduce two types of criteria here. One of them is the mass flow rate of the product per unit active area of the ion-selective membrane, $K_1$ (unit: kg m$^{-2}$ s$^{-1}$):

$$K_1 = \frac{\dot{m}_{\mathrm{P}}}{nWL} = \frac{\dot{m}_{\mathrm{P}}}{A_{\mathrm{eff}}} \tag{7.123}$$

where $n$ is the number of cell pairs and $W$ is the width of the flow chamber. Optimum configuration of ED line as well as the operating mode correspond to the maximum of $K_1$ criterion, which, among others, tells how to design ED systems for specific applications.

Another criterion is the mass flow rate of the product (kg s$^{-1}$) per unit of investments $K_2$:

$$K_2 = \frac{\dot{m}_P}{Q_C} \tag{7.124}$$

where $Q_C$ represents the investments into ED technology.

The optimum design of the technology corresponds, in this case, again to the maximum of $K_2$ criterion. This criterion is closely followed by ED systems and technologies suppliers in order to increase their competitiveness and to maximize their profit.

Possible requirements for continuous or discontinuous operation must be further considered when optimizing an ED technology.

Examples of ED technology optimization, based on $K_1$ criterion, for the case of desalting of aqueous solution of $Na_2SO_4$ with different initial and final electrolyte concentrations are shown in Fig. 7.25. The basis for the ED technology design was a test with the pilot electrodialysis stack ED-Y, available from MemBrain s.r.o. The test was carried out in the batch operating mode, according to instructions in Section 7.4.2, while a constant circulation flow rate of the media was maintained. The batch volume and the media conductivity at the inlet to electrodialysis stack were subject to monitoring. The dependence of the solution density and conductivity on $Na_2SO_4$ concentration was known from the literature. Data from the pilot test were processed using eqs. (7.107) and (7.109), which resulted in dependencies $\dot{m}_{DIL,out}(w_{i,DIL,in})$ and $w_{i,DIL,out}(w_{i,DIL,in})$. Using the methodology described in Section 7.4.5, dependencies $J_{DIL}(w_{i,DIL})$ and $J_i(w_{i,DIL})$ were obtained and using the mathematical tasks described in Section 7.4.6, dependencies $\dot{m}_P$ on the flow path length and the diluate operating mode were obtained as well.

In this task, for simplicity, all standardized electrodialysis stacks from the portfolio of companies MEGA a.s. and MemBrain s.r.o. were considered, that is, electrodialysis stacks ED-Z, ED-Y, ED-X and ED-II/ED-III with flow path lengths of 0.16, 0.4, 0.59 and 1.3 m successively, whereas electrodialysis stacks ED-II and ED-III can be connected into up to two-stage (ED-II) and four-stage (EDR-III) lines, respectively, which corresponds to flow path lengths of 2.6 and 5.2 m, respectively. The ED-II electrodialysis stack is used for partial demineralization of more concentrated solutions, while the field of use of the EDR-III electrodialysis stack is limited to partial demineralization of waters and aqueous solutions with total dissolved solids (TDS) of up to approx. 3,000 mg dm$^{-3}$. The laboratory and pilot electrodialysis stacks ED-Z, ED-Y and ED-X were included in these optimization tasks only to demonstrate the interesting impact of the flow path length on the specific performance

(a)

(b)

(c)

- - - - - - batch
- · - - - modified batch
- - - - - - feed and bleed
———— single-pass (with bypass)

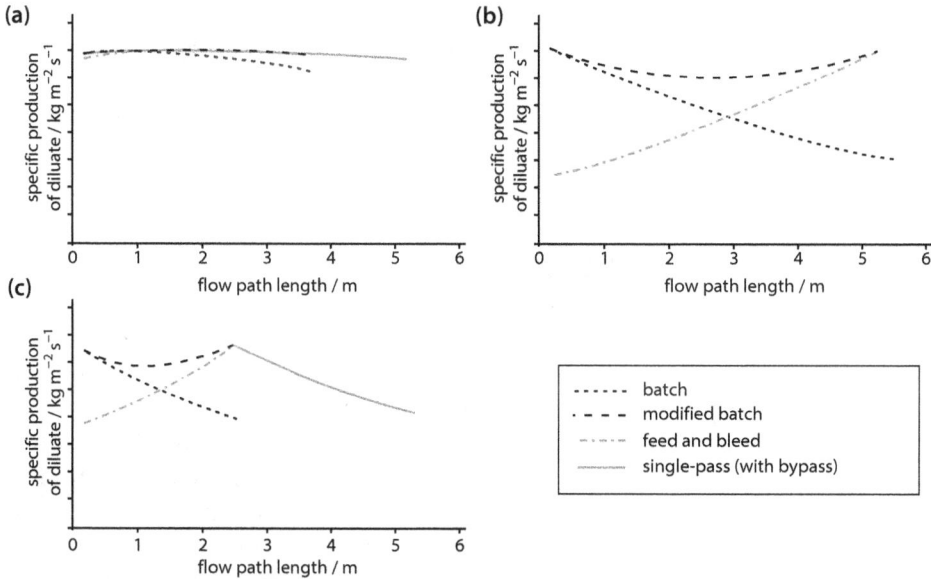

**Fig. 7.25:** Optimization of ED technology at desalination of $Na_2SO_4$ solution: (a) from 20 to 10 kg m$^{-3}$, (b) from 10 to 1 kg m$^{-3}$, (c) from 2.0 to 0.5 kg m$^{-3}$.

of the technology; the realization of industrial ED technologies with these electrodialysis stacks would obviously be uneconomical. The task solution is, therefore, limited to determination of optimum number of hydraulic and electric stages of the line with electrodialysis stacks ED-II/EDR-III as well as the diluate operating mode.

When demineralizing the solution from 20 to 10 kg m$^{-3}$ $Na_2SO_4$, there are slight differences in the specific performance of ED technology (criterion $K_1$) between individual configurations of the line (flow path lengths) and the diluate operating modes, see Fig. 7.25a. As for the application field of ED-II and EDR-III electrodialysis stacks, we, therefore, choose as optimal, the two-stage line of electrodialysis stacks ED-II operated in the feed and bleed mode of diluate (link to $K_2$ criterion). As for demineralization from 10 to 1 kg m$^{-3}$ $Na_2SO_4$, which still falls in the application area of ED-II electrodialysis stack, we choose the single-stage lines with this electrodialysis stack operated in a batch mode of diluate, see Fig. 7.25b. If we manage to avoid, by selecting a suitable geometry, see Section 7.4.6.1, the mixing of the diluate batch in the tank, we can also use the two-stage lines of ED-II electrodialysis stacks and the specific performance of ED technology will be very close to a continuous single-pass operating mode. Demineralization from 2 to 0.5 kg m$^{-3}$ $Na_2SO_4$ falls in the application area of EDR-III electrodialysis stacks. Based on the results of calculations, see Fig. 7.25c, the obvious choice would be the two-stage line of EDR-III electrodialysis stacks operated in a continuous single-pass operating mode.

From the model calculations presented in Fig. 7.25, it is evident that by choosing an inappropriate configuration of the line or the operating mode, the specific performance of ED technology may easily decrease, even by dozens of percent. It is, therefore, necessary to pay considerable attention to processing and evaluation of data from laboratory and pilot tests.

## 7.5 Electrodialysis stack design

Conventional electrodialysis stacks are, most often, devices which conceptually remind a plate-and-frame filter press. Spiral-wound electrodialysis stacks are considerably less widespread than the analogue type of EDI modules, see Section 8.8, and in terms of their practical use, they are unsuitable due to uneven distribution of current densities in radial direction and rather difficult or even impossible rebuild. According to the working orientation of the membrane stack, plate-and-frame electrodialysis stacks can be further divided into horizontal and vertical. An example of the typical horizontal industrial electrodialysis stack of a plate-and-frame type is shown in Fig. 7.26. The key advantage of the horizontal design is higher mechanical stability of the membrane stack, which is particularly important in devices with a large number of cell pairs. Another reason for choosing the horizontal arrangement can

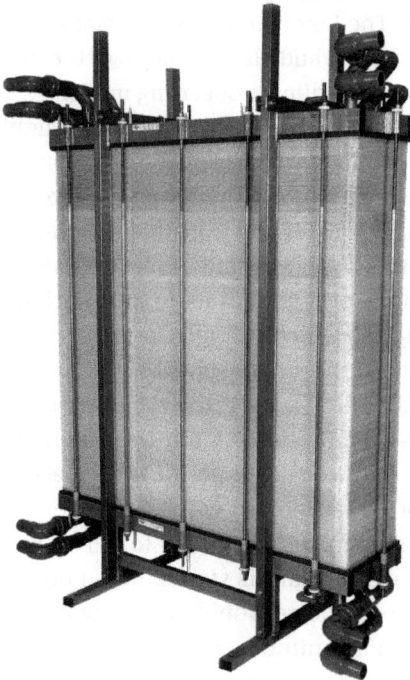

**Fig. 7.26:** Horizontal industrial electrodialysis stack EDR-III/600-0.8 by MEGA a.s. (photo MEGA a.s.).

be the design of working spacers (spacers with U-flow, tortuous path spacers, see hereinafter).

Some parts of a typical industrial electrodialysis stack of a plate-and-frame type are shown in Fig. 7.27. The electrodialysis stack consists of **end plates** in between which is a membrane stack made of regularly alternating cation- and anion-selective membranes interlaced with **spacers**. The entire assembly is – in order to ensure the necessary internal and external tightness – clamped by tie rods or bolts. The clamping force or torque depends on the material of the working spacers, the surface texture of the working spacers frame and mechanical properties of ion-selective membranes; it normally ranges from 40 to 100 N m. Tie rods are usually electrically insulated by means of heat-shrink tubing. This prevents electrical short-circuit, for example, at the incrustation of the membrane stack surface by salts. As a result of the membrane stack clamping by tie rods and because of working pressures, the end plates are subject to significant mechanical stress. The required stiffness is, therefore, ensured by frame or plate made of stainless steel or cast iron that forms the outer part of the end plate, usually with a non-conductive surface coating. The metal frame or plate is electrically insulated from the membrane stack by means of plastic plate,

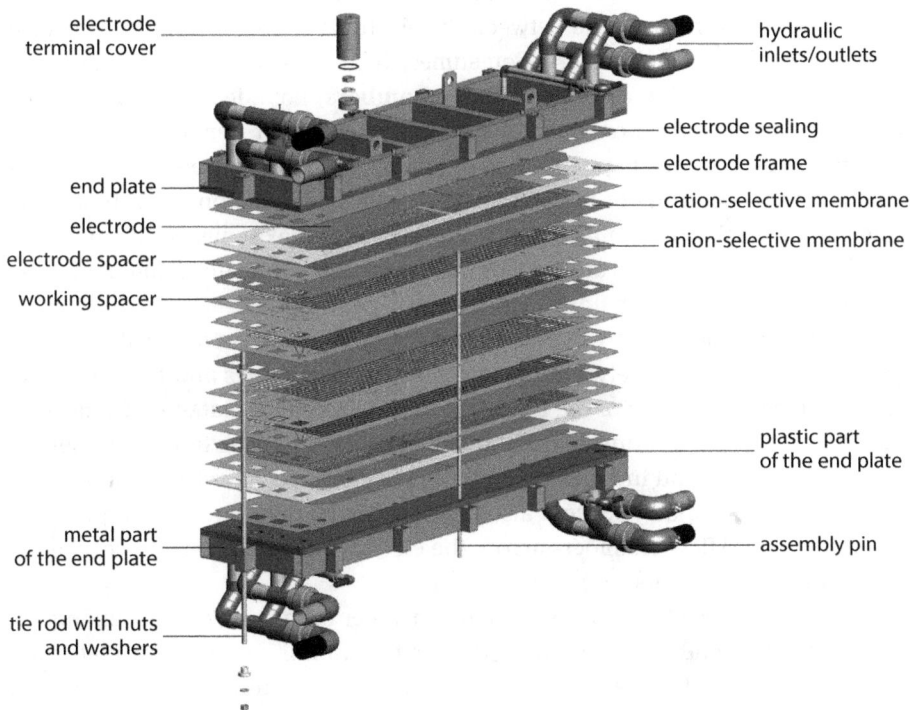

**Fig. 7.27:** Parts of an industrial electrodialysis stack.

which forms the inner part of the end plate. The plastic plate is usually made of PP or PVC and on the inner side, it is fitted with an electrode including electrical wiring, while on the outer side, it is provided with hydraulic inlets and outlets for working media (diluate, concentrate, electrode solution). In the case of horizontally oriented electrodialysis stacks, the plastic part of the bottom end plate can also fulfil the function of collecting the leaks from the membrane stack and their discharge. Large-format electrodialysis stacks may require the use of so-called assembly pins, which maintain the alignment of the membrane stack parts during assembly and prevent from deflection of the side walls of working spacers' frames from the membrane stack during clamping or operation. These assembly pins are usually made of plastic (PP) or stainless steel with plastic coating. Additionally, for the same purpose, the electrodialysis stack may be reinforced by metal square tubes from the outside.

The structure of the membrane stack from the anode to the cathode may be of the type CM-AM-CM, see Fig. 7.1, or AM-CM-AM. Priority is given to the first variant since, in this case, $Cl^-$ ions, which are involved in electrode reactions with the formation of dangerous products ($Cl_2$), shorten the service lifetime of the first and last membranes and penetrate the electrode solution at a significantly lower rate. Another reason is, generally, better chemical stability of CM. The first and last CM are usually thicker than other membranes in the stack as they are mechanically stressed by the difference of the pressure between the electrode chambers and the adjacent diluate and concentrate chambers. Sometimes, the edges of the membrane stack may consist of **protective** or **neutral flow chambers**, bounded by only one type of ion-selective membrane on both sides, in which, neither demineralization nor concentration of solutions happens. This reduces negative consequences of leakage electric currents – burning of membrane stack edges, see Section 7.3.4, protecting the membrane stack from damage due to current density extremes occurring when electrodes in the form of rods are used or further reducing the rate of contamination of the electrode solution with unwanted ions such as chlorides.

The primary function of working spacers is to create a geometrically-defined space for the flow of fluid between the ion-selective membranes and to ensure adequate turbulence promotion of the flow. Due to the concept of plate-and-frame electrodialysis stacks, the working spacers' frames also assure the external tightness of electrodialysis stacks and internal tightness against leakages between hydraulic circuits. In addition, working spacers also provide for the distribution of liquid flow between individual flow chambers across the electrodialysis stack and over the entire width of flow chambers.

Depending on the shape of the working chamber or, more likely depending on the way how the liquid flows through the chamber, working spacers for plate-and-frame electrodialysis stacks are either tortuous path or with a rectangular chamber, see Fig. 7.28. According to the method of supporting of the ion-selective membrane and ensuring the liquid flow turbulence promotion, working spacers are either net-like or with cross-straps. Threads of the net or cross-straps ensure transverse mixing

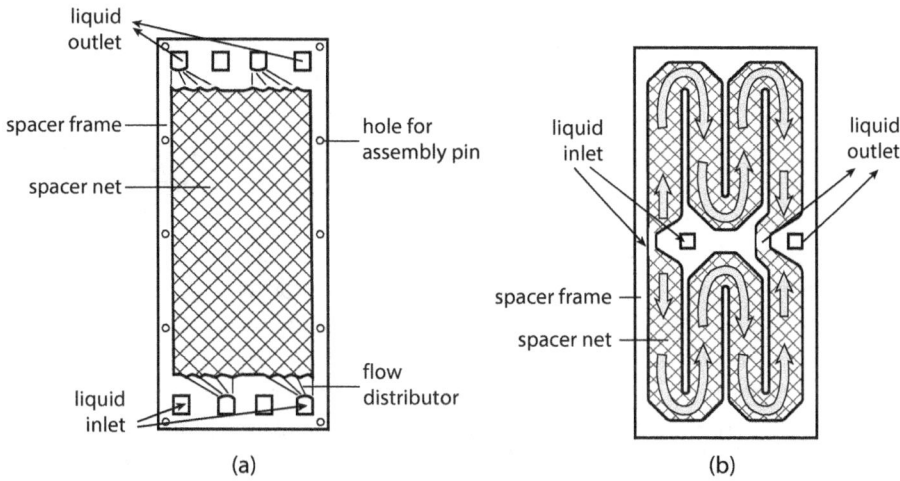

**Fig. 7.28:** Scheme of the working ED spacer of net-like type (a) and tortuous path type (b).

of the liquid layers, thus facilitating mass transfer – this is especially important when demineralizing diluted solutions. The tortuous flow is achieved by dividing the working chambers by barriers having the same thickness as the spacer frame into a couple of narrower and parallel-oriented channels through which the liquid flows in series. This narrows the flow-through channel and extends the flow path length (see Fig. 7.28 on the right), with only a slight reduction of the active area of the ion-selective membrane due to the presence of dividing barriers. This results in the increase of linear velocity of the liquid while maintaining the residence time of the liquid in the flow chamber. One of the practical limitations of tortuous path spacers is the high pressure loss of liquid, especially when combined with a spacer net. Spacer nets were, previously, only available in limited quantities, their quality was rather poor (variable thickness) and were expensive. Therefore, the transverse mixing of the liquid flow in some older spacers was preferentially ensured by the system of cross-straps perpendicular to the direction of liquid flow, thinner than the spacer frame (spacers were produced by gluing two parts with half thickness together). To ensure that the active area of the ion-selective membrane is not covered too much, these barriers were placed farther from each other [17]. However, at steady laminar flow of liquid through a smooth rectangular channel, a characteristic parabolic profile of linear fluid velocity developed across the channel. By analysing the convective-diffusion model of mass transfer in electrodialysis stack, in the case of processing of diluted electrolyte solutions, we can show the negative impact of such liquid flow character through the flow chamber on mass transfer [11]. The desired effect was, therefore, only possible with greater linear velocity of liquid (20–40 cm s$^{-1}$). To achieve the required demineralization of the liquid while passing through the diluate chamber, a certain residence time of the fluid in the chamber was necessary, which in combination with the requirement for greater linear velocity of liquid necessitated the use of tortuous path spacers only. A significant disadvantage of tortuous path spacers with turbulence promoters in the form of cross-straps was the high portion of inefficiently used area of the ion-selective membrane.

However, in connection with the progress of pressure membrane separation processes, especially the reverse osmosis process, suitable spacer nets were developed, which were then applied in ED as well. Spacer net ensures sufficient turbulence promotion of liquid flow at linear velocity of 5–10 cm s$^{-1}$. Therefore, most working spacers are net-like spacers today, although the tortuous path of liquid has been maintained in many cases, often in order to assure the backward compatibility with older types of spacers.

Working spacers frames are usually made of elastomers, for example, natural, synthetic or silicone rubber, polyolefins (PE, PP), PVC, EVA, etc., mostly by cutting them out of a particular sheet. The formatting device may take the form of discontinuous die-cutting press or a continuous roll design. The second option is considerably more expensive in terms of investment costs, but production costs are lower, so it can be more advantageous in continuous production. In some cases, especially when small formats are involved, cast or injection moulds can be used. To maximize the efficient use of the ion-selective membrane area, the spacer frame, including the barriers, should occupy as small an area as possible. The portion of an efficiently used area of the ion-selective membrane varies in the case of working spacers for the current industrial electrodialysis stacks in the range of 65–85 %.

Non-woven nets made of chemically inert polymers, typically PE or PP, with mesh size from 1 to 6 mm, are mostly used as spacer nets. Both sets of fibres must not lie in a single layer to avoid plugging of the working chambers. Examples of spacer nets are shown in Fig. 7.29. The spacer net is thermally welded to the spacer frame or more frequently pressed into the spacer frame in the pressing machine at higher temperatures or both parts are joined during cast or injection moulding.

**Fig. 7.29:** Examples of spacer nets.

The spacer-net surface may be modified by a material with ion exchange properties [18, 19]. Such working spacers can, preferably, be used at demineralization of low conductive electrolyte solutions, in which an anion has a considerably lower mobility than a cation or vice versa. A typical example would be the demineralization of organic acids salts or diluted solutions of inorganic acids and hydroxides. In these cases, the net with ion exchange properties extends the area of the relevant ion-selective membrane into the space of flow chamber, thereby facilitating mass transfer.

The thickness of working spacers for ED usually varies from 0.5 to 1.0 mm. Thinner spacers improve ED process kinetics at processing of diluted aqueous solutions as they have lower ohmic resistance and, at the same volumetric flow rate, the liquid flows through these spacers at a higher linear velocity. At the same time, however, they exhibit higher pressure loss. Due to the limited external tightness of electrodialysis stacks (usually up to a pressure of 350 kPa) and the increase in energy consumption on pumping, there seems to be a certain optimum thickness of working spacers.

Diluate and concentrate spacers are mostly of the same design, but due to the system of inlets and outlets, they are inserted into electrodialysis stack relatively horizontally or vertically rotated by 180°. In Fig. 7.2, depending on the position and orientation of the flow distributors on two adjacent working spacers, the relative orientation of both spacers is evident (they are vertically overturned). The same figure further illustrates the way in which the channels are created to distribute and collect the liquid flow across the electrodialysis stack through appropriate openings in the working spacers' frames and in the same positions of ion-selective membranes as well as the way the diluate and concentrate are distributed into relevant working chambers.

The minimum flow rate through the flow chambers is determined by the need to ensure uniform distribution of the fluid flow along the working chamber width, adequate transverse mixing of liquid, removal of heat generated by dissipation of electrical energy and by prevention of ion-selective membranes fouling.

To use the active area of ion-selective membranes efficiently and to avoid the risks of the existence of so-called **hydrodynamic shadows** (see Fig. 6.7), it is necessary to ensure uniform distribution of liquid flow along the entire width of the working chamber. For this purpose, **flow distributors** are used. The most common types are shown in Fig. 7.30. The simplest is the situation with tortuous path spacers characterized by small width of the flow channel. As for wide spacers (relative to the size of openings forming the flow distribution manifolds), inlets into the working chamber and outlets from it need to be solved by several parallel openings for flow distribution manifolds and the same number of flow distributors or sets of flow distributors. Flow distributors in the form of nets (Fig. 7.30a) and in the form of narrow channels or slits (Fig. 7.30b) can be characteristic by problematic tightness, which may cause internal leakage between main hydraulic circuits (in the first case, due to the thickness difference between the spacer frame and the spacer net at the flow distributor or due to pressing of ion-selective membrane into the spacer net; in the latter case, due to elastic or plastic deformation of ion-selective membrane at the flow distributor channels). The latter case is shown in Fig. 7.31. Therefore, the distributor channel is sometimes filled with a net with a thickness that slightly exceeds the thickness of the spacer frame which is – by deformation at clamping of the membrane stack – compressed to the thickness corresponding to spacer frame. But it can also have a more complex structure, see Fig. 7.32.

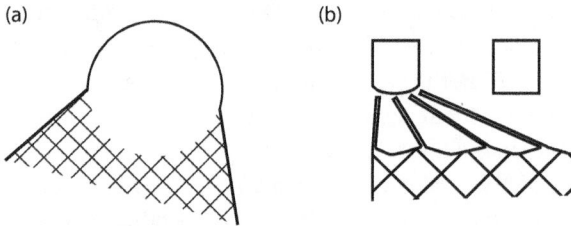

**Fig. 7.30:** Common types of flow distributors: (a) net-like, (b) in the form of narrow channels.

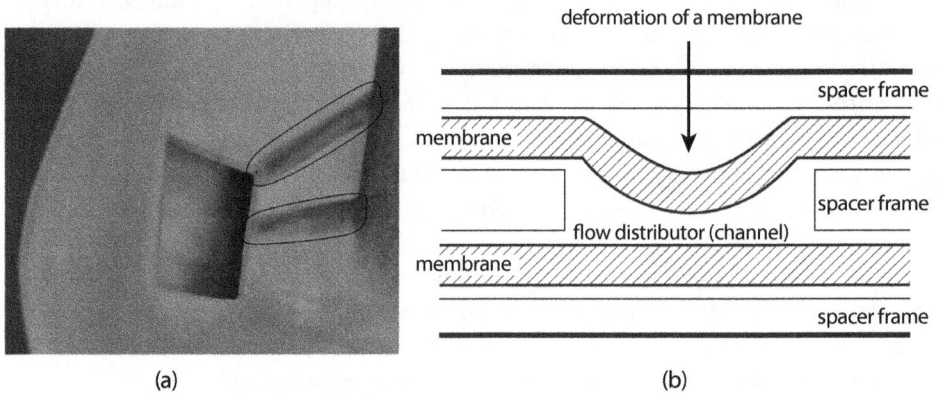

**Fig. 7.31:** Example (on the left) and mechanism (on the right) of ion-selective membrane deformation in the area of liquid flow distributor (according to [13]).

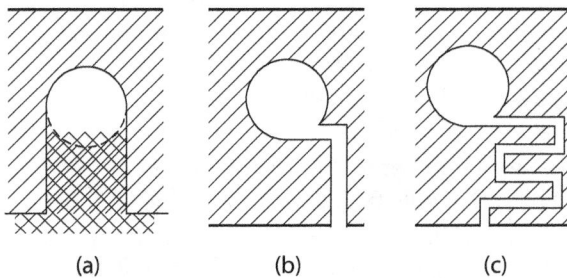

**Fig. 7.32:** Special types of liquid flow distributors [20].

Electrode spacers have a design similar to that of working spacers. They are often thicker (approx. twice as much), which is sometimes explained by the effort to achieve higher flow rate through the electrode chamber and thus to reduce the risk of scaling or fouling of the electrode while maintaining low pressure loss of the electrode solution [7]. Other reasons for using thicker electrode spacers include the concerns for spacer net breaking by membrane stack and plugging of electrode chamber or direct contact of the edge ion-selective membrane with the electrode. Electrode solution inlets and outlets on end plates are usually routed directly to the working space of the electrode spacer, so the electrode solution is not led across the electrodialysis stack as diluate and concentrate. This simplifies the design of the working spacers and reduces the risk of internal leakage between individual hydraulic circuits.

The minimum linear velocity of the electrode solution is determined by the necessity to reliably remove gases generated at electrode reactions, see eqs. (7.1) to (7.3), which is particularly important in the case of horizontal electrodialysis stacks, and also to remove the heat generated by the dissipation of electrical energy on the electrode and at the interface between the electrode and the electrode solution. In the case of horizontal industrial electrodialysis stack, EDR-III, available from MEGA a.s., featuring a 2 mm thick electrode chamber, the minimum superficial velocity of electrode solution of 8 cm s$^{-1}$ was determined to be safe.

As for anode, DSA/ATA/MMO type is usually used, that is, anode made of titanium (Ti) and activated by a coating of platinum (Pt), transition metals oxides ($RuO_2$, $TiO_2$ etc.) or their mixtures, see Section 13.1.3, while cathode is mostly made of stainless steel. In some cases, both electrodes are of Ti/Pt type, which allows the electrode polarity to be reversed – the process is then referred to as **electrodialysis (polarity) reversal (EDR)**. When changing the polarity of electrodes with respect to the membrane stack symmetry, see Fig. 7.1, the function of flow chambers is reversed, that is, originally diluate chambers are changed to concentrate chambers and vice versa. The polarity reversal is usually performed once every 15–60 min and it helps to increase the resistance of membranes to scaling (formation of inorganic deposits) and fouling (membrane surface contamination by organic substances, microorganisms, etc.).

The new trend in EDR process for partial demineralization of brackish water is the use of porous carbon electrodes. The electrodes themselves operate on the principle of electrodes for capacitive deionization (CDI), see Section 9.1, that is, on the principle of adsorption and desorption of ions in the electrical double layer. In combination with the ion exchange resin (ion-selective membrane) coating, these electrodes replace the conventional metal electrodes and electrode chamber system. This eliminates the need to work with electrode solution and the risk of cathode scaling. Also, no electrode gases are generated. In addition, these electrodes are cheaper than conventional electrodes.

The membrane stack of the electrodialyser, as the live part of the electrical equipment, poses a significant risk of electric shock to operators. Therefore, large industrial electrodialysis stacks that are currently equipped with plastic covers or other means are used to prevent operators from touching the membrane stack (fencing, light curtains, etc.). Examples of solution are shown in Fig. 7.33. Membrane stack casing also protects the operator in cases when, due to a mechanical defect, water starts to squirt from the membrane stack.

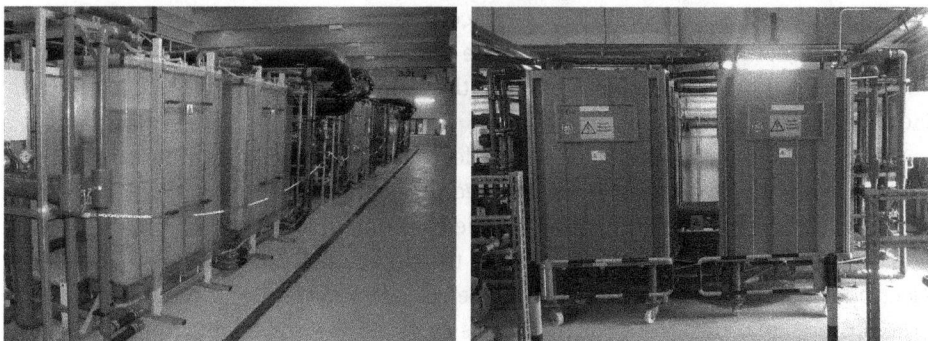

**Fig. 7.33:** Membrane stack casing options – on the left EDR-III by MEGA a.s., on the right EDR 2020 by GE Water & Process Technologies (now Suez Water Technologies & Solutions) (photo MEGA a.s.).

## 7.6 Operation of electrodialysis stacks

The current ED systems work with modern elements of measurement and regulation, that is, with pneumatically actuated valves in combination with solenoid valves, analogue/digital sensors of temperature, pressure, flow rate, pH, conductivity and level, programmable DC power supplies and variable frequency drives, various switches, etc. These electro-components allow direct or transmitter-mediated communication with programmable logic controller unit (PLC). Thanks to this, today's industrial ED technologies are characterized by high level of automation.

Figure 7.34 shows the diagrammatic example of a simple batch ED technology with single electrodialysis stack and a pair of tanks for the processed solution. One tank always contains the currently processed batch, while the other one is emptied and subsequently filled with a fresh batch. Switching between these tanks allows running the batch ED process continuously. No tanks are needed for hydraulic circuits operated in any of the continuous modes.

The advantage of ED in comparison with pressure-driven membrane-separation processes are rather low demands for pre-treatment. For example, in water treatment applications, sand filtration followed by candle or bag filters (10–50 µm) is usually sufficient. In some cases, however, dechlorination or softening of feed water or the use of various precipitation procedures, is necessary.

The transport or circulation of working media is almost exclusively ensured by centrifugal pumps with a three-phase motor controlled by a variable frequency drive. Thanks to this, it is possible to optimize the electricity consumption on pumping while ensuring the design margin of the pump.

In ED systems, usually only the circulation flow rate of the diluate is directly controlled and, possibly, also the make-up into the diluate and concentrate loop. However, more often, the make-up to the diluate and concentrate loop is controlled on the basis of the target diluate or concentrate conductivity, respectively. Zero pressure difference is maintained between the diluate and concentrate loops in ED. Therefore, the circulation flow rate of the concentrate and electrode solution is usually controlled on the basis of pressure difference between the diluate and these hydraulic circuits at the inlet. In water applications, the diluate and concentrate flow rates are practically the same. However, this is no longer the case in applications aimed at demineralization of whey, where diluate is a more viscous medium due to the high lactose content. In this case, the flow rate of the concentrate must be twice or even three times higher than the flow rate of the diluate in order to meet the zero pressure difference requirement. To prevent the risk of demineralized product contamination by the concentrate, it is sometimes recommended to maintain a slight overpressure in the diluate circuit, in comparison with the concentrate circuit, which eliminates potential internal leakage of concentrate into diluate. Generally, the overpressure up to approx. 5 kPa is sufficient. Such an operating mode requires the use of sufficiently sensitive differential pressure sensors.

In Fig. 7.34, note the method of implementation of the so-called **diversion valves** before the electrodialysis stack (line of electrodialysis stacks) and behind it. These valves are used in the EDR process. In continuous EDR technologies for water treatment, electrode polarity reversal is carried out during operation. Due to the mixing of diluate with concentrate, for a certain amount of time after electrode polarity reversal, EDR system produces the demineralized product, which does not meet the requirements, the so-called **off-spec product** or **OSP**. The off-spec product is either returned back to the feed or discharged to drain.

Electrode solution is fed into the anode as well as the cathode chamber from one tank and after passing through the electrodialysis stack, both electrode streams return back to the tank. The electrode solution tank must be vented to prevent accumulation of the $H_2$ generated on the cathode and to eliminate the risk of explosion. The lower explosive limit of hydrogen in a mixture with air is 4.1 vol %. Electrode gases are removed from the building by means of a special piping and a blower. The blower performance must be chosen to dilute the hydrogen to approx. one-tenth of the lower explosion limit or lower. The exception are continuous single-pass ED technologies processing water with very low concentration of electrolyte (up to approx. 3,000 mg $dm^{-3}$), where the electrode solution is branched directly from the

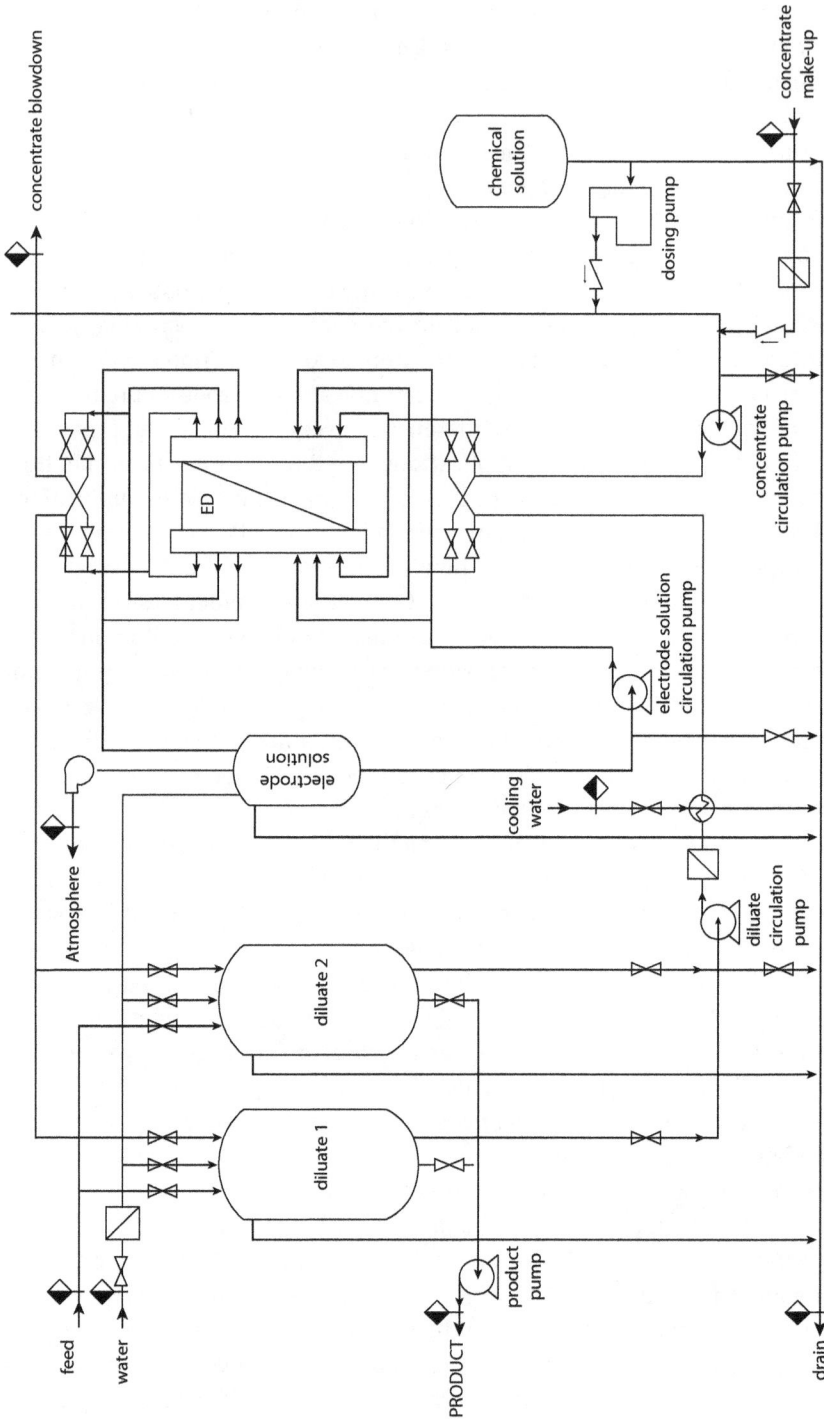

**Fig. 7.34:** Example of P&ID diagram of batch ED technology.

processed water and after passing through the electrode chambers, it is either led directly to drain (waste) or mixed with concentrate make-up.

For safety reasons, it is preferable to operate electrodialysis stacks in the constant voltage mode (potentiostatic mode) compared to the constant current mode (galvanostatic mode). During the ED process, the electrolyte concentration decreases in the diluate, which results in the reduction of diluate conductivity and increase of the electrodialysis stack electrical resistance. In galvanostatic mode, an uncontrolled increase in voltage in the electrodialysis stack or voltage overload of power supplies could occur and in extreme cases, even the complete destruction of the membrane stack due to overheating (Joule heating).

Most ED systems require regular **chemical cleaning** (**CIP**, *cleaning in place*) due to fouling and scaling. In practice, cleaning is most often carried out using acids (HCl and $HNO_3$) or hydroxides (NaOH and $NH_4OH$) at concentrations of up to 3–5 mass %. Each chemical cleaning cycle is followed by single or multiple rinsing of ED system with fresh water. The possibility of chemical cleaning may be – especially in case of hydroxides – limited by the chemical stability of the ion-selective membranes used.

A typical ED system usually consists of three basic parts – electrodialysis stacks, hydraulics covering most piping, fittings, measurement and regulation and one or more switchboards with transformers, DC power supplies, variable frequency drives, and PLC, functioning simultaneously as a control panel with touch panel, buttons and switches. ED technology control system usually communicates with the end-user's parent control system, thus allowing for its remote control from the control room.

## 7.7 Electric power consumption in the electrodialysis process

The consumption of electricity in ED is generally composed of two items:
- power supply of modules (by direct current),
- pumping power.

The DC power input, $P_Z$ (unit: W) is,

$$P_Z = \sum_{i=1}^{n} \frac{U_{Z,i} I_{Z,i}}{\eta_{Z,i}} \qquad (7.125)$$

where $U_{Z,i}$ is voltage, $I_{Z,i}$ is electric current and $\eta_{Z,i}$ is the efficiency of $i$th DC power supply. The corresponding consumption of electricity, $W_Z$ (unit: J) is

$$W_Z = \int_0^{\tau_P} P_Z(\tau)\, d\tau \tag{7.126}$$

where $\tau_P$ is the operating time. The input power of pumps $P_P$ can be expressed by the equation

$$P_P = \sum_{i=1}^{n} \frac{\Delta p_i \dot{V}_i}{\eta_{P,i}} \tag{7.127}$$

where $\Delta p_i$ is the differential pressure between the suction and the discharge side of the pump and $\dot{V}_i$ is the volume flow rate in the $i$th hydraulic circuit and $\eta_{P,i}$ is the efficiency of the relevant pump. The electricity consumption for pumping $W_P$ is analogically

$$W_P = \int_0^{\tau_P} P_P(\tau)\, d\tau \tag{7.128}$$

Total power input of the ED technology is

$$P_{tot} = P_Z + P_P \tag{7.129}$$

and total power consumption $W_{tot}$ is

$$W_{tot} = W_Z + W_P \tag{7.130}$$

In addition to the DC power supplies and pumps, the real ED technology also features other electrical equipment and appliances such as blower for venting of electrode gases, dosing pumps, transmitters and sensors, control system (PLC), electrically actuated valves, lighting, etc.

In practice, especially in batch ED technologies, the value of the maximum power input is quite important due to the correct dimensioning of the local electrical grid. In continuous ED technologies, we work with power input, while in batch ED technologies, we rather work with consumption of electricity. In both cases, the specific consumption of electricity per unit volume or mass of processed feed water or one of the products (diluate or concentrate) or per unit mass of electrolyte transferred through ion-selective membranes is often also indicated.

# 7.8 Economics of electrodialysis

While calculating the ED process economy, we consider:
- capital (investment) costs, so-called CAPEX,
- operating costs, so-called OPEX.

CAPEX enter into calculations in the form of amortization (depreciations) of investments in relation to the planned life of the equipment or ED technology. The sum of both CAPEX and OPEX items represents the total costs.

Operating expenses consist of replacement of ion-selective membranes and electrodes, consumption of electricity, see Section 7.7, water and chemicals, costs on operators, lighting, etc. The costs for the replacement of ion-selective membranes and electrodes, operation and lighting are fixed, while the consumption of electricity, water and chemicals are variable costs, which depend on the performance of technology. The consumption of water and chemicals increases in direct proportion to technology output. This, however, does not apply to consumption of electricity, which increases faster. Higher technology output means operation at higher current load, that is, higher voltage. In EDR applications for partial demineralization of brackish and river waters, an annual replacement of ion-selective membranes of approx. 15 % is commonly taken into account.

By relating the total costs to technology output, we get average costs. It is obvious that a part of average costs including depreciations, costs for replacement of ion-selective membranes and electrodes, operators and lighting will show a downward trend with increasing output of technology, while another part of average costs such as consumption of electricity will increase with increasing output of technology. The existence of these two conflicting trends means that there is an extreme (minimum) of average costs for a specific output of technology. This extreme corresponds to the optimum output of technology [21].

# References

[1]  R. W. Baker, Membrane Technology and Applications. 2nd Edition. John Wiley & Sons Ltd, Chichester, 2004. ISBN 0-07-135440-9.
[2]  RU1262900.
[3]  LV5046.
[4]  K. Nagasubramanian, F. P. Chlanda, and K. Liu, Use of bipolar membranes for generation of acid and base – an engineering and economic analysis. J. Membr. Sci. 2: 109–124, 1977.
[5]  K. N. Mani, F. P. Chlanda, and C. H. Byszewski, Aquatech membrane technology for recovery acid/base values from salt streams. Desalination 68: 149–166, 1988.
[6]  H. Strathmann and H. Chmiel, Die Elektrodialyse – ein Membranverfahren mit vielen Anwendungsmöglichkeiten. Chem. Ing. Tech. 56(3): 214–220, 1984.

[7]    F. H. Meller, et al. Electrodialysis (ED) & Electrodialysis Reversal (EDR) Technology. Corporate literature of Ionics Inc, 1984.

[8]    V. A. Shaposhnik, V. A. Kuzminykh, and O. V. Grigorchuk, Analytical model of laminar flow electrodialysis with ion-exchange membranes. J. Membr. Sci. 133: 27–37, 1997.

[9]    V. V. Nikonenko, A. G. Istoshina, and M. K. Urtenova, Analysis of electrodialysis water desalination costs by convective-diffusion model. Desalination 126: 207–211, 1999.

[10]   V. V. Nikonenko, V. I. Zabolotsky, and C. Larchet, Mathematical description of ion transport in membrane systems. Desalination 147: 369–374, 2002.

[11]   D. Tvrzník, L. Machuča, and A. Černín, Mathematical model of mass transfer in an electrodialyzer with net-like spacers. Desalin. Water Treat. 14: 1–5, 2010.

[12]   N. Káňavová, L. Machuča, and D. Tvrzník, Determination of limiting current density for different electrodialysis modules. Chemical Pap. 2013. DOI: 10.2478/s11696-013-0456-z.

[13]   Y. Tanaka, Ion Exchange Membranes – Fundamentals and Applications. In Membrane Science and Technology Series 12. Elsevier, Amsterdam, 2007. ISBN 978-0-444-51982-5.

[14]   R. P. Allison, High water recovery with electrodialysis reversal. Proceedings of 1993 AWWA Membrane Conference, April 1 to 4, 1993, Baltimore, USA.

[15]   R. Kodým, P. Pánek, D. Šnita, D. Tvrzník, and K. Bouzek, Macrohomogeneous approach to a two-dimensional mathematical model of an industrial-scale electrodialysis unit. J. Appl. Electrochem. 42: 645–666, 2012.

[16]   H. Lee, F. Sarfert, and H. Strathmann, Designing of an electrodialysis desalination plant. Desalination 142: 267–286, 2002.

[17]   A. Gottberg, New high-performance spacers in Electro-Dialysis Reversal (EDR) systems, Proceedings of 1998 AWWA Annual Conference, June 21 to 25, 1998, Dallas, USA.

[18]   E. Korngold, L. Aronov, and O. Kedem, Novel ion-exchange spacer for improving electrodialysis – I. Reacted spacer. J. Membr. Sci. 138: 165–170, 1998.

[19]   R. Messalem, Y. Mirsky, and N. Daltrophe, Novel ion-exchange spacer for improving electrodialysis – II. Coated spacer. J. Membr. Sci. 138: 171–180, 1998.

[20]   US Patent 3,284,335.

[21]   R. E. Lacey and S. Loeb, Industrial Processing with Membranes. Wiley-Interscience, New York, 1972. ISBN 0471511366.

David Tvrzník, Aleš Černín, Luboš Novák

# 8 Electrodeionization

## 8.1 Basic principles

In electrodialysis, the rate of electrolyte separation in the treatment of dilute solutions is limited, so there are high demands on the installed area of ion-selective membranes. The method is also unable to remove ionisable (dissociable) neutral components from the water. However, when processing solutions with low conductivity, the mass transfer in electrodialysis stack may be significantly enhanced if diluate chambers of the electrodialysis stack are filled with mixed bed, a mixture of strong acid cation-exchange resin and strong base anion-exchange resin in the form of spherical beads.

The described hybrid process combining electrodialysis (ED) with ion-exchange resins in a single device is called **electrodeionization** (EDI) or **continuous (electro)deionization** (CEDI). It is used for a deep demineralization of low conductivity solutions. Its typical application is the preparation of high purity and ultra-pure water with conductivity from 0.055 to 0.2 μS cm$^{-1}$ and with residual content of sodium, chlorides, dissolved (reactive) silica and other contaminants at levels meeting the requirements of most industrial applications. The technical implementation of the process is an **electrodeionization module** (hereinafter referred to as **EDI module**).

The advantage of this hybrid separation process was first shown by Kollsman [1] and Walters [2] in the 1950s. Until the second half of the 1980s, the EDI process was only subject to scientific works and academic studies, for example, [3–5]. The very first commercial EDI modules and systems were introduced to the market under the trade name Ionpure by Millipore Corporation only in 1987. The rise of EDI was linked to the development of another membrane separation process – reverse osmosis (RO), which gradually became the exclusive step of pre-treatment of feed water for EDI.

The mass transfer intensification mechanism in EDI, as compared to ED, can be explained as follows. The type of ion-exchange resin bead in the given position of the mixed bed is only expressed by the probability given by the volume fraction of the relevant type of ion-exchange resin in the mixed bed. However, from the idealized image of the monodisperse ion-exchange resin bed, with individual beads organized in the closest hexagonal arrangement, it can be deduced that the vast majority of cation-exchange resin and anion-exchange resin beads form the relevant continuous zones in mutual contact across the diluate chamber. At statistical processing of the mixed bed issue, the reality rather corresponds to the situation indicated in Fig. 8.1. As ionic conductivity of ion-exchange resins is higher compared to diluate, the quality of which varies from ultra-pure water to RO-permeate, there is an alternative path for electric current passage through the ion-exchange resin bed (solid phase) in diluate chambers, instead of diluate, as also shown in Fig. 8.1. The electric current enters through the anion-selective membrane into the continuous anion-exchange resin zone across

https://doi.org/10.1515/9783110739466-010

the diluate chamber, passes through this zone and gradually passes through bipolar anion-exchange resin/cation-exchange resin interface into the continuous cation-exchange resin zone across the diluate chamber, from which it enters – through the cation-selective membrane – into the adjacent concentrate chamber. In fact, electric current also passes through the anion-selective membrane/cation-exchange resin and anion-exchange resin/cation-selective membrane bipolar interfaces. The electrolyte separation not only takes place in the diluate/ion-selective membrane interface, as in the case of ED, but also on the surface of ion-exchange resin beads, thus increasing the rate of separation. The separation of components from the diluate consists in the transport of ions and ionisable components in the diluate by diffusion and migration to the solid phase surface, their adsorption by solid phase with the conversion of neutral components to their conjugate ionic forms, their transport by migration towards the relevant ion-selective membrane and, finally, their transfer into the adjacent concentrate chamber.

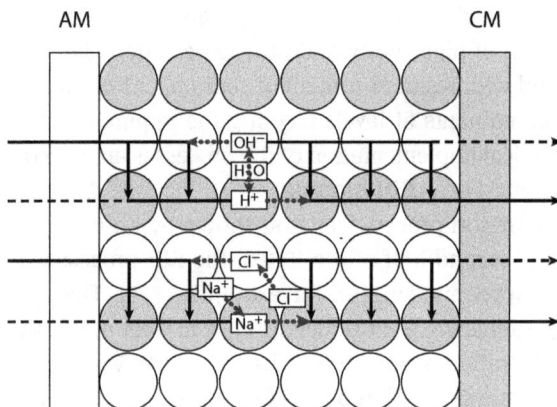

**Fig. 8.1:** Simplified and idealized diagram showing the arrangement of cation-exchange resin and anion-exchange resin beads and the passage of electric current (⟶) and components ( ---▸ ) through the mixed bed in diluate chambers of EDI module.

The process can be operated in two modes: ED and EDI. The second mode is also known as electro-regeneration. In ED mode, the electrolyte is separated from the liquid phase especially in the vicinity of bipolar contacts, in between of ion-exchange resin beads, due to the flow of electric current through these interfaces. However, it has been found that even in this mode, it is possible to achieve high degree of demineralization and – with high current efficiency – to prepare a product with conductivity in the order of $10^{-1}$ μS cm$^{-1}$. In terms of product quality, it is more advantageous to operate the module in EDI mode, where water is split on bipolar interfaces into H$^+$ and OH$^-$ ions, which regenerate cation-exchange resin and anion-exchange resin and allow the ion-exchange resin packing of diluate chambers to act as a continuously electrochemically regenerated column filled with a mixed bed. From the thermodynamics point of view, the operation in EDI mode requires a difference of electrical potential between both continuous ion-exchange resin zones of at least 0.83 V. Another advantage of the EDI mode is that the separation of electrolyte from the flow phase runs over the entire active surface of ion-exchange resin beads, not just in the vicinity of bipolar contacts in between ion-exchange resin beads. This significantly increases the rate of mass transfer from liquid to solid phase. Moreover, the regenerated ion-exchange resin beds separate neutral conjugate forms of some ions from the liquid phase, in particular $CO_2$, $SiO_2$, $H_3BO_3$, $NH_3$, etc.

The water-splitting phenomenon was studied by a number of authors on ion-selective, especially anion-selective membranes subject to overlimiting current loads. The existence of this phenomenon is, among others, the basis of the function of bipolar membranes. From theories explaining this phenomenon, the most frequently mentioned is the Second Wien Effect or the catalytic effect of protonation and deprotonation reactions of the following type:

$$B + H_2O \rightleftharpoons BH^+ + OH^-$$
$$BH^+ + H_2O \rightleftharpoons B + H_3O^+$$

$$(8.1)$$

This mechanism was discussed, for example, by Simons [6] and today, it seems to be the most likely explanation of the reality observed. From the work of the above mentioned author, an important conclusion can be drawn – that the quaternary ammonium groups of strong base anion-exchange resins ($-N^+R_3$) do not catalyse the water-splitting. Only the tertiary ammonium groups ($-N^+HR_2$) resulting from their reduction, or hydroxy groups ($-OH$) are capable of doing that. As for the cation-selective membranes, water splitting was only observed under certain conditions, for example, in the presence of alkaline earth metal ions ($Ca^{2+}$, $Mg^{2+}$) or $NH_4^+$ ions, which is, considering the possible reactions, in conformance with the mechanism, according to the following equations:

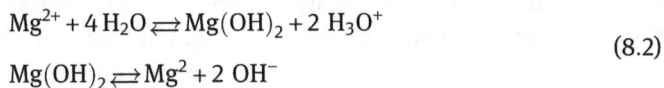

$$Mg^{2+} + 4\,H_2O \rightleftharpoons Mg(OH)_2 + 2\,H_3O^+$$
$$Mg(OH)_2 \rightleftharpoons Mg^2 + 2\,OH^-$$

$$(8.2)$$

Ion-selective membranes and ion-exchange resins are of the same chemical nature, so the conclusions can be generalized for both types of materials. While only strong electrolytes are removed from water in the ED mode, the EDI mode allows removal of the weak electrolytes ($NH_3$, $CO_2$, $SiO_2$, $H_3BO_3$, humic substances, etc.) too. The principle consists in the weak electrolyte ionization by reaction with ion-exchange resin in regenerated state, for example:

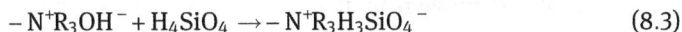

$$-N^+R_3OH^- + H_4SiO_4 \rightarrow -N^+R_3H_3SiO_4^-$$

$$(8.3)$$

and the migration of the resulting ion through the ion-exchange resin bed, just as the strong electrolyte ion. The recent EDI modules are generally capable of removing about 95 to 99 % $SiO_2$ from the feed water in a single pass through the diluate chambers.

## 8.2 Modules with layered ion-exchange resin bed in diluate chambers

The main disadvantage of EDI systems with mixed bed in diluate chambers is the fact that the fraction of active beads decreases with the increasing number of ion-exchange resin layers between the membranes. In order to achieve a high degree of demineralization of the processed liquid, relatively thin diluate chambers (thickness from 2 to 3 mm, max.) had to be used in the past. Due to the characteristically thin diluate chambers, in literature, this EDI conception is sometimes referred to as **thin cell**. The limited size of the active ion-exchange surface area is the cause of rather low output of this type of equipment per unit of the membrane active area. Considering the high price of the ion-selective membranes and spacers, this can make the EDI technology significantly more expensive.

Due to the above reasons, systems were developed in which cation-exchange resin and anion-exchange resin form separate layers or zones in the diluate chamber, along the membrane [7, 8] – Fig. 8.2. An ion adsorbed by an ion-exchange resin bead in a given layer or zone reaches the relevant ion-selective membrane without the risk of being released into the water again, and it is therefore possible to work with a thicker diluate chamber and, at the same time, with a proportionally higher flow rate per unit membrane active area, while maintaining a high level of demineralization of the processed liquid.

In systems where cation-exchange resin and anion-exchange resin form separate layers or zones in the diluate chamber, the ion transport in the ED mode – as in the case of systems with a mixed bed – is simply impossible, because the conditions in the given layer or zone (ion-exchange surface area, low resistance of the environment) for transport of both types of ions, cations and anions, are not equivalent. The only exceptions are bipolar interfaces between the adjacent layers (zones) of ion-exchange resins or bipolar interfaces between the layer of ion-exchange resin and the ion-selective membrane. However, these interfaces represent only a minimum portion of the total number of beads in the diluate chamber. The removal of ions and weakly dissociated components from the water is thus almost exclusively done by the ion-exchange mechanism between the cation-exchange resin in $H^+$ form or anion-exchange resin in $OH^-$ form in the given layer or zone and the flowing liquid, with the occurrence of characteristic pH shifts between individual layers or zones. This is advantageously used to ionize weak electrolytes (especially $SiO_2$) at

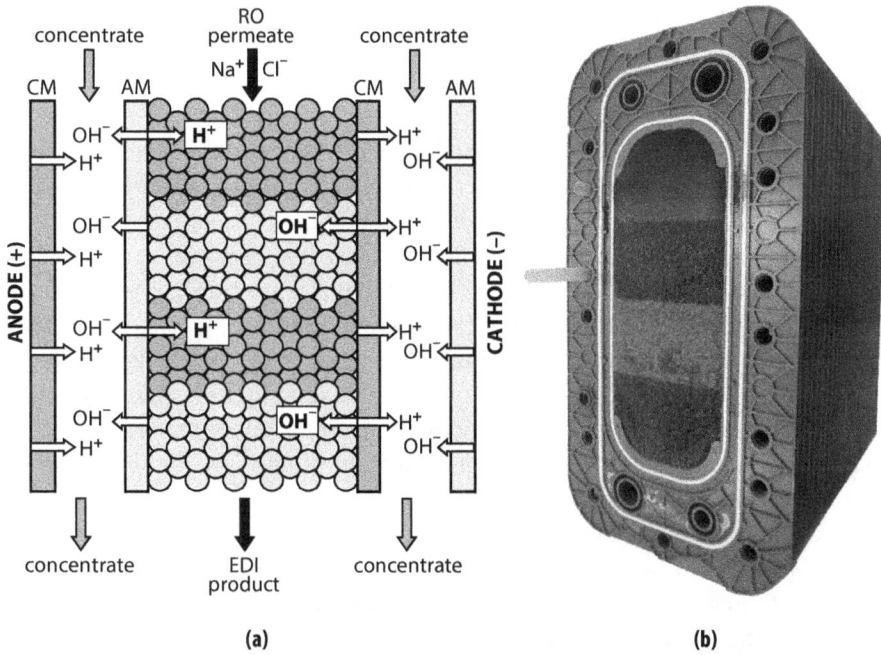

**Fig. 8.2:** (a) EDI principle with a layered bed of ion-exchange resins in diluate chambers. (b) Diluate chamber with alternating layers of cation-exchange resin (dark) and anion-exchange resin (light) in EDI module of Ionpure LX series by Evoqua Water Technologies [9].

their removal. The trapped ions are transported by migration towards the respective membrane and subsequently transferred to the concentrate by the action of the electrical potential gradient. The volume ratios between the cation-exchange resin and anion-exchange resin layers or zones as well as the type of ion-exchange resin should be chosen after taking into account the required level of demineralization, pH neutrality of water coming out of diluate chambers or removal of specific components.

The regeneration of ion-exchange resins in diluate chambers of EDI systems with layered or separated cation-exchange resin and anion-exchange resin beds is most often achieved by water splitting to $H^+$ and $OH^-$ ions due to electric current passing through the bipolar interfaces of anion-selective membrane/cation-exchange resin and anion-exchange resin/cation-selective membrane. The problem of this method of regeneration was originally in the excessive electrical resistance of the aforementioned bipolar interfaces, using the common strong acid cation-exchange resins and strong base anion-exchange resins of type I and heterogeneous ion-selective membranes, based on these ion-exchange resins. A voltage gradient of tens of volts was required per each cell pair to achieve the necessary electric current. Furthermore, a part of the current tended to pass through the bipolar interface between the cation-exchange resin and anion-exchange resin layers or zones, so that in the area of these interfaces, there was a local increase in the current density in the ion-selective membrane, at the expense of its uniform distribution through the membrane and demineralization capabilities of these systems. Later, the strong base anion-exchange resins of type I were at least partially replaced by strong base anion-exchange resins of type II or weak base anion-exchange resins with better capability to catalyse the reaction of water splitting, which resulted in a reduction of the voltage loss in the aforementioned bipolar interfaces [10].

In the case of a layered bed, the distribution of current densities is often influenced by a different electrical resistance of individual layers or cation-exchange resin and anion-exchange resin zones in the relevant ion forms, which, as a consequence, slows down the rate of ion exchange in the layer or zone with higher electrical resistance. This problem is usually solved by adjusting the electrical resistance of one of the layers by adding a suitable dopant – ion-conductive materials such as weak or strong acid cation-exchange resins, weak or strong base anion-exchange resins or inert materials, such as glass beads and inert ion-exchange resins [11, 12].

Alternatively, the EDI module can be divided into two hydraulic or possibly also electrical stages or two EDI modules can be hydraulically connected in series. In this case, the cation-exchange resin and anion-exchange resin beds can be completely separated, and each ion-exchange resin type placed in different hydraulic stage. Other variants are also possible, such as the use of layered ion-exchange resin bed or mixed bed in one of the stages. The liquid is completely demineralized only after it passes through both stages. Each of them can be optimized for separation of different types of components, for example, silica or boron, and contain different number of cell pairs. That is how even the most demanding requirements for the quality of product water can be met, even in industries such as semiconductor manufacture. A precisely defined electrical current passes through each stage, thus avoiding the problems mentioned in the previous paragraph [13, 14].

The only sources of $H^+$ and $OH^-$ ions for the regeneration of relevant cation-exchange resin and anion-exchange resin layers or zones in the diluate chamber are the anion-selective membrane/cation-exchange resin and anion-exchange resin/cation-selective membrane bipolar interfaces. The level of ion-exchange resin regeneration thus decreases towards the opposite membrane, which slows down the ion-exchange rate between solid and liquid phases (separation rate). Such a decrease may be only compensated by applying a higher electrical current, which however increases the power consumption of the EDI process and may even result in degradation of the bipolar interfaces mentioned above. It again points to a certain limitation of the diluate chamber thickness with regard to the increasing degree of ion-exchange resin conversion into salt form, resulting in slower rate of ion-exchange and higher electrical resistance towards the relevant membrane. In the literature, the most often mentioned thicknesses of diluate chambers are 8 to 12 mm for these types of equipment. Due to the relatively large thickness of diluate chambers compared to systems with mixed bed in diluate chambers, in literature, these systems are referred to as **thick cell**.

The larger thickness of diluate chambers allows the use of sealing O-rings and the choice of spacer material only from a mechanical point of view [15–17]. For instance, the Ionpure spacers made of polysulphone allow operating temperatures up to 45 °C and periodic sanitization of the module using hot water at a temperature up to 80 °C [8], while in case of spacers made of polyethylene, undesirable deformation of the material due to mechanical and viscoelastic properties such as creep would be experienced even at temperatures ranging from 35 to 38 °C.

The disadvantage of systems with layered or separated ion-exchange resin bed is the limited area of bipolar contacts between ion-exchange resin and ion-selective membranes. In order to achieve the required current density, in comparison with mixed bed, higher operating voltage would be required, despite the innovations mentioned above. Therefore, the chemical degradation of these interfaces is considerably faster, and the voltage required for reaching the necessary current density may increase over time.

## 8.3 Approaches to mathematical modelling of mass transfer in electrodeionization process

There is still no mathematical model that satisfactorily describes the mass transfer in the EDI process with mixed bed in diluate chambers, especially due to very complex system geometry (polydisperse bead size distribution, imperfect sphericity of ion-exchange resin beads, existence of two solid phases and one liquid phase in the diluate chamber). For the purposes of process modelling, it is necessary to adopt an idealized perception of system geometry, that is, to consider for example, cubic or the closest hexagonal arrangement of ion-exchange resin beads with monodisperse particle size distribution in the space of the diluate chamber. The position of a particular type of bead (cation-exchange resin or anion-exchange resin) is expressed only by a probability. The situation is even more complicated by the ongoing reaction of water splitting on the oriented bipolar interfaces. It is clear from this, that the problem cannot be solved by simply combining the ion-exchange-based mass transfer model with ED.

Considerably simpler is the situation with layered or separated ion-exchange resin beds in diluate chamber, where ions are removed from the solution almost exclusively by the ion-exchange mechanism, which is easy to describe mathematically for the case of single ion-exchange resin bead and film kinetics. Examples of such models are presented, for example, in [18]. To extend this model from a single bead to a volume element of ion-exchange resin bed in diluate chamber, the solid phase specific surface area, $a_P$ (dimension $m^{-1}$) must be introduced. This quantity is defined as the total surface area of solid phase (ion-exchange resin) in the unit of bed volume. The following can be deduced:

$$a_P = 6(1-\varepsilon)/d_p \qquad (8.4)$$

where $\varepsilon$ is the bed void fraction and $d_p$ is the ion-exchange resin bead diameter.

Let us suppose that the flow of ions through the solid phase is only going in the direction of $x$-axis, ion exchange is controlled by film kinetics and the liquid in the diluate chamber is ideally mixed in the direction of $x$-axis. The balance of $i$-th ion removed from the liquid phase in the volume element of the solid phase is defined as follows:

$$\frac{dJ_{i,P,x}(x,y)}{dx} - a_P J_{i,r}(x,y) = 0 \qquad (8.5)$$

where $J_{i,P,x}$ is ion flux in the solid phase in $x$ direction and $J_{i,r}$ is ion flux from the liquid phase to solid phase (ion-exchange rate) in radial direction of the ion-exchange resin beads.

Let us consider the case of $i^{z+}$ cation separation from the liquid phase by ion-exchange mechanism with electrochemically regenerated cation-exchange resin bed in diluate chamber; see Fig. 8.3. Let us assume that the migration is the exclusive ion transport mechanism in the solid phase. The regeneration of the cation-exchange resin bed occurs by water splitting on the anion-selective membrane/cation-exchange resin bipolar interface and by migration of $H^+$ ion through the cation-exchange resin bed, together with $i^{z+}$ ions absorbed from the liquid phase. At the point $x = 0$ (anion-selective membrane/cation-exchange resin bipolar interface, where water splitting occurs), the concentration of $i^{z+}$ ion in the solid phase is zero, which gives us – at the known local current density $j_x$ and considering the eqs. (3.47) and (7.11) – the initial condition for the fluxes of $H^+$ ion and $i^{z+}$ ion in the solid phase $J_{H^+,P,x} = j_x/F$ and $J_{i,P,x} = 0$, respectively. We further define the conversion of cation-exchange resin into the regenerated form as

$$X_{H^+} = -\frac{\bar{c}_{H^+}}{z_M \bar{c}_M} \qquad (8.6)$$

where $\bar{c}$ denotes concentration in the solid phase and subscript M denotes a functional group in the solid phase.

Based on the Nernst–Planck equation, the following shall apply to $X_{H^+}$:

$$X_{H^+} = \frac{\bar{D}_i J_{H^+,P,x}}{\bar{D}_i J_{H^+,P,x} + \bar{D}_{H^+} J_{i,P,x}} \qquad (8.7)$$

where $\bar{D}_{H^+}$ and $\bar{D}_i$ are diffusion coefficients of components in the solid phase. Equation (8.7) holds in the solution where only cations $H^+$ and one another cation (marked with the subscript $i$) are present. At the point $x = 0$, the conversion of cation-exchange resin into $H^+$ form is complete, however it is decreasing in the direction of $x$-axis due to the ongoing ion exchange with the liquid phase.

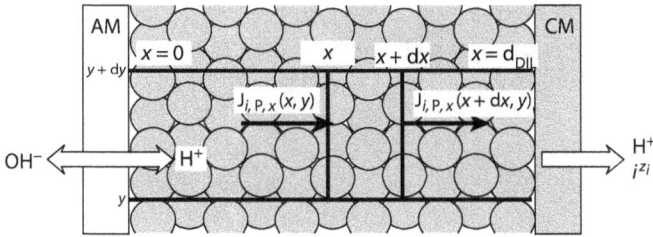

**Fig. 8.3:** Mass balance in ion (cation) exchange resin bed volume element.

As we can express the ion-exchange resin bed conductivity, $\kappa_L$ as the function of conversion into the form with the ions exchanged, the voltage drop in the diluate chamber can be expressed as follows:

$$\Delta\varphi_{\text{DIL}} = -\int_{x=0}^{d_{\text{DIL}}} \frac{j_x(y)}{\kappa_L(x,y)}\,dx \tag{8.8}$$

The same applies to other cell pair components, that is, to the anion-selective and cation-selective membranes and to the concentrate chamber. The local current density, $j_x(y)$ is searched in the same way as described in Section 7.3.2.

To change the ion concentration in the liquid phase of diluate chamber along the membrane (in the direction of $y$-axis), we get the following relation from the balance:

$$\frac{dc_{i,S}}{dy} = -\frac{Wa_P}{\dot{V}_{\text{DIL}}}\int_{x=0}^{x=d_{\text{DIL}}} J_{i,r}(x,y)\,dx \tag{8.9}$$

where $W$ is the diluate chamber width and $\dot{V}_{\text{DIL}}$ is the flow rate through the diluate chamber.

The mass transfer model in the EDI process with separate cation-exchange resin bed and anion-exchange resin bed in the diluate chamber is discussed in the literature, for example, [19–21]. The authors even extended their model by diffusion in liquid phase in the direction of the liquid flow. However, at standard linear velocities of liquid in the order of cm s$^{-1}$, this mass transfer mechanism is rather irrelevant.

Considering the statistical representation of the mixed bed (see e.g. Fig. 8.1), the above-mentioned approach to modelling of systems with layered or separated cation-exchange resin and anion-exchange resin beds can be easily applied to such a case, as well.

For the purposes of modelling the ion-exchange resin bed conductivity, the **porous-plug model** [22, 23] was developed, which assumes the passage of an electric current by three ways: through ion-exchange resin beads separated by a layer of solution (1), through ion-exchange resin beads in mutual contact (2) and through the liquid phase (3).

For the ion-exchange resin bed conductivity, the following relation was proposed:

$$\kappa_L = \frac{a\kappa\bar{\kappa}}{d\kappa + e\bar{\kappa}} + b\bar{\kappa} + c\kappa \tag{8.10}$$

where $\kappa$ and $\bar{\kappa}$ are conductivities of liquid and solid phases, respectively, and $a$, $b$, $c$, $d$ and $e$ are empirical constants to be determined on the basis of experimental tests.

## 8.4 EDI technology designing

EDI technology is usually a part of some larger technological entity. However, this chapter is focused exclusively on EDI technology. EDI technology designing means the selection of suitable types of modules, their arrangement and determination of appropriate operating conditions. To do this, the following input information must be known:
-   feed water temperature and composition,
-   required product flow rate,
-   required product quality (conductivity, residual concentration of contaminants).

Some manufacturers have specialized modules for certain branches of industry, for example, hot water sanitizable (HWS) modules (see Section 8.7) for pharmaceutical industry or modules assuring high removal of selected species or impurities (silica and boron), which is especially important for applications in power generation industry or in manufacture of semiconductors. While selecting the type of module, it is also necessary to thoroughly consider the application field.

For the designing itself, the software tools provided by manufacturers together with EDI modules are used (Evoqua Water Technologies/Ionpure: IP-PRO, Suez Water Technologies & Solutions/E-Cell: E-CALC, SnowPure: EDICAD). Some of them have the form of a self-executable program or a web application, while others are available as Excel spreadsheet. The basic functions of these tools include the implementation of the general water chemistry (balance of cations and anions, dissociation equilibria, estimations of conductivity, etc.), prediction of product quality (conductivity and residual concentrations of certain ions), pressure losses, maximum water recovery, electrical voltage and current and power consumption. Some designing tools allow the choice between co-current and counter-current flow mode for the concentrate, while others are capable of offering the user a suitable type of the module based on

the user-specific requirements as well as to recommend changes in the feed water pre-
treatment for EDI.

The knowledge of the feed water composition is necessary due to EDI process
limits and also because of its impact on the EDI product quality. Table 8.1 shows the
typical specification of the feed water composition for EDI. However, it is a common
practice to describe the feed water by a suitable summary characteristic. In practice,
we mainly see these three values:

- FCE (**feed conductivity equivalent**) (unit: $\mu S\ cm^{-1}$) defined by the following
equation:

$$FCE = \kappa + 2.66\ c_{CO_2} + 1.94\ c_{SiO_2} \tag{8.11}$$

- TEA (**total exchangeable anions**) (unit: $mmol\ dm^{-3}$) defined as a sum of equivalent concentrations (normalities) of anions and those components whose conjugate deprotonated (base) forms are anions by the equation:

$$TEA = \sum_{\substack{i=1 \\ z_i < 0}}^{n} \frac{|z_{i,ef}| c_i}{M_i} \tag{8.12}$$

- TEC (**total exchangeable cations**) (unit: $mmol\ dm^{-3}$) defined as a sum of equivalent concentrations (normalities) of cations and those components whose conjugate protonated (acid) forms are cations by the equation:

$$TEC = \sum_{\substack{i=1 \\ z_i > 0}}^{n} \frac{|z_{i,ef}| c_i}{M_i} \tag{8.13}$$

where $z_i$ is a charge of ion $i$, $z_{i,eff}$ is its effective charge, $\kappa$ is conductivity expressed
in $\mu S\ cm^{-1}$, $c_i$ is concentration of a ion $i$ expressed in $mg\ dm^{-3}$ and $M_i$ is molar mass
expressed in $g\ mol^{-1}$. In the case of ions, $z_i$ corresponds with their actual charge.
For neutral components such as $NH_3$, whose conjugate protonated (acid) form is
a cation, $z_i = 1$ is taken. For neutral components such as $CO_2$, $SiO_2$, $H_3BO_3$, whose
conjugate deprotonated (base) form is an anion, $z_i = -1$ is taken. $z_{i,eff} = z_i$ for all components except for $CO_2$ forms. The value of $z_{i,eff}$ for $CO_2$ forms is determined by the
manufacturer. Evoqua (Ionpure modules) or DOW use $z_{i,eff} = 1.7$ for $CO_2$ forms, while
Suez (E-CELL modules) uses $z_{i,eff} = 1.1$. Both TEA and TEC can be alternatively expressed in the calcium carbonate equivalent units (unit: $mg\ dm^{-3}$ as $CaCO_3$) by
using the following equation:

$$TEx_{CCE} = \frac{TEx\ M_{CaCO_3}}{2} \tag{8.14}$$

where $x$ represents anions (A) or cations (C) and $M_{CaCO_3}$ is the molar mass of $CaCO_3$ (100 g mol$^{-1}$). The key parameters that determine the quality and residual concentrations of contaminants in the product are:
- feed water TEA/TEC/FCE,
- concentration of total $CO_2$ in feed water,
- $SiO_2$ content in feed water,
- boron content in feed water,
- $NH_4^+/NH_3$ content in feed water,
- feed water temperature,
- product flow rate through the module,
- electrical current,
- water recovery.

In the subsequent calculations, the maximum values of TEA and TEC are used, here identified as TEM:

$$TEM = \max\ (TEA, TEC) \tag{8.15}$$

Since the feed water for EDI is almost exclusively RO-permeate, in which the essential components are $CO_2$ forms, including free $CO_2$ in most cases, in the vast majority of cases, TEA value is used for the calculations.

The feed water temperature has a considerable influence on the product quality. It affects the mobility of ions or the diffusion coefficients of individual components, in general, which is related to the temperature dependence of water viscosity. At low temperatures, the mobility of ions and the rate of separation of individual components from the diluate, as well, are rather low. If EDI module produces water with resistivity over 16 MΩ cm (conductivity below 0.0625 µS cm$^{-1}$) at a feed water temperature of 25 °C under certain conditions, the same module may produce water with resistivity in the order of MΩ cm units, if the temperature of water is in the range between 5 and 10 °C. The negative impact of low feed water temperature can be partially compensated by applying a higher electric current, thus increasing the degree of regeneration of ion-exchange resins in the diluate chambers. At higher temperatures, the rate of separation of components from the liquid running through the diluate chambers increases.

The product quality prediction using software tools is not based on the mathematical model of mass transfer in EDI, but uses empirical procedures, instead. It draws information from the extensive database of operational data available to module manufacturers. Over the years, for example, $E$-factor [24] turned out to be the parameter suitable for the product quality prediction:

$$E\text{-factor} = \frac{n\,I\cdot 3\,600}{F\,\dot{V}_P\,TEM} = \frac{n\,I\cdot 3\,600\,M_{CaCO_3}}{2\,F\,\dot{V}_P\,TEM_{CCE}} \tag{8.16}$$

where $n$ is the number of cell pairs, $I$ is electric current in A, $F$ is the Faraday constant in C mol$^{-1}$ and $\dot{V}_P$ is the product flow rate in m$^3$ h$^{-1}$. This parameter expresses the ratio between the applied electric current and the electric current that is theoretically needed to remove 100 % all ions and ionisable components present in the feed water. Considering the typical EDI product quality, this parameter expresses the reciprocal value of the current efficiency (fraction).

The EDI module's product quality is controlled by performance curves similar to those in Fig. 8.4a. Analogical dependences are also used for the removal of selected components (e.g. SiO$_2$), voltage and pressure loss. For completeness, it is necessary to add that, although manufacturers of EDI modules use similar procedures in practice, summary characteristics such as FCE or TEA cannot properly express the actual influence of individual components of feed water on the product quality (e.g. they do not take into account the back diffusion, negative effect of SiO$_2$ or boron concentration in the feed water on product conductivity, especially in the presence of CO$_2$ and HCO$_3^-$).

To describe the hydraulic characteristics of EDI modules, the following well-known equation may be used, considering the thickness of diluate and concentrate chambers and their ion-exchange resin filling:

$$\frac{\Delta p}{L} = \frac{150 \eta v}{d_p^2} \frac{(1-\varepsilon)^2}{\varepsilon^3} + \frac{1.75 \rho v^2}{d_p} \frac{(1-\varepsilon)}{\varepsilon^3} \tag{8.17}$$

where $p$ is pressure, $L$ is the flow path length, $\eta$ is dynamic viscosity of water, $v$ is superficial velocity of the liquid, $d_p$ is equivalent diameter of ion-exchange resin beads, $\varepsilon$ void fraction of ion-exchange resin bed and $\rho$ is water density.

From the linear dependence of the pressure loss within the range of the diluate working flow rates (Fig. 8.4b), it is obvious that the hydraulic characteristics depend on the first member in the eq. (8.17). It is therefore easy to take into account the influence of the feed water temperature on the pressure loss, as the temperature dependence of $\eta$ is known. The example of the dependence of the pressure loss of diluate on E-factor in the EDI module is shown in Fig. 8.4c.

From the scheme in Fig. 8.5, it is obvious that, from the technological point of view, the EDI module works with single inlet (feed water) and three outlets (product, concentrate, electrode solution). The water recovery is defined by the following relation:

$$R_w = \frac{\dot{V}_P}{\dot{V}_F} = \frac{\dot{V}_P}{\dot{V}_P + \dot{V}_C + \dot{V}_E} \tag{8.18}$$

where $\dot{V}_F$, $\dot{V}_P$, $\dot{V}_C$ and $\dot{V}_E$ represent flow rates of feed water, product, concentrate (concentrate blowdown in older EDI modules with concentrate recycle loop) and electrode solution, respectively. Water recovery is often expressed in percent and

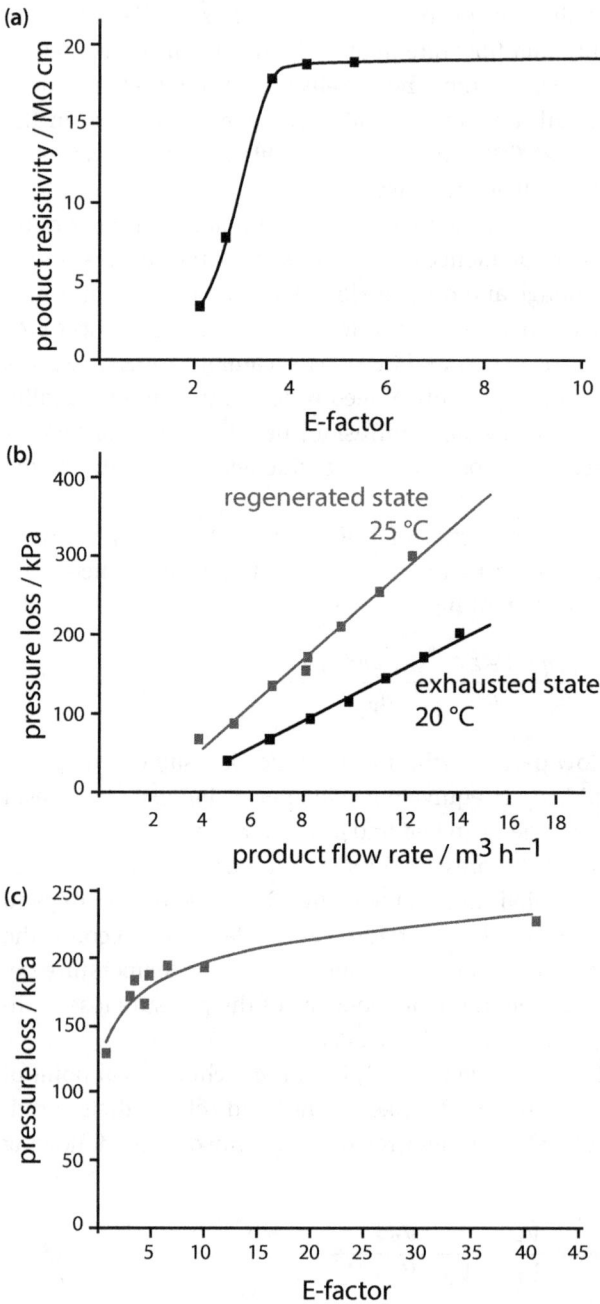

**Fig. 8.4:** Example of demineralization characteristics (a) and hydraulic characteristics (b) of EDI module, dependence of the pressure loss of diluate in EDI module on *E*-factor (c).

limited by the minimum flow rate of electrode solution and concentrate as well as the feed water hardness and concentration of $CO_2$ forms.

## 8.5 EDI modules operation

EDI process is extremely sensitive to feed water composition. Due to the use of ion-selective membranes and ion-exchange resins and the method of fixation of the ion-exchange resins in the flow chambers and the process mechanism, the EDI process is particularly susceptible to plugging by solid and colloidal particles, fouling, scaling and poisoning of ion-exchange materials. In terms of product quality, the process is sensitive to the total content of ionic and ionisable components and particularly to the concentration of $CO_2$ forms in the feed water. Therefore, the use of current EDI modules is almost exclusively restricted to the treatment of RO-permeate, usually from the double pass RO, or demineralized water. Permeate from single pass RO has higher hardness and concentration of $CO_2$ forms, which are usually its main anionic components. This fact increases the risk of scaling and decreases the quality of EDI product. Nevertheless, the EDI process can be also used to process the permeate from the single pass RO. In addition to RO, the pre-treatment may further comprise other processes such as softening (elimination of $Ca^{2+}$ + $Mg^{2+}$). Modern EDI modules, especially HH type modules, (see Section 8.8), are capable of processing the permeate from single pass RO directly, as they are characterized by higher tolerance to the feed water hardness compared to previous modules (2–4 mg dm$^{-3}$ as $CaCO_3$ versus 0.1–1 mg dm$^{-3}$ as $CaCO_3$). Processing of other media, for example, condensates in power or chemical industry, is generally not recommended because the EDI module could easily be contaminated by solid and colloidal particles, organic substances, corrosion products, microorganisms, etc.

In order to avoid scaling, feed water for RO is usually subject to acidification and therefore, the produced permeate is often contaminated with high free $CO_2$ concentrations. In order to reduce the $CO_2$ content in permeate as the feed water to EDI, LIQUI-CEL™ membrane contactors are sometimes used. The pre-treatment of feed water for EDI may further include UV-oxidation or mechanical filtration using filters with absolute efficiency for particles above 0.5 or 1 μm.

Due to the above reasons, strict feed water composition limits are defined for each EDI module. An example is given in Tab. 8.1.

Each EDI module is also characterized by detailed specification of the operating conditions range. A typical example is given in Tab. 8.2. The operating range of the product flow rate varies mostly within the limits of 50 % to 150 % of the nominal value. The selection of the operating product flow rate depends on the feed water composition and the required product quality.

**Tab. 8.1:** Example of feed water specification for EDI module.

| Parameter | Value |
| --- | --- |
| Source | RO permeate |
| TEA (mg dm$^{-3}$ as CaCO$_3$) | < 25 |
| TEC (mg dm$^{-3}$ as CaCO$_3$) | < 25 |
| Temperature (°C) | 5–40 |
| Oxidizing agents (chlorine or chloramine, ozone) (mg dm$^{-3}$) | < 0.05 |
| Heavy metals (Fe, Mn) (mg dm$^{-3}$) | < 0.01 |
| (Hydrogen)sulphides (mg dm$^{-3}$) | < 0.01 |
| Hardness (mg dm$^{-3}$ as CaCO$_3$) | < 1 |
| TOC (mg dm$^{-3}$) | < 0.5 |
| Silica (mg dm$^{-3}$ as SiO$_2$) | < 1 |

**Tab. 8.2:** Example of EDI module operational parameters.

| Parameter | Value |
| --- | --- |
| Product flow rate, minimum–maximum (nominal) (m$^3$ h$^{-1}$) | 5–15 (10) |
| Concentrate flow rate, minimum–maximum (m$^3$ h$^{-1}$) | 0.3–2.0 |
| Flow rate through electrode chamber, minimum–maximum (m$^3$ h$^{-1}$) | 0.1–0.2 |
| Feed water pressure (kPa) | < 700 |
| Product pressure loss[1] (kPa) | 110–250 |
| Product outlet pressure | > Concentrate outlet pressure[2]<br>> Concentrate inlet pressure[3] |
| DC voltage (V) | < 300 |
| Electrical current (A) | < 16 |
| Product quality[4] (as resistivity in MΩ cm) | > 16 |

[1] At nominal product flow rate and 25 °C.
[2] At co-current operation of concentrate.
[3] At counter-current operation of concentrate.
[4] Under conditions determined by the design tool.

The P&ID diagram of a simple system with an EDI module is shown in Fig. 8.5. The permeate is usually fed into the EDI system directly from the RO system. The RO system design must therefore take into account the pressure on the permeate side of RO membrane typically reaching 200–500 kPa. The feed water may alternatively be supplied to the EDI system from the RO-permeate tank using the feed water pump (it is recommended, in this case, to place a safety filter with absolute efficiency for particles above 0.5–1.0 μm before EDI modules). Feed water is then divided as 80 to 95 % for the feed flow to diluate chambers and the rest for the feed flow to concentrate (and electrode) chambers. The product outlet pressure in EDI modules must be maintained at a higher level than the pressure of concentrate, in order to avoid product contamination due to internal concentrate leakage. The EDI system must be provided with interlocks that will turn off DC power supplies if the flow rate in any circuit falls below the minimum value. This is because modules operated under current load with no flow are quickly irreversibly damaged or destroyed. Where magnetic float rotameters are used, this is achieved by means of limit switches. The system is also equipped with a product conductivity measurement, which is the main indicator of quality. In order to avoid the end-user system contamination by a product of poor quality, the EDI system is equipped with two branches for the product pipe, both of which are fitted with automatic ball or butterfly valves, and the output from one or the other branch is controlled on the basis of product conductivity.

Each EDI module is powered by its own DC power supply. EDI modules are typically operated in galvanostatic mode.

Similar to ED, EDI modules may be combined into parallel lines in order to increase the process capacity. These lines are installed on a single frame (skid). By connecting several skids in parallel, even higher product flow rates can be achieved.

The literature often discusses the possibility of recycling the concentrate from EDI back to the feed of RO, thus achieving a certain increase in water recovery in the RO + EDI system. However, the practical limitations of this method result from the permeability of RO membranes for $CO_2$ and boron, which leads to accumulation of these substances in the system and has a negative impact on EDI product quality.

As follows from Fig. 8.5, EDI systems are equipped with connections for **chemical cleaning** (CIP). EDI operation requires occasional chemical cleaning in relation to operating conditions (water recovery), feed water composition (design of RO and pre-treatment – softening) and technological discipline of operators. However, compared to ED systems, EDI requires less regular and less frequent chemical cleaning; the need for cleaning is indicated, for example, by deterioration of product quality at constant operating conditions and feed water temperature and composition, increase in the module electrical resistance, increase of pressure loss, etc. The need for chemical cleaning mainly results from (bio)fouling and scaling (most frequently, $CaCO_3$) on the concentrate side of anion-selective membranes, in the concentrate chambers, on the cathode or in the cathode chamber. The following chemicals are used for the chemical cleaning of EDI modules:

- 2 % hydrochloric acid (scaling removal),
- 1 % sodium hydroxide mixed with 3–5 % sodium chloride solution (fouling removal),
- 0.04 % peracetic acid stabilized by hydrogen peroxide (chemical sterilization).

Sodium percarbonate is sometimes used for the removal of organic fouling and sterilization. In addition to that, there are some preliminary chemical cycles including flushing of EDI module with NaCl solution, to prevent scaling on alkaline cleaning or thermal degradation of the resins on hot water sanitization. Each cycle of chemical cleaning is followed by flushing with water that meets the requirements for feed water. During operation after chemical cleaning, the product must be temporarily

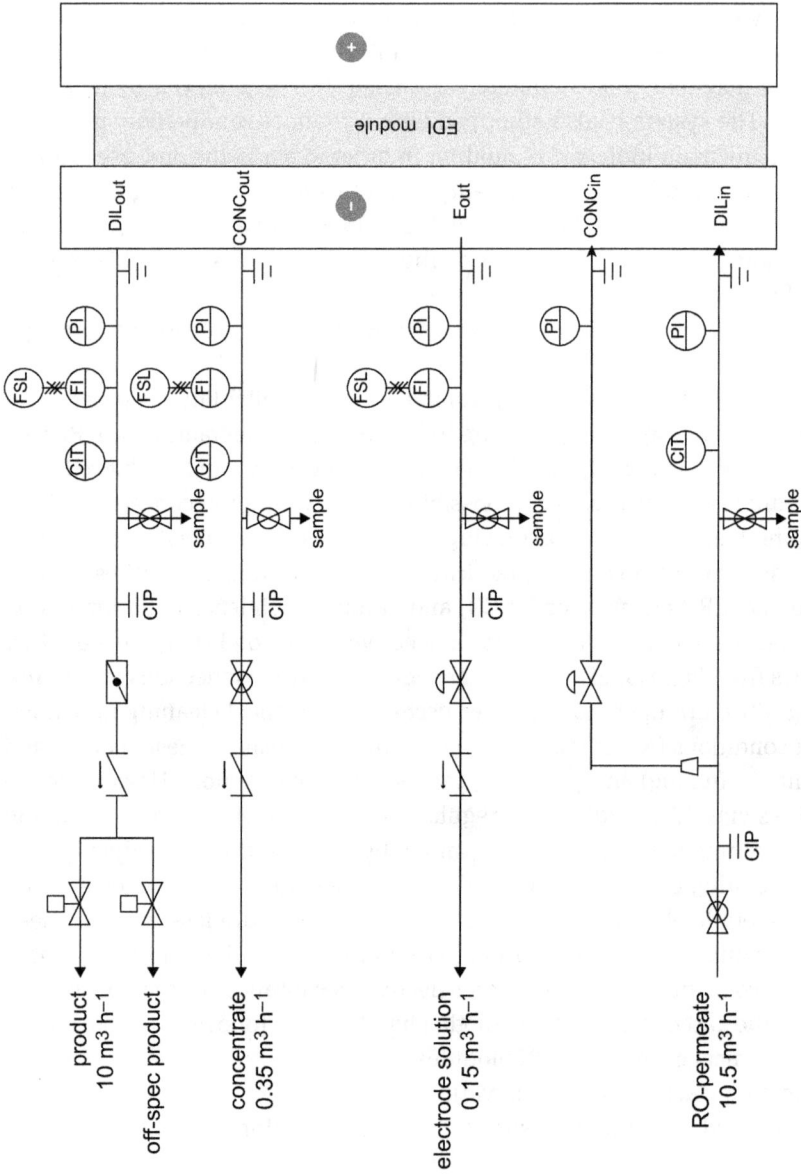

**Fig. 8.5:** Example of P&ID diagram of a system with single EDI module with nominal product flow rate of 10 m$^3$ h$^{-1}$.

drained to waste or recycled back to the feed of RO, until the required product quality is reached again. In modules of HWS type, which are intended for pharmaceutical industry, sterilization is carried out using warm water at temperatures from 80 to 90 °C.

## 8.6 Operating costs

EDI operating costs primarily include the costs for replacement of modules after reaching the end of their service lifetime. In applications, the estimated lifetime of EDI modules varies from 3 to 5 years. Quite often, it is also necessary to replace unreliable DC power supplies for EDI modules. Other operating costs are mainly generated by consumption of electricity needed to power EDI modules. Power consumption of the current EDI modules is less than 0.2 kWh per 1 m$^3$ of product water. Usually it is even lower than 0.1 kWh m$^{-3}$ of product water. With MPure™ 36 module available from MEGA a. s., it is possible to prepare the product with conductivity of 0.0625 μS cm$^{-1}$, at nominal product flow rate of 10 m$^3$ h$^{-1}$, from the model solution of NaCl with conductivity of 20 μS cm$^{-1}$, at the power consumption of only about 0.04 kWh m$^{-3}$ of product water. This holds only if no feed water pump is used, that is, the module is connected directly after RO or no concentrate recirculation pump is used. Otherwise, due to the pressure loss in EDI modules, electricity consumption is significantly higher. The literature [7] says that the typical electricity consumption of a system with EDI modules and mixed bed in diluate chambers ranges around 0.25 kWh m$^{-3}$ of product water, of which two-thirds are the energy consumed by the concentrate recirculation pump. Operating costs are relatively low because EDI systems usually work reliably and require minimum attention.

## 8.7 Plate-and-frame design of electrodeionization module

The design of EDI modules is, in principle, similar to electrodialysis stacks and most industrial EDI modules are of a plate-and-frame type, just as electrodialysis stacks. Compared to ED, the emphasis is on higher mechanical reliability and leak-free operation. EDI modules are designed for much higher operating pressures (now commonly 500–700 kPa) than electrodialysis stacks, which are reflected in their robust construction. The need for EDI modules operation at higher pressure is related to higher pressure loss in EDI modules due to ion-exchange resin filling of flow chambers as well as to the potential need for transporting the demineralized product to a site of consumption without the use of product pump which is – especially in the power industry – often equivalent to high pump heads.

Examples of plate-and-frame modules by leading manufacturers are shown in Fig. 8.6.

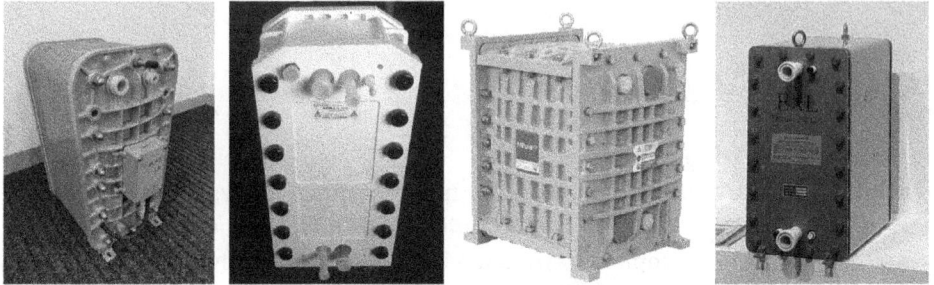

**Fig. 8.6:** Plate-and-frame EDI modules (photo: MemBrain).

The necessary stiffness of spacer frames is usually assured by electrically insulated stainless steel bolts passing directly through the spacer frames. If the design does not provide sufficient rigidity (stiffness) for spacer frames, the EDI module may be mechanically reinforced from outside using metal plates or square tubes or placed in a cylindrical housing or vessel made of fibreglass reinforced plastic (FRP). Today, the prevailing materials of both end and side plates are aluminium alloys or, less often, stainless steel.

As in ED, the plastic parts of the end plates are made of PP or PVC and fitted with electrodes on the inside. The anode is commonly of Ti/Pt or MMO type, while the cathode is made of stainless steel. The space for electrode chambers is created within the end plate or the electrode spacer is used. If the concentrate and electrode solution form a single hydraulic circuit, a concentrate spacer is used for this purpose.

Nowadays, internal and external tightness of the EDI module is not always ensured by the classical sealing effect between the spacer frame and ion-selective membrane, as in ED. In that case, ion-selective membranes are attached to the diluate or concentrate spacer either by suitable adhesives or by welding. However, the progress in module design, especially thick-cell design, allows for the use of sealing O-rings and selection of spacer materials purely from a mechanical point of view. The sealing elements are usually only located in the diluate spacer, while the surface of concentrate spacers is smooth. As an alternative to sealing O-rings, sealing structures are created on the surface of the diluate spacer by applying a layer of thermoplastic elastomer (TPE). Another way is to use raised features on the diluate spacer in combination with a smooth concentrate spacer made of TPE.

Diluate spacers are usually made of natural PP, PP filled with glass fibres or with an admixture of blowing agent – in the latter case, in order to limit the deformation at cooling of the parts in the injection mould. In case of modules sanitizable by hot water at temperatures from 80 to 90 °C, so-called HWS (*hot water sanitizable*) modules, which are intended for pharmaceutical industry, polysulfone (PSU) or polyphenylenoxide (PPO) is used. The material of concentrate spacers may be the same as in the case of diluate spacers. In addition, polycarbonate, polyester, acetal or TPE is used. The selection of material for the concentrate spacer depends on the method of securing the internal and external tightness of the membrane stack.

Ion-selective membranes for EDI are expected to offer especially excellent mechanical properties, low permeability for water and high resistance against back diffusion of electrolytes and neutral components. The most commonly used ion-selective membranes are of a heterogeneous type, based on a thermoplastic blend of ground ion-exchange resin with polymer binder (polyethylene, polypropylene), in most cases, without any reinforcing fabrics. To improve mechanical properties and reduce the effect of back diffusion, the active component (ion-exchange resin) content in the membrane is reduced approximately to a half. Giuffrida [25] presented the benefits of using LLDPE as a binder in ion-selective membranes of this type. However, homogeneous types of ion-selective membranes based on copolymer of styrene with divinylbenzene are also used. Ion-selective membranes are loosely inserted, glued or welded to the spacers in EDI modules.

# 8.8 Classification of modules for electrodeionization

EDI modules may be classified on the basis of the following criteria:
- module design,
- type of ion-exchange resin filling(packing) of diluate chambers,
- diluate chamber thickness,
- type of concentrate chamber filling,
- number of the liquid passes through diluate or concentrate chambers,
- method of electrode solution operation.

Most EDI modules are – similar to electrodialysis stacks – designed as a plate-and-frame type (see Section 7.5). These plate-and-frame EDI modules have proven their usefulness in practice, but they are also criticized because of some disadvantages, particularly the complex and costly construction of working spacers (inlets and outlets including liquid flow distributors in each spacer, sealing elements, the method of fixation of ion-selective membranes and packing ion-exchange resin beds), possibility of internal leakages, complex system of inlets and outlets, the necessity for individual filling of working chambers with ion-exchange resins, impossibility of changing the ion-exchange resin bed without disassembling the module (in most cases, taking apart the module means destruction of the module), larger size, higher weight and loss of electrical power due to current leakage [26–32].

Therefore, **spiral wound EDI modules** [26–32] were developed (see Fig. 8.7). This variant of EDI process is sometimes referred to by the abbreviation **SWEDI (spiral wound EDI)**. A typical spiral wound EDI module consists of a central cathode in the form of a stainless steel tube, which also serves as an inlet and outlet for the concentrate. A membrane stack consisting of a concentrate spacer of a mesh type, to the frame of which the cation-selective membrane is attached on one side and the anion-selective membrane on the other side is wound around the cathode. The space of diluate chambers is defined by longitudinal ribs. The membrane stack spiral assembly is housed in a tubular stainless steel or FRP housing or vessel, which gives the module the necessary mechanical stability and eliminates external leaks of the module. The tubular housing or vessel is fitted with an anode on the inside. In contrast to plate-and-frame modules, SWEDI modules are filled with ion-exchange resins only after the assembly is finished. To fix the bed, there are strainers above and below the stack, which also serve as flow distributors. The occurrence of internal leakages is less common than in plate-and-frame EDI modules. They also require a lower footprint.

The advantages and disadvantages of both EDI module concepts are widely discussed. However, it is evident that the disadvantage of spiral wound EDI modules is particularly the uneven distribution of current density in radial direction of the membrane stack (lower in the peripheral part of the module and higher in the centre of the module).

Diluate chambers of EDI modules are filled with mixed-bed or layers of cation-exchange resins and anion-exchange resins arranged in the direction of the liquid flow. Other arrangements are rare and have no practical meaning. The thickness of diluate chambers is related to the mechanism of separation, based on the type of ion-exchange resin filling used in diluate chambers, and was discussed in Sections 8.1 and 8.2. It needs to be noted that the issue of using the mixed-bed in diluate chambers of greater thickness (8 mm) was later successfully resolved by Asahi Glass Company and Glegg Water Conditioning [33].

All EDI modules used had originally in their concentrate chambers the filling in the form of a mesh, similarly to ED. The thickness of these spacers was similar to working spacers in ED (e.g. SnowPure uses in its EDI modules the concentrate spacers with a thickness of 0.7 mm). In such a case, the electrical resistance of EDI module is controlled by the conditions in the concentrate chambers. The application of EDI technology is almost exclusively limited to the processing of RO-permeate, in practice, usually from double pass RO, with a low conductivity (4 to 10 $\mu S\ cm^{-1}$). Due to the limitation of water recovery in EDI modules because of minimum flow rates of concentrate and electrode solution or the risk of scaling, the steady-state conductivity of the concentrate may be very low. The concentrate conductivity may also be very low after a longer shutdown. It was therefore necessary to operate the concentrate in the feed and bleed mode, and also to inject NaCl into the concentrate in order to achieve the required conductivity of 100–1,000 $\mu S\ cm^{-1}$. The need to add chemicals is however in

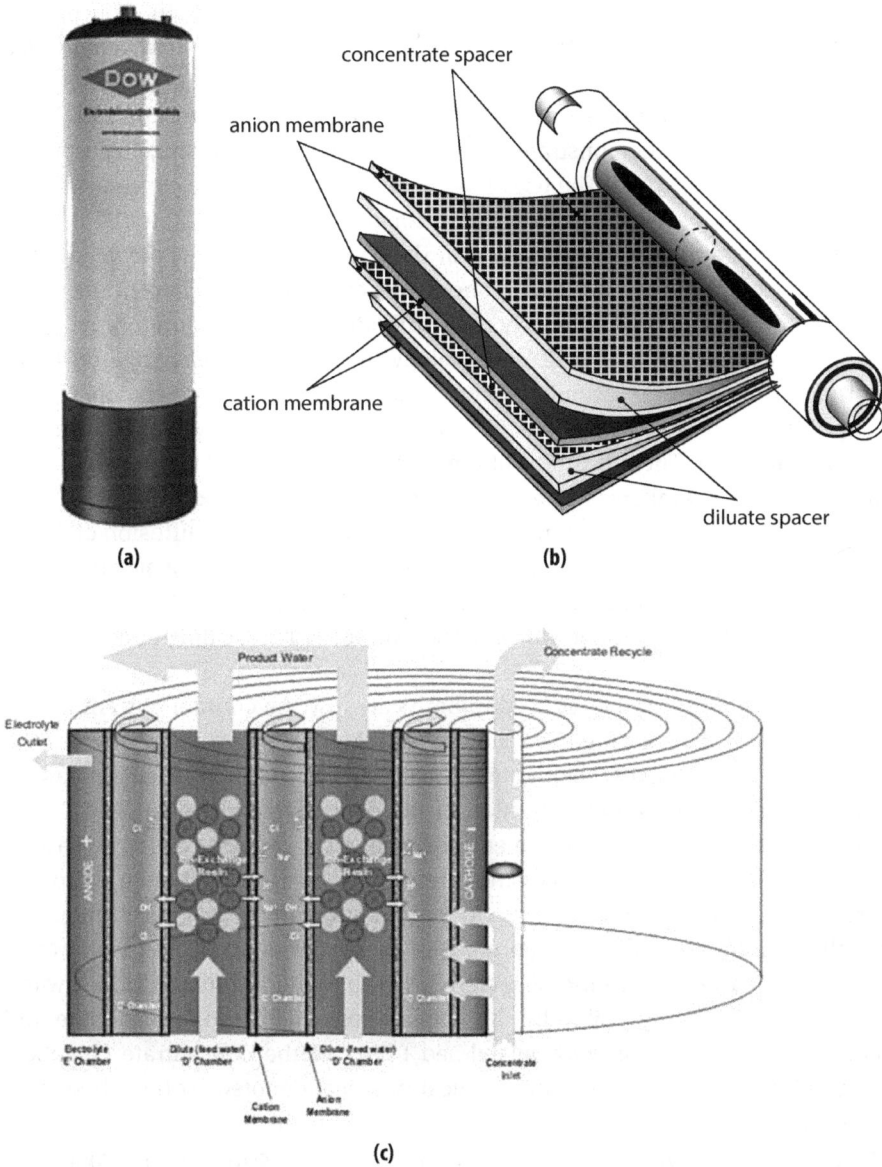

**Fig. 8.7:** Spiral wound EDI module by DuPont™ (Omexell) company (a), its internal structure (b), and diluate and concentrate flow through the spiral wound EDI module (c).

direct conflict with the philosophy of EDI technology as a process that does not require any chemicals for operation, in contrast to conventional ion-exchange technologies. The requirement for the use of recirculation pump and injection of NaCl makes the operation more expensive, reduces the reliability of EDI systems and increases the consumption of electricity (most of the energy is consumed by the concentrate recirculation pump). The concentrate stream with high salinity and conductivity was also a source of leaks and incrustation on the module surface, thus increasing the potential risk of module burning due to leakage electric current (shunt current).

In 1994, Ganzi et al. demonstrated the advantages of ion-exchange resin filling of concentrate chambers [34]. The use of recirculation pump for the concentrate and NaCl injection into the concentrate were no longer necessary, as the EDI module electrical resistance became independent of the concentrate conductivity. In modules with ion-exchange resin filling of concentrate chambers, the concentrate is operated in a single pass mode, similarly to the diluate. The vast majority of modules available from world leaders in EDI technology today use ion-exchange resins in concentrate (and electrode) chambers. The thickness of concentrate chambers usually ranges from 3 to 4.5 mm. In practice, mixed-bed or cation-exchange resin are most commonly used. The advantage of the first variant is higher resistance to scaling in concentrate chambers, while the disadvantage is the back diffusion of $CO_2$, which may result in product contamination. The problem of back diffusion of $CO_2$ is significantly suppressed by using a pure cation-exchange resin filling of concentrate chambers. However, at co-current operation of diluate and concentrate or at a higher feed water hardness, this variant brings in the extreme risk of scaling in concentrate chambers.

The risk of scaling can be reduced in both cases by the counter-current operation of diluate and concentrate [35]. The $Ca^{2+}$ and $Mg^{2+}$ ions are removed from the diluate in the inlet portion of the diluate chamber, and at counter-current operation, they are transferred to the outlet portion of the concentrate chamber. The risk of scaling affects only this part, not the entire concentrate chamber as in the case of co-current operation. The EDI process requires higher pressure of diluate compared to concentrate, in order to avoid the product contamination by the concentrate. Due to this, the counter-current operation requires higher pressure of feed water compared to the co-current operation. EDI modules with higher tolerance to feed water hardness, usually made possible by the counter-current operation of diluate and concentrate, in combination with mixed-bed filling of the concentrate chambers and reduced water recovery, are, in technical practice, denoted by the HH symbol (high hardness).

Most of the current EDI modules are operated in a continuous single-pass mode, both in terms of diluate flow and concentrate flow. However, the liquid flows through the given type of the flow chamber multiple times in some cases, that is, these are multi-pass systems. As for the diluate, there are basically two reasons for this solution. One is the possibility to control the operation mode (ED versus EDI regime). In such a case, the module is divided into two hydraulic and electric stages, where the possibility of independent control of the electrical current in the first stage allows reduction of the risk of scaling in concentrate chambers [36]. The second reason was mentioned in Section 8.2 – it consists in the possibility of optimizing the

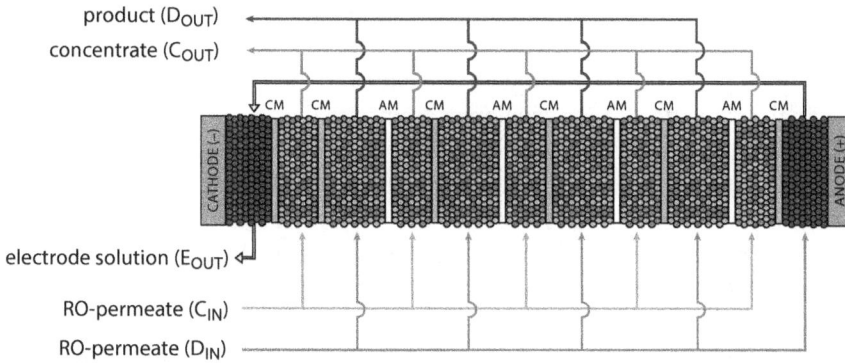

**Fig. 8.8:** Example of internal hydraulic arrangement of EDI module. Diluate chambers are delimited by AM on the anode side and CM on the cathode side, concentrate chambers are delimited by CM on the anode side and AM on the cathode side, electrode chambers are delimited by electrodes and the adjacent membranes.

ion-exchange resin bed in each stage for the separation of certain components [13]. In the case of diluate, the application of multi-pass arrangement is limited by large pressure loss of the liquid passing through the diluate chamber. The reason for the multi-pass operation of the concentrate is – in consideration of the requirement for minimum flow rate of concentrate through each chamber (kinetics, heat dissipation) – the effort to increase the total water recovery. It is obvious that the applicability of such a solution is significantly limited by the feed water hardness.

EDI modules from various manufacturers differ in the way the stream passing through the electrode chambers is treated. Some modules have an electrode circuit completely hydraulically separated, some have the electrode chambers feed flow branched from the feed flow to diluate or concentrate chambers (Fig. 8.8), while others have the only hydraulic circuit with the concentrate (the electrode and concentrate chambers are hydraulically connected in parallel). The specific solution has impact on the number of hydraulic connections of the module (four to six). The concentrate and the electrode chambers have different demands on the liquid flow rate, not only because of their different thickness, but also due to the fact that the flow through the electrode chamber must also secure a reliable removal of gases generated on electrodes at electrode reactions. Where the electrode solution is at least partially separated from diluate or concentrate, it is possible to choose between the parallel inflow to electrode chambers or their connection in series. In the latter case, the liquid flows through the anode chamber first, where it is acidified, and then through the cathode chamber, where it is neutralized (this variant is again illustrated in Fig. 8.8). This solution helps to avoid scaling on the cathode. Due to the requirement for the minimum flow rate through each electrode chamber, the second solution increases the water recovery, as the total flow rate of electrode solution is only half the parallel inflow.

## 8.9 Electrodeionization benefits and competitiveness

The disadvantage of mixed-bed columns in conventional ion-exchange technologies is gradual exhaustion of ion-exchange resins during the operation. Once the undesirable ions have broken through the bed, the water treatment process must be interrupted, cation-exchange resins separated from anion-exchange resins, and ion-exchange resins regenerated. This entails the necessity of handling some aggressive chemicals – hydrochloric or sulphuric acid and sodium hydroxide, neutralization of their excess volumes and subsequent disposal of high-concentration salt waste water. Moreover, handling these substances increases the risks in the workplace.

Industries requiring continuous supply of demineralized water and using conventional ion-exchange columns must have their demineralization lines duplicated, so at least one of them remains always in operation, while the other one is regenerated. Duplication of demineralization lines obviously increases costs, complexity and size of such systems. Moreover, conventional systems require extra space for storage of chemicals, handling, neutralization and disposal of waste.

In comparison with conventional water demineralization using chemically regenerated ion-exchange columns, EDI provides a number of benefits. In particular, it is a continuous process in which a steady state is maintained by electrical current passing through it (exhaustion and regeneration of ion-exchange resins occurs simultaneously), which eliminates the need for interrupting the operation and using chemicals for regeneration of ion-exchange resins. In EDI, the costs of regeneration and regeneration chemicals are replaced by only small consumption of electricity (see Section 8.6). The elimination of regeneration entails the overall recovery of the entire working environment as well as higher operational safety. At present, when the conditions in workplaces and waste treatment methods are subject to increasingly stringent standards and regulations in the field of safety and environmental protection, storage, use, neutralization and disposal of hazardous chemicals add to hidden costs associated with monitoring and administration. Besides that, vapours, especially from acids, may often cause corrosion of structures and equipment.

As a continuous process, EDI provides constant quality of water produced. Contrary to ion-exchange technologies, no duplication of demineralization lines is required to ensure continuous operation. However, the duplication is often done following a conservative approach, especially in the power generation industry. Due to the separation mechanism, EDI works with several times higher (approx. by an order of magnitude) flow rate per unit volume of the used ion-exchange resin bed, compared to chemically regenerated ion-exchange columns. The EDI process is thus, considering the elimination of infrastructure for storage of chemicals, neutralization and disposal of waste, characterized by significantly less demands on footprint, which results in additional savings. These are the most common reasons for end users switching from the existing ion-exchange technologies to EDI [37]. Waste water from EDI can be, in many cases, recycled back to the feed of RO or reused in other processes, without neutralization or any additional treatment.

EDI is an alternative to conventional demineralization systems using ion-exchange technologies. The competitiveness of EDI in relation to these technologies in terms of investment and operating costs is sometimes subject to discussions in specialized literature (see e.g. [38, 39]). Model cases were used to compare the investment and operating costs of EDI systems and mixed-bed technology for three different system capacities and for three different TDS of processed water [38], and the benefits of EDI technology over the mixed-bed technology were demonstrated in this comparison, especially for systems with lower capacity, whereas in case of higher operating capacity, the costs were comparable. Economic indicators must be however assessed comprehensively in terms of the entire water treatment technology, which does not include only EDI. Thus the situation may change quickly depending on which technologies are compared and the system capacity. Indeed, the following technologies can be considered as competitive:
- ion-exchange technologies (single- to four-stage water demineralization, possibly in combination with $CO_2$ degassing),
- technologies combining RO with mixed bed,
- technologies combining RO and EDI.

Single- or double-pass RO may be used as a source of the feed to EDI. In case the EDI technology is the final step in the water treatment process, it may include softening in the column filled with strong acid cation-exchange resin in $Na^+$ form, NaOH dosing into RO permeate from the first pass, etc. In many cases, water treatment in EDI is, for safety reasons, followed by a mixed-bed polisher or microfilters. Qualified assessment of EDI technology advantages compared to other alternatives would therefore require the development of sophisticated tools that quantify the individual items of investment and operating costs in details. Similar aspirations can be found, for example, in [40].

RO + EDI systems with lower capacity (approx. up to 20 $m^3\ h^{-1}$) are often integrated into one skid only, in order to save costs and simplify the operation management (Fig. 8.9a).

EDI technology is constantly evolving towards higher product flow rate through individual modules, modularity enabling easier integration of modules into modular systems offering higher capacity, mechanical and process reliability, higher product quality and lower costs. Evoqua's Ionpure® VNX module (Fig. 8.9b) has nominal capacity 11.4 or 12.5 $m^3\ h^{-1}$ and VNX-Max module, even 15 $m^3\ h^{-1}$. The costs associated with integration of this module into EDI systems were reduced by placing the membrane stack in a tubular FRP housing, which in addition ensures leak-free operation, by choosing cheap materials for spacers (polypropylene filled with glass fibres and covered by a layer of thermoplastic elastomer) and by smart module design, which reduces footprint and allows easy integration into the system, similar to RO modules [15] (see Fig. 8.9b).

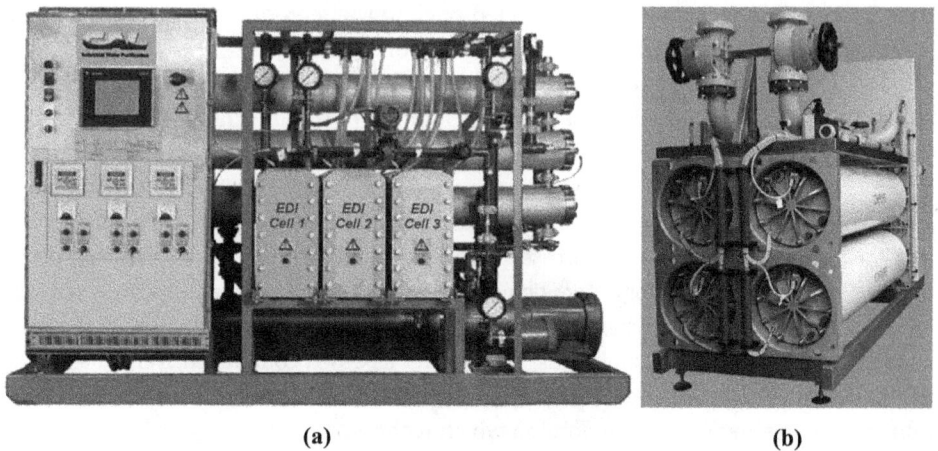

(a)                                        (b)

**Fig. 8.9:** (a) Integrated RO + EDI system by CAL WATER. (b) System assembled from four VNX50 modules by Evoqua with nominal product flow rate of 45.6 m$^3$ h$^{-1}$. Photo author.

# References

[1]   US Patent 2815320.
[2]   W. R. Walters, D. W. Weiser, and I. L. Marek, Concentration of radioactive aqueous wastes – electromigration through ion-exchange membranes. Ind. Eng. Chem. 47: 61, 1955.
[3]   N. P. Gnusin, V. D. Grebenyuk, and S. N. Bazhenova. Zh. Fiz. Chim. 41(5): 1177–1179, 1967.
[4]   Z. Matějka. J. Appl. Chem. Biotechnol. 21: 117–120, 1971.
[5]   V. A. Shaposhnik, A. K. Reshetnikova, and R. I. Zolotareva. Zh. Prikl. Chim. 46(12): 2659–2663, 1973.
[6]   R. Simons, Electric field effects on proton transfer between ionizable groups and water in ion exchange membranes. Electrochim. Acta 29(2): 151–158, 1984.
[7]   J. D. Gifford and D. Atnoor, An innovative approach to continuous electrodeionization module and system design for power applications. Official Proceedings of the 61st International Water Conference, 479–485 (October 2000).
[8]   L. Liang and L. Wang, Continuous electrodeionization process for production of ultrapure water. Semiconductors Pure Water and Chemicals Conference, 2001.
[9]   Service Training Master. Ionpure, 2006.
[10]  US Patent 5858191.
[11]  US Patent 6284124.
[12]  US Patent 6514398.
[13]  US Patent 6649037.
[14]  A. D. Jha and J. D. Gifford, Ultrapure CEDI for microelectronics applications: A cost effective alternative to mixed bed polishers. Ionpure corporate literature.
[15]  A. D. Jha, L. Liang, and J. D. Gifford, Advances in CEDI module construction and performance. Ionpure corporate literature.
[16]  J. Wood, Field experience with a new CEDI module design. International Water Conference, 2003.

[17]  L. Liang, Deionization – Evolution in Design of CEDI Systems. Ultrapure Water 2003
      (13 – 17 October).
[18]  F. Helfferich, Ion Exchange. McGraw-Hill, New York, 1962.
[19]  H. Neumeister, L. Fürst, R. Flucht, and V. D. Nguyen. Ultrapure Water 13(7): 60–64, 1996.
[20]  H. N. Verbeek, L. Fürst, and H. Neumeister. Comput. Chem. Eng. 22(Suppl.): 913–916, 1998.
[21]  H. Neumeister, L. Fürst, and R. Flucht, Ultrapure Water 2000 (April), 22–30.
[22]  M. R. J. Wyllie and P. F. Southwick. J. Petroleum Technol. 6: 44, 1954.
[23]  A. Mahmoud, L. Muhr, and F. Lapicque, Analysis of Electrical Phenomena Occurring in
      Ion-Exchange Assisted Electrodialysis for Treatment of Rinsing Solutions. CHISA, 2004.
[24]  US Patent 5762774.
[25]  US Patent 5346924.
[26]  US Patent 5292422.
[27]  US Patent 5376253.
[28]  US Patent 6190528.
[29]  A. Dey and G. Li, Introduction of spiral wound EDI – exclusive design and its application.
      International Water Conference, 2004.
[30]  A. Dey, SWEDI – a more forgiving electrodeionization technology with higher feed water
      hardness tolerance. Water Condition. Purif. (June):32–40, 2005.
[31]  A. Dey and J. Tate, Deionization – Part 1: A review of spiral-wound electrodeionization
      technology. Ultrapure Water (July – August): 20–26, 2005.
[32]  A. Dey and J. Tate, Deionization – Part 2: A comparison between spiral-wound and
      plate-and-frame EDI technologies. Ultrapure Water (September): 47–51, 2005.
[33]  US Patent 5961805.
[34]  US Patent 5308466.
[35]  US Patent 6149788.
[36]  US Patent 4925541.
[37]  D. Beattie, Using RO/CEDI to Meet USP 24 on Chloraminated Feed Water. Ultrapure Water
      Expo 2001.
[38]  C. Edmonds and E. Salem, Demineralization – An economic comparison between EDI and
      mixed-bed ion exchange. Ultrapure Water (November): 43–47, 1998.
[39]  E. Matzan, P. Maitino, and J. Tate, Deionization – cost reduction and operating results of an
      RO/EDI treatment system. Ultrapure Water (October): 20–23, 2001.
[40]  R. Laflamme and R. Gerard, Technology selection tools for boiler feed water applications
      (High Purity Water). International Water Conference, 2009.

David Tvrzník, Aleš Černín, Luboš Novák

# 9 Capacitive deionization

## 9.1 Basic principle

**Capacitive deionization** (**CDI**, **CapDI**) is an alternative separation process for partial demineralization of water with low power consumption, which has recently been receiving considerable attention [1].

In the simplest arrangement, CDI cell consists of two porous carbon electrodes under direct current voltage from external DC power supply, with water flowing in between them; see Fig. 9.1a. The principle of separation by CDI process is based on the "adsorption" of ions in the electrical double layer at charging of the electrical double layer (see Section 3.5.5), while electrode reactions are considered here as parasitic. This limits the voltage in CDI cells to 1.2 V. In terms of the principle, the process is characterized by two operating cycles: adsorption and desorption. In the adsorption cycle, electrolyte ions in the solution flowing between the electrodes are attracted to the relevant electrode, that is cations to negatively charged electrode and anions to positively charged electrode, and adsorbed in the electrical double layer. In this way, the flowing solution is demineralized. In potentiostatic mode, a characteristic drop of electrical current due to charging of the electrical double layer could be observed. After the electrical double layer is charged, demineralization stops and in the subsequent desorption cycle, the inserted voltage is switched off or the electrical potential of electrodes is reversed – this results in desorption of counter-ions which are released into the solution flowing in between both electrodes to recover them. At the same time, it is possible to recover some of the electricity that was previously supplied to the process during the adsorption cycle. By periodic alternation of these cycles, we get alternating periods of production of partially demineralized feed water and concentrate, as shown in Fig. 9.2.

The problem of charging the electrical double layer in the adsorption cycle is that it occurs not only by adsorption of counter-ions, but also at desorption of co-ions, which reduces the capacity of electrode and slows down the demineralization of the liquid. The problem can be solved by covering the electrode by an ion-selective membrane which will ensure in both cycles the transport of only one type of ions (counter-ions) between the flow phase and the electrode; see Fig. 9.1b. Such a modified process is known as **membrane capacitive deionization** (**MCDI**). Among others, the ion-selective membrane holds co-ions in electrode pores, thus increasing the electrical double layer capacity.

Due to the limitation of the voltage as the main driving force of CDI process, attention is focused on the development of electrodes with maximum specific surface area and capacity of the electrical double layer. Today, mainly porous carbon electrodes are used for these purposes.

https://doi.org/10.1515/9783110739466-011

a)

b)

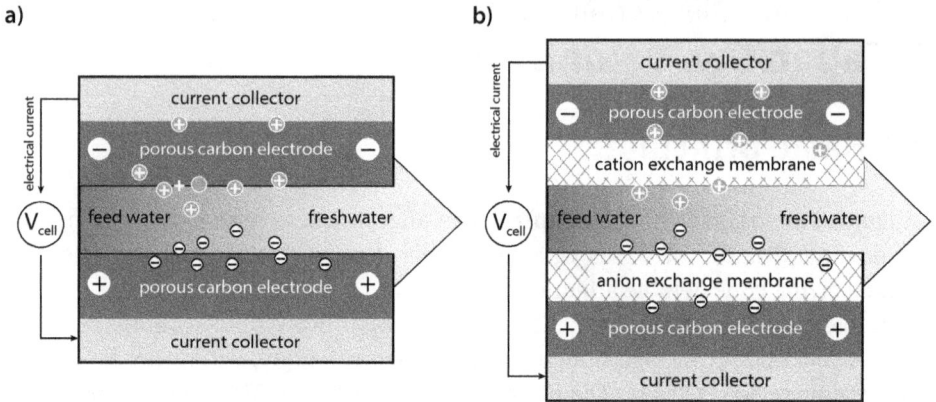

**Fig. 9.1:** Capacitive deionization (CDI) (a) and membrane capacitive deionization (MCDI) (b) [1].

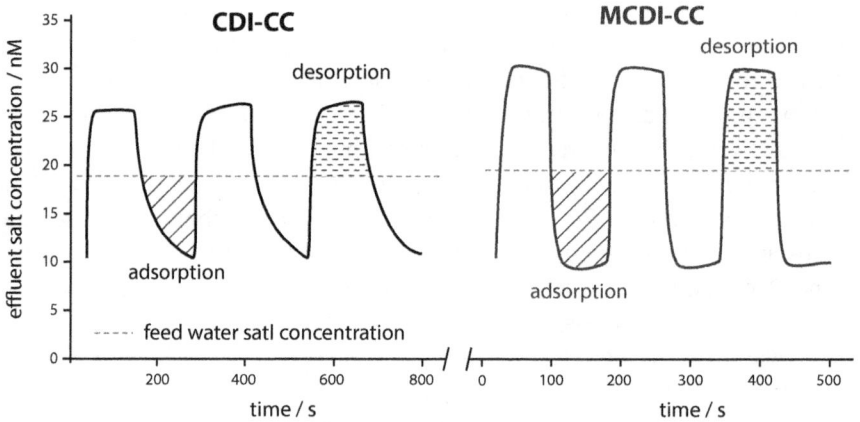

**Fig. 9.2:** Effluent salt concentration for constant-current (CC) operation of CDI and MCDI [1].

# 9.2 Application of capacitive deionization

Over the last two decades, the function of (M)CDI process has been demonstrated under laboratory conditions and at pilot scale, followed by small-scale industrial applications. At present, there are applications with the capacity of hundreds to thousands of $m^3\ h^{-1}$. However, wider utilization of the process in industry is hindered by problems with the manufacture of large-size equipment, problems with insufficient capacity and mechanical stability of porous carbon electrodes, their high price, operating limitations arising from the process principle, open issues with scaling and fouling of ion-selective membranes and inability to clean the electrodes. The (M)CDI process cannot currently compete with conventional membrane separation processes (RO, ED) in terms of investment and operating costs as well as capacity. In addition, another disadvantage is its discontinuous operation. The future extension of the (M)CDI process in a larger scale is therefore conditional on resolving of the problems mentioned above.

The efforts to innovate the (M)CDI process and solve the above problems resulted, for example, in the variant of the CDI process with the flow through electrodes. The substantial improvement of demineralization ability has been achieved in this way compared to conventional flow-by cells [2]. Another innovation of MCDI process uses a suspension of activated carbon microparticles, circulating through "electrode chambers" (space between the power collector and the adjacent ion-selective membrane) instead of solid carbon electrodes. This process is called flow electrode capacitive deionization (FCDI) [3]. The FCDI process allows, at the cost of a certain increase of complexity, achieving continuous operation in steady state. The carbon particles from both "electrode" circuits are mixed at the outlet from the cell, get discharged by their mutual contact, with the electrolyte being desorbed to form a concentrate, and then recycled to the inlet to the cell.

Despite that, (M)CDI systems are commercially offered by many companies, for example, Voltea, Atlantis Technologies, ECO EWP and Current Water Technologies. Most of them have a patented custom solution and use a unique identification of the (M)CDI process. Voltea used MCDI technology to purify cooling tower blowdown and succeeded to reduce the feed water consumption by 22 %, the volume of liquid waste by 40 % and the consumption of chemicals by 77 % at a power consumption only 0.4 kWh m$^{-3}$. As another potential application, the company described the use of MCDI technology for waste water recycling, softening of water for households, preparation of drinking water or irrigation water for agriculture. The company says that MCDI technology is more economical for waste water recycling than RO [4].

Atlantis Technologies refers to its MCDI solution as **radial deionizing** (RDI™) and compared to other solutions, MCDI technology is characteristic by faster demineralization process, lower investment costs and broader range of application. RDI technology is primarily used for treatment of waste water from oil and mining industries with total dissolved solids (TDS) in the order of tens of g dm$^{-3}$. The company says that in case of treatment of water with TDS of 1.6 g dm$^{-3}$, its RDI technology is more economical than RO. Among other possible applications, softening and demineralization of make-up water for steam boilers, demineralization of waste water from flue gas desulfurization and demineralization of retentate from RO are mentioned [5].

ECO EWP (formerly known as Aqua EWP) refers to its CDI technology as **electronic water purifier** (EWP) and its application is again focused on the treatment of waste water with high TDS from oil industry. The company presents EWP technology in container arrangement as the new zero liquid discharge (ZLD) process [6].

Current Water Technologies (formerly known as Enpar Technologies) refers to its CDI technology as **electrostatic deionization** (ESD) and sees the potential application of the process among others in removal of arsenic, nitrates and fluorides from groundwater or ammonium ions from waste water [7].

Manufacturers of cells for (M)CDI seek to profile the process in particular as an alternative to RO for partial demineralization of brackish water with TDS of up to 10 g dm$^{-3}$. However, the perspective is also seen in the treatment of waste water with high TDS (higher than seawater) from oil industry, where conventional technologies such as RO experience problems or cannot be used at all, and also the treatment of retentate from RO, as a part of ZLD technologies (see Section 12.3 or 12.4). Due to undemanding operation and easy control, the use in households (softening, removal of nitrates from water etc.) is also under consideration.

# References

[1]   S. Porada, R. Zhao, and A. Van Der Wal, Review on the science and technology of water desalination by capacitive deionization. Prog. Mater. Sci. 58: 1388–1442, 2013.
[2]   M. E. Suss, T. F. Baumann, and W. L. Bourcier, Capacitive desalination with flow-through electrodes. Energy Environ. Sci. 5: 9511–9519, 2012.
[3]   S. Jeon, H. Park, and J. Yeo, Desalination via a new membrane capacitive deionization process utilizing flow-electrodes. Energy Environ. Sci. 6: 1471–1475, 2013.
[4]   www.voltea.com.
[5]   www.atlantiswater.com.
[6]   http://www.ecoewp.com.
[7]   www.currentwatertechnologies.com.

Zdeněk Palatý
# 10 Diffusion dialysis

Strictly speaking, diffusion dialysis does not belong among electromembrane processes as the ionic transport is not accelerated by an electric field here. Nevertheless, it was decided to include this process among them. The logical reason is that all these processes apply ion-selective membranes, and are suitable for the treatment of solutions containing ions.

## 10.1 Introduction

Dialysis and diffusion dialysis are membrane separation processes based on different diffusivities (diffusion coefficients) and solubilities of components transported through a polymeric membrane. The only difference between them is that diffusion dialysis utilizes an ion-selective membrane (anion-selective or cation-selective). The principle of diffusion dialysis is explained in Fig. 10.1, where an anion-selective membrane separates two electrolyte solutions having different concentrations of components, for example aqueous solutions of hydrochloric acid and sodium chloride. If the concentrations of both components on side 1 are higher than those on side 2, then transport of both components occurs in the direction from side 1 to side 2. Hydrochloric acid and sodium chloride are completely dissociated so that $H^+$, $Cl^-$ and $Na^+$ ions are transported through the membrane. The transport of $Cl^-$ ions is accelerated due to attractive forces in the membrane phase, while the transport of $Na^+$ ions is decelerated.

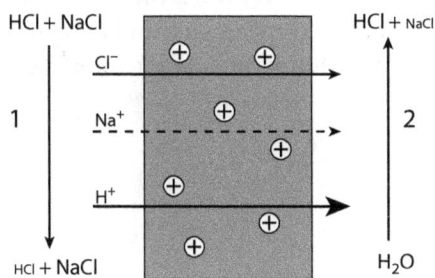

**Fig. 10.1:** Principle of diffusion dialysis with anion-selective membrane.

Hydrogen ions exhibit high mobility not only in aqueous solutions but also in the membrane and are also transported through the membrane. As a result, the acid concentration on side 2 will be higher than that of salt on the same side, whereas the solution on side 1 will be enriched with salt. Thus, diffusion dialysis can be

https://doi.org/10.1515/9783110739466-012

applied in the separation of mixtures containing acids and their salts [1–14]. In a similar way, if a cation-selective membrane is used instead of an anion-selective, this technique can be used to separate mixtures of alkalis and their salts [15, 16].

## 10.2 Transport of one component through membrane

Although diffusion dialysis is used for the separation of liquid mixtures, the studies focused on the transport of one component through the membrane are worth because reference data are obtained for dialysis of multicomponent mixtures. The transport of one component through the membrane can be quantified by an overall mass transfer coefficient, that is, dialysis coefficient which reflects not only properties of the component transported but also properties of liquid solutions on both sides of the membrane. According to the film theory, diffusion boundary layers (DBLs) are created at each liquid–membrane interface. In these layers, a part of the total mass transfer resistance is concentrated – the main part of resistance is in the membrane. Based on this theory, the concentration profiles of the component in the membrane and layers are established under steady state (see Fig. 10.2). Under these conditions, the flux of the component is expressed by the following equations:

$$J = k_{L1}(c_1 - c_{1,if}) \tag{10.1}$$

$$J = k_M(c_{1M} - c_{2M}) \tag{10.2}$$

$$J = k_{L2}(c_{2,if} - c_2) \tag{10.3}$$

where $k_{L1}$ and $k_{L2}$ are the mass transfer coefficients and $k_M$ ($= D_M/\delta_M$) is the membrane mass transfer coefficient ($D_M$ is diffusion coefficient in the membrane and $\delta_M$ is the membrane thickness). The subscript "if" means the solution/membrane interface.

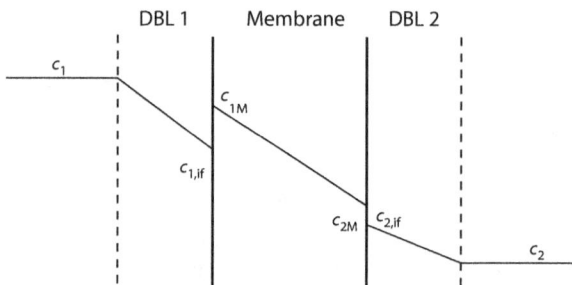

**Fig. 10.2:** Concentration profiles of component in the membrane and diffusion boundary layers.

Equations (10.1) and (10.3) describe the transport through the diffusion boundary layers (DBLs), while eq. (10.2) concerns the transport through the membrane. On both the interfaces, the membrane–solution equilibria are established, which are characterized by the solubility coefficients, $S$

$$c_{1M} = S_1 c_{1,\text{if}} \tag{10.4}$$

$$c_{2M} = S_2 c_{2,\text{if}} \tag{10.5}$$

From eqs. (10.1) to (10.3), it is possible to express the driving forces, that is

$$c_1 - c_{1,\text{if}} = \frac{1}{k_{L1}} J \tag{10.6}$$

$$c_{1M} - c_{2M} = \frac{1}{k_M} J \tag{10.7}$$

$$c_{2,\text{if}} - c_2 = \frac{1}{k_{L2}} J \tag{10.8}$$

After multiplying eqs. (10.6) and (10.8) by the solubility coefficient and subsequent summation of all equations, the component concentrations in the liquid and membrane at the interfaces ($c_{1,\text{if}}$, $c_{1M}$, $c_{2M}$, $c_{2,\text{if}}$) are eliminated, and the following relation for the component flux is obtained

$$J = \frac{S_1 c_1 - S_2 c_2}{\dfrac{S_1}{k_{L1}} + \dfrac{1}{k_M} + \dfrac{S_2}{k_{L2}}} \tag{10.9}$$

If the solubility coefficients are identical ($S_1 = S_2 = S$), then eq. (10.9) becomes

$$J = \frac{c_1 - c_2}{\dfrac{1}{k_{L1}} + \dfrac{1}{S\,k_M} + \dfrac{1}{k_{L2}}} = K(c_1 - c_2) \tag{10.10}$$

where $K$ is the **dialysis coefficient**.

$$K = \frac{1}{\dfrac{1}{k_{L1}} + \dfrac{1}{S\,k_M} + \dfrac{1}{k_{L2}}} \tag{10.11}$$

The product of the membrane mass transfer coefficient and the solubility coefficient is **permeability of the membrane** to the component

$$P = S\,k_M \tag{10.12}$$

## 10.3 Batch cell

The basic information on the course of diffusion dialysis can be obtained from experiments carried out in a two-compartment cell [2, 17–21] (Fig. 10.3). At the beginning of an experiment, compartment 1 is filled with a solution of known composition and volume, while compartment 2 is filled with a stripping solution (mostly deionized water). Both compartments are mixed – the intensity of mixing being the same. Due to the concentration difference of component across the membrane, the unsteady-state transport of the component exists in the direction from compartment 1 to compartment 2 until the equilibrium, characterized by equal concentrations, is reached (Fig. 10.4). At the same time, the transport of stripping solution can occur in both directions.

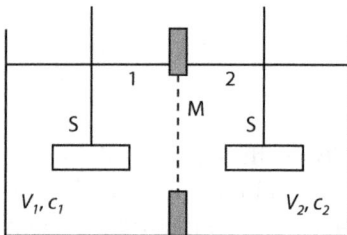

Fig. 10.3: Two-compartment dialysis cell: M – membrane, S – stirrer, 1, 2 – compartments.

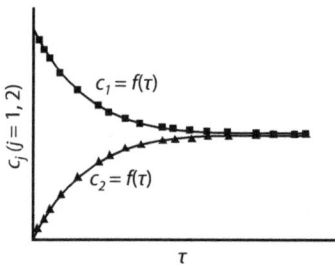

Fig. 10.4: Dependence of component concentration upon time in individual compartments.

From the balance of molar amount of the component in compartments 1 and 2, the following basic differential equations can be derived

$$-\frac{dn_1}{d\tau} = -\frac{d(V_1 c_1)}{d\tau} = KA(c_1 - c_2) \tag{10.13}$$

$$\frac{dn_2}{d\tau} = \frac{d(V_2 c_2)}{d\tau} = KA(c_1 - c_2) \tag{10.14}$$

The initial conditions are

$$\tau = 0; \ c_1 = c_{10}; \ c_2 = c_{20} = 0 \tag{10.15}$$

If the changes in the liquid volumes are negligible, then eqs. (10.13) and (10.14) can be rewritten into eqs. (10.16) and (10.17), which describe dependences of the component concentrations in both compartments upon time:

$$\frac{dc_1}{d\tau} = -K\frac{A}{V_1}(c_1 - c_2) \tag{10.16}$$

$$\frac{dc_2}{d\tau} = K\frac{A}{V_2}(c_1 - c_2) \tag{10.17}$$

By subtracting eq. (10.17) from eq. (10.16), we can obtain new differential equation

$$\frac{d(c_1 - c_2)}{c_1 - c_2} = -KA\left(\frac{1}{V_1} + \frac{1}{V_2}\right)d\tau \tag{10.18}$$

with the initial condition

$$\tau = 0; \; c_1 - c_2 = c_{10} \tag{10.19}$$

The integration of eq. (10.18) in the limits from $c_{10}$ to $c_1 - c_2$ and from 0 to $\tau$ leads to

$$\ln\frac{c_{10}}{c_1 - c_2} = KA\left(\frac{1}{V_1} + \frac{1}{V_2}\right)\tau \tag{10.20}$$

The dialysis coefficient can be obtained by a linear regression of experimental data as shown in Fig. 10.5.

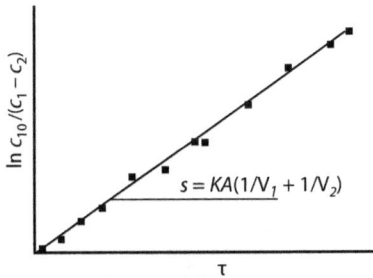

**Fig. 10.5:** Linearization of eq. (10.20).

Generally, the dialysis coefficient is dependent on the intensity of mixing and the initial component concentration in compartment 1. A better characteristic of the dialysis process, which is not affected by mixing, is the **permeability of the membrane** to the component. Its value can be obtained from a series of experiments carried out at a constant initial component concentration in compartment 1 and various rotational speeds of stirrers ($N_1 = N_2$). From eqs. (10.11) and (10.12), it follows that

$$\frac{1}{K} = \frac{1}{k_{L1}} + \frac{1}{P} + \frac{1}{k_{L2}} \tag{10.21}$$

If both compartments are identical and $N_1 = N_2$, then $k_{L1} = k_{L2} = k_L$ and eq. (10.21) can be rewritten as

$$\frac{1}{K} = \frac{1}{P} + \frac{2}{k_L}$$

(10.22)

The liquid mass transfer coefficient, $k_L$, can be estimated from a suitable correlation, for example

$$\text{Sh} = C \, \text{Re}_{\text{m}}^p \, \text{Sc}^q$$

(10.23)

In eq. (20.23), Sh is the Sherwood number, $\text{Re}_{\text{m}}$ is the Reynolds number for mixing, $C$ is a constant, Sc is Schmidt number and $p$ and $q$ are exponents. Assuming average values of kinematic viscosity and diffusion coefficient, the mass transfer coefficient can be expressed as

$$k_L = C' \, N^p$$

(10.24)

where

$$C' = C \, d^{2p-1} \, v^{q-p} \, D^{1-q}$$

(10.25)

and $N$ is the rotational speed of stirrers. In eq. (10.25), $d$ is the diameter of a stirrer, $D$ is the diffusion coefficient and $v$ is the kinematic viscosity.

The permeability coefficient is then obtained by an extrapolation of the dependence $1/K = f(N^{-p})$ to $N^{-p} = 0$. This situation is depicted in Fig. 10.6.

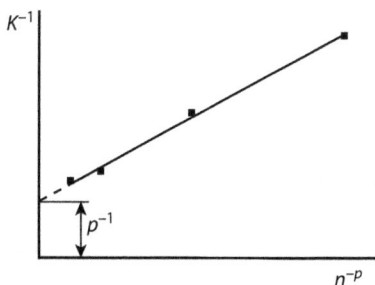

Fig. 10.6: Determination of permeability coefficient.

## 10.4 Continuous dialyser

Figure 10.7 presents the balance scheme of a counter-current continuous dialyser where differential volumes $dV_j$ ($j = 1, 2$) are highlighted. The membrane area in each volume is $dA$. At steady state, the balance of the molar amount of the component in the individual differential volumes is given by

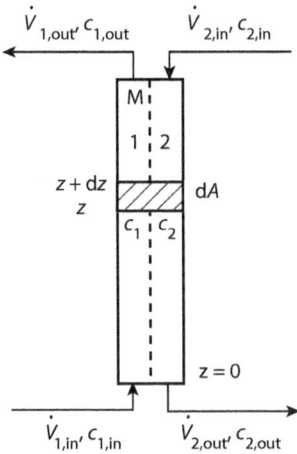

**Fig. 10.7:** Balance scheme of counter-current dialyser.

$$\left(\dot{V}_1 c_1\right)\big|_z - \left(\dot{V}_1 c_1\right)\big|_{z+dz} - K\,(c_1 - c_2)\mathrm{d}A = 0 \tag{10.26}$$

$$\left(\dot{V}_2 c_2\right)\big|_{z+dz} - \left(\dot{V}_2 c_2\right)\big|_z + K\,(c_1 - c_2)\mathrm{d}A = 0 \tag{10.27}$$

Assuming constant volumetric liquid flow rates in the compartments, eqs. (10.26) and (10.27) can be rearranged into the following forms:

$$\frac{\mathrm{d}c_1}{\mathrm{d}A} = -\frac{K}{\dot{V}_1}\,(c_1 - c_2) \tag{10.28}$$

$$\frac{\mathrm{d}c_2}{\mathrm{d}A} = -\frac{K}{\dot{V}_2}\,(c_1 - c_2) \tag{10.29}$$

By subtracting eq. (10.29) from eq. (10.28), it is possible to obtain a new differential equation

$$\frac{\mathrm{d}(c_1 - c_2)}{c_1 - c_2} = -K\left(\frac{1}{\dot{V}_1} - \frac{1}{\dot{V}_2}\right)\mathrm{d}A \tag{10.30}$$

that can be integrated in the limits from $(c_{1,in} - c_{2,out})$ to $(c_{1,out} - c_{2,in})$ and from 0 to $A$. This step results in

$$\ln \frac{c_{1,in} - c_{2,out}}{c_{1,out} - c_{2,in}} = K\left(\frac{1}{\dot{V}_1} - \frac{1}{\dot{V}_2}\right) A \tag{10.31}$$

From the balance equation written over the dialyser

$$\dot{n} = JA = \dot{V}_1\,(c_{1,in} - c_{1,out}) = \dot{V}_2\,(c_{2,out} - c_{2,in}) \tag{10.32}$$

it is possible to express the reciprocal value of $\dot{V}_2$

$$\frac{1}{\dot{V}_2} = \frac{1}{\dot{V}_1} \frac{c_{2,\text{out}} - c_{2,\text{in}}}{c_{1,\text{in}} - c_{1,\text{out}}} \qquad (10.33)$$

and these reciprocal values substitute it into eq. (10.31):

$$\frac{1}{\dot{V}_1} - \frac{1}{\dot{V}_2} = \frac{1}{\dot{V}_1} \frac{(c_{1,\text{in}} - c_{1,\text{out}}) - (c_{2,\text{out}} - c_{2,\text{in}})}{c_{1,\text{in}} - c_{1,\text{out}}} = \frac{(c_{1,\text{in}} - c_{2,\text{out}}) - (c_{1,\text{out}} - c_{2,\text{in}})}{\dot{n}} \qquad (10.34)$$

After a rearrangement, a combination of eqs. (10.31) and (10.34) leads to eq. (10.35) for the total molar amount of the component transferred in the dialyser

$$\dot{n} = JA = K \frac{(c_{1,\text{in}} - c_{2,\text{out}}) - (c_{1,\text{out}} - c_{2,\text{in}})}{\ln \frac{c_{1,\text{in}} - c_{2,\text{out}}}{c_{1,\text{out}} - c_{2,\text{in}}}} A = K (\Delta c)_{\text{m}} A \qquad (10.35)$$

where $(\Delta c)_{\text{m}}$ is the logarithmic mean of driving forces on both boundaries of the dialyser

$$(\Delta c)_{\text{m}} = \frac{(c_{1,\text{in}} - c_{2,\text{out}}) - (c_{1,\text{out}} - c_{2,\text{in}})}{\ln \frac{c_{1,\text{in}} - c_{2,\text{out}}}{c_{1,\text{out}} - c_{2,\text{in}}}} \qquad (10.36)$$

Equations (10.32) and (10.35) are the basic equations for the design of a dialyser. Moreover, they can be used in the treatment of experimental data to calculate the dialysis coefficient.

If a solution containing two components, that is acid (A) and its salt (B), is treated, then the dialysis coefficient for each component is determined. Ratio (10.37), which depends upon the concentration of both components in the feed, is called the **separation factor of a membrane**, $\alpha$. For several systems, the separation factors are summarized in Tab. 10.1 [22].

$$\alpha = K_B / K_A \qquad (10.37)$$

**Tab. 10.1:** Separation factors for selected acid–salt systems.

| System | $c_A \times 10^3$ mol m$^{-3}$ | $c_A \times 10^3$ mol m$^{-3}$ | $\alpha \times 10^2$ |
|---|---|---|---|
| HCl–MgCl$_2$ | 2.0 | 1.0 | 1.26 |
| HNO$_3$–Mg(NO$_3$)$_2$ | 2.0 | 2.0 | 1.72 |
| H$_2$SO$_4$–MgSO$_4$ | 4.0 | 1.0 | 0.67 |
| HClO$_4$–Mg(ClO$_4$)$_2$ | 2.0 | 2.0 | 1.23 |

In the industrial practice, an apparatus for diffusion dialysis is very similar to a conventional filter press or plate heat exchanger. The basic unit is a membrane stack, in which gaskets and ion-selective membranes of the same type are alternately assembled and clamped. In this way, a series of compartments separated from each other by membranes are created. In order to maintain a large concentration difference of components across the membrane, a counter-current flow system is exclusively applied. The membrane stack is schematically depicted in Fig. 10.8.

Diffusion dialysis finds its application in the recovery of acids (mainly HCl, $H_2SO_4$, $HNO_3$, $H_3PO_4$ and HF) from waste waters containing acids and their salts. Several examples of applications are given in Tab. 10.2. Others can be found in review [23]. The applications of the diffusion dialysis process exhibit a series of advantages: (i) low energy consumption; (ii) continuous mode; (iii) easy maintenance; (iv) stable operation; (v) short payback period and (vi) low operating costs.

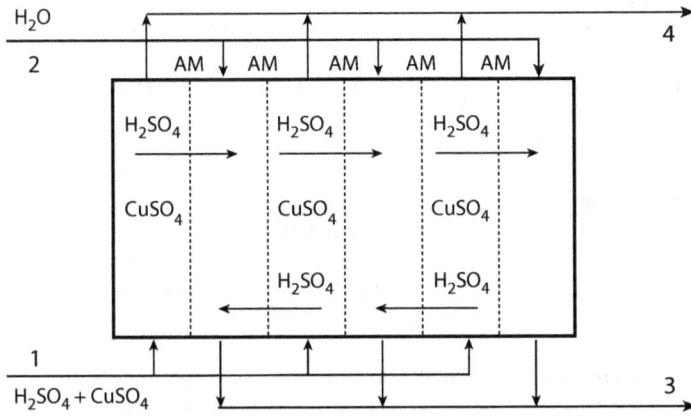

Fig. 10.8: Scheme of continuous dialyser. 1 – feed; 2 – stripping agent (water); 3 – regenerated acid; 4 – waste; AM – anion-selective membrane.

Tab. 10.2: Applications of diffusion dialysis.

| Source | Recovered component |
| --- | --- |
| Spent pickling baths | HCl, $H_2SO_4$, $HNO_3$, HF |
| Refining of waste acid from car batteries | $H_2SO_4$ |
| Waste acids in metal refining | HCl, $H_2SO_4$ |
| Spent aluminium etching baths | NaOH |
| Spent aluminium anodizing baths | $H_2SO_4$ |
| Waste acids in organic technologies | HCl, $H_2SO_4$ |
| Waste alkalis | NaOH |

# References

[1]   Y. Kobuchi, H. Motomura, Y. Noma, and F. Hanada, Application of ion-exchange membranes to the recovery of acids by diffusion dialysis, J. Membr. Sci. 27(2): 173–179, 1986.

[2]   A. Narebska and A. Warszawski, Diffusion dialysis – effect of membrane composition on acid salt separation, Sep. Sci. Technol. 27(6): 703–715, 1992.

[3]   E. Elmidaoui, A. T. Cherif, J. Molenat, and C. Gavach, Transfer of $H_2SO_4$, $Na_2SO_4$ and $ZnSO_4$ by dialysis though an anion-exchange membrane, Desalination 101(1): 39–46, 1995.

[4]   A. Narebska and M. Staniszewski, Separation of fermentation products by membrane techniques. 1. Separation of lactic acid lactates by diffusion dialysis, Sep. Sci. Technol. 32(10): 1669–1682, 1997.

[5]   S. J. Oh, S. H. Moon, and T. Davis, Effects of metal ions on diffusion dialysis of inorganic acids, J. Membr. Sci. 169(1): 95–105, 2000.

[6]   T. W. Xu and W. H. Yang, Sulfuric acid recovery from titanium white (pigment) waste liquor using diffusion dialysis with a new series of anion exchange membranes – static runs, J. Membr. Sci. 183(2): 193–200, 2001.

[7]   T. W. Xu and W. H. Yang, Industrial recovery of mixed acid (HF + $HNO_3$) from the titanium spent leaching solutions by diffusion dialysis with a new series of anion exchange membranes, J. Membr. Sci. 220(1–2): 89–95, 2003.

[8]   T. W. Xu and W. H. Yang, Tuning the diffusion dialysis performance by surface cross-linking of PPO anion exchange membranes – simultaneous recovery of sulfuric acid and nickel from electrolysis spent liquor of relatively low acid concentration, J. Hazard. Mater. 109(1–3): 157–164, 2004.

[9]   J. Jeong, M. S. Kim, B. S. Kim, S. K. Kim, W. B. Kim, and J. C. Lee, Recovery of $H_2SO_4$ from waste acid solution by a diffusion dialysis method, J. Hazard. Mater. 124(1–3): 230–235, 2005.

[10]  J. Xu, D. Fu, and S. G. Lu, The recovery of sulphuric acid from the waste anodic aluminum oxidation solution by diffusion dialysis, Sep. Purif. Technol. 69(2): 168–173, 2009.

[11]  J. Xu, D. Fu, and S. G. Lu, Recovery of hydrochloric acid from the waste acid solution by diffusion dialysis, J. Hazard. Mater. 165(1–3): 832–837, 2009.

[12]  C. Wei, X. B. Li, Z. G. Deng, G. Fan, M. T. Li, and C. X. Li, Recovery of $H_2SO_4$ from an acid leach solution by diffusion dialysis, J. Hazard. Mater. 176(1–3): 226–230, 2010.

[13]  J. Y. Luo, C. M. Wu, Y. H. Wu, and T. W. Xu, Diffusion dialysis of hydrochloride acid at different temperatures using PPO-$SiO_2$ hybrid anion exchange membranes, J. Membr. Sci. 347(1–2): 240–249, 2010.

[14]  J. Y. Luo, C. M. Wu, Y. H. Wu, and T. W. Xu, Diffusion dialysis processes of inorganic acids and their salts: The permeability of different acidic anions, Sep. Purif. Technol. 78(1): 97–102, 2011.

[15]  R. Wycisk and W. M. Trochimczuk, Salt and base separation with interpolymer type carboxylic ion-exchange membrane, J. Membr. Sci. 65(1–2): 141–146, 1992.

[16]  H. Wang, C. M. Wu, Y. H. Wu, J. Y. Luo, and T. W. Xu, Cation exchange hybrid membranes based on PVA for alkali recovery through diffusion dialysis, J. Membr. Sci. 376(1–2): 233–240, 2011.

[17]  Z. Palatý and A. Žáková, Diffusion dialysis of sulfuric acid in a batch cell, Collect. Czech. Chem. Commun. 59(9): 1971–1982, 1994.

[18]  Z. Palatý, A. Žáková, and P. Doleček, Modelling the transport of $Cl^-$ ions through the anion-exchange membrane Neosepta-AFN. Systems HCl/membrane/$H_2O$ and HCl-$FeCl_3$/membrane/$H_2O$, J. Membr. Sci. 165(2): 237–249, 2000.

[19]  Z. Palatý, J. Kaláb, and H. Bendová, Transport properties of propionic acid in anion-exchange membrane Neosepta-AFN, J. Membr. Sci. 349(1–2): 90–96, 2010.
[20]  Z. Palatý and H. Bendová, Dialysis of aqueous solutions of nitric acid and ferric nitrate, Chem. Eng. Process. 50(2): 160–166, 2011.
[21]  Z. Palatý and H. Bendová, Transport of nitric acid through anion-exchange membrane in the presence of sodium nitrate, J. Membr. Sci. 372(1–2): 277–284, 2011.
[22]  M. Nishimura, Dialysis (Chapter 10), in Membrane Science and Technology (Y. Osada and T. Nakagawa Eds.) Marcel Dekker, New York, 1992, 361–376.
[23]  J. Y. Luo, C. M. Wu, T. W. Xu, and Y. M. Wu, Diffusion dialysis — concept, principle and applications, J. Membr. Sci. 366(1–2): 1–16, 2011.

Luboš Novák, Aleš Černín

# 11 Electrophoresis

## 11.1 Introduction

The term electrophoresis is used for description of physical phenomena or effect. Dealing with surface treatment, we call the procedure electrocoating, electrophoretic painting, electrophoretic deposition or electropainting [1–3].

The application potential of membrane technologies in the field of surface treatment is mainly in the process of electrophoretic application of thin organic coatings for anticorrosive protection of steel products or metal products in general. Depending on whether the polymer has a cation or anion form, we distinguish **cataphoresis** and **anaphoresis**. In this technology, the use of ion-selective membranes is crucial for the automatic adjustment of pH, that is, to achieve the required stability of the electrophoretic coating bath.

Electrophoresis is a relatively new and modern technology that belongs to the category of electromembrane synthesis processes. The first industrial use dates back to 1963, when the company Ford used anaphoresis to paint small car body parts. In Europe, it was used for the first time by the company Peugeot in 1967. The intensive development of the organic resins used and the process optimization brought about a relatively short period of fundamental change in technology, in particular the shift to cataphoretic deposition. Cathodic paints were developed by the company PPG in the USA in the early 1970s and industrially applied for the first time in 1975. In Europe, they appeared three years later, again in France, where the company Chrysler–France (current PSA Group) used them to car bodies. At present, the cataphoresis and cataphoretic technology is important for the anticorrosive protection of metal products by organic coatings. The reason is the extraordinary adhesion of such coatings, as well as chemical and climatic resistance of the layer, achieved at the thickness of only a few tens of micrometres. A major advantage is the possibility of application to products with complicated shapes or in cavities and a low sensitivity to mutual interaction or shielding of products treated in one batch. The method allows bulk coating with the high productivity of electrocoat paint application and stable reproducibility of the result with a relatively easy and effectively controlled process and required parameters of the applied protective layer. Also, the workplace and environment pollution with organic solvents and other pollutants is marginal. Due to its properties, the cataphoretic paint is applied as a highly resistant anticorrosive primer. It is usually covered by one or two layers of the final electrocoat paint to achieve the required functional and visual characteristics. In many cases, however, especially on non-visible parts, it remains the final protective layer of the product. Besides its dominant place in the automotive industry, it has its irreplaceable position in the production of other transport technology, construction and agricultural machinery, radiators and consumer goods.

https://doi.org/10.1515/9783110739466-013

## 11.2 Principle and basic description of cataphoretic technology

Cataphoresis uses controlled deposition of organic resin, ion solved in demineralized water. The deposited layer of resin is then cross-linked by high-temperature polymerization. The protective layer can be gained by dipping of the coated product. The process principle is schematically shown in Fig. 11.1.

The basic conditions for the implementation of this technology are:
– organic resin in the cation form in a water solution,
– electrical conductivity of the coated object,
– ability to dip the object into an application bath,
– resistance of the coated products against temperature up to 160–200 °C.

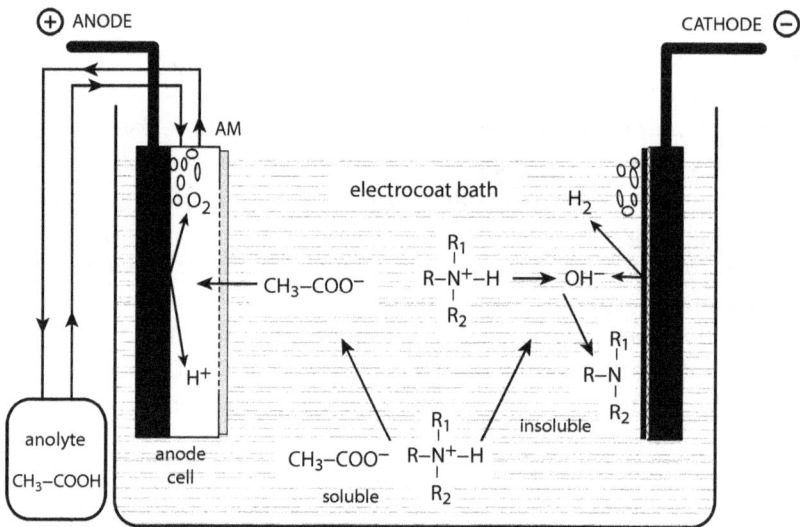

**Fig. 11.1:** Principle of cataphoretic deposition of organic resin.

The water-coating bath consists of the cation form of organic resin itself. The counterion is usually acetate, or ion of sulphamic acid in new systems. After the external direct current voltage (typically from 150 to 350 V) is connected, two processes occur in parallel between the auxiliary electrode and the coated object:
– migration of the both ions to the respective electrode; cation moves to the cathode (coated object) and anion to the auxiliary anode,
– electrolysis of water, producing gaseous hydrogen on the cathode and oxygen on the anode.

For deposition, the process is crucial occurring on the cathode, where, besides gaseous hydrogen, a hydroxide anion is also generated which neutralizes the resin cation immediately at the coated surface and transfers it into a neutral molecular form, insoluble in water. Similarly, hydrogen ions are generated on the anode as another product to oxygen formed by water decomposition. If these were not removed from the functional bath or otherwise neutralized, the pH of the solution would change due to the maintaining of solution electroneutrality. As the ion solution of the resin in water has a limited stability, depending on pH value, the dissolved resin would coagulate eventually and the coating bath would be irreversibly degraded. The problem is effectively resolved by ion-selective membranes, in this case of the anion-selective type, which can only transfer anions and are practically impermeable for water. The acetate therefore migrates through this membrane, which also prevents the transfer of the resulting hydrogen cation to the coating solution, to the auxiliary anode placed in a hydraulically separated anode cell. The concentration of organic acids in the anolyte (or the conductivity of the anolyte) is maintained by automated addition of make-up deionized water. In practice, this system is called the automatic control of the pH of the electrocoat bath, or in short the anolyte circuit.

## 11.3 Advantages of cataphoretic surface treatment

The current dominant position of this technology results from a number of positive technical, environmental and process aspects. The most important ones include:
- electrochemical process working at the molecular level, which substantially affects the properties of the protective film, adhesion, penetrability and so on;
- automatically controlled technological process;
- excellent reproducibility of results;
- relatively simple and objective change of the output quality, especially the thickness of the applied protective layer;
- very small thicknesses of layer (10–35 μm) sufficient for surface protection;
- high anticorrosive resistance at a low thickness of the electrocoat film (approximately 25 μm per 1,000 h in salt mist);
- excellent penetration of the electrocoat primer and protection even in cavities and hardly accessible parts of products;
- extraordinary uniformity of the deposited layer – deviations in the order of several micrometres;
- high productivity – the deposition of paint usually takes 180 s;
- yield close to 100 %;
- minimum share of manual work;

– minimum content of organic solvents in the functional bath, routinely under 1 % (m/m);
– favourable working conditions and very good environmental safety.

Besides the advantages mentioned above, the cataphoretic technology has its limitations. It is only applicable for electrically conductive products, which must be resistant up to the temperature of 200 °C. There is also a very small range of final colours and relatively high power consumption of electrophoresis. Moreover, the operating conditions of majority coating baths require the permanent circulation of process solutions, which additionally increases power consumption even outside the operating time.

## 11.4 Operating application

The operating application of the cataphoretic technology has three basic technological steps:
– surface pre-treatment,
– cataphoretic part itself,
– firing in a baking oven.

The complete delivery of a paint shop must be complemented with other devices and technologies, the most important of which is the conveyor system, neutralization station, ultrafiltration technology and technology of reverse osmosis to prepare demineralized water. The deposition section is shown in Fig. 11.2.

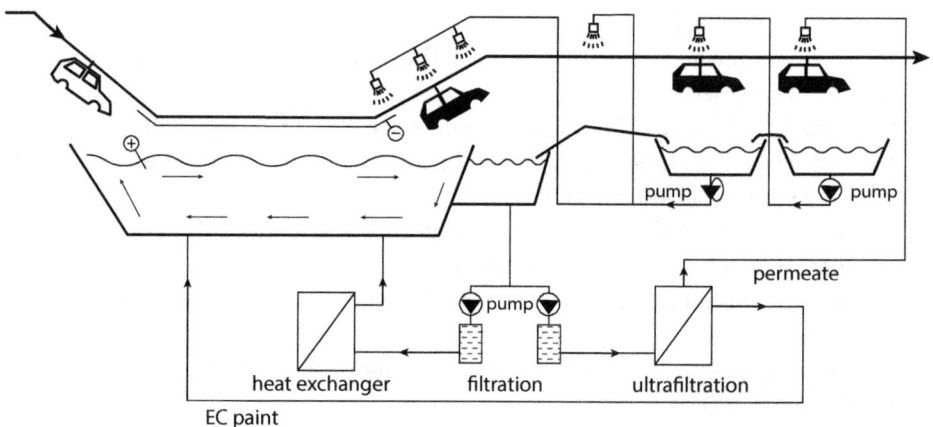

Fig. 11.2: Simplified block diagram of the circuit of cataphoretic paint coating.

## 11.4.1 Pre-treatment of the surface

Cataphoresis would not ensure the expected protection of products without proper pre-treatment, especially degreasing and application of the interlayer to increase the adhesion of the electrocoat paint. Usually, chemical processes are used, in which amorphous ferric phosphate or better-quality crystalline zinc phosphate is formed finally. These treatments secure the best quality of the cataphoretic coating and are required by car manufacturers practically everywhere. More recently, nano-ceramic zirconium interlayers have been used, but their flexibility does not reach yet the level of zinc phosphate treatment mentioned above.

## 11.4.2 Cataphoretic part

This part includes:
- coating of paint – current baths contain 14–17 % of paint (per 100 % of dry matter), up to 1 % (m/m) of solvent and pH of painting bath is maintained in the range of 5.4 to 5.8;
- circulation system (bath and other process solutions) under required temperature conditions;
- automatic adjustment of pH of the painting bath by the use of anode cells with ion-exchange membranes;
- recycling system for paint re-use and rinsing system for washing and stabilization of the product (by ultrafiltration permeate);
- DC power supply – driving force of the electrophoretic process.

## 11.4.3 Baking oven

It is used for the subsequent polymerization of the applied organic resin. It is a commercial product made on the basis of the requirements and nature of the coated object.

## 11.4.4 Circuit of the paint

To avoid sedimentation and maintain homogeneity, the functional bath must be stirred permanently. The maximum interruption of circulation is about 2–4 h. The circuit of the paint consists of the coating bath, storage tank for pumping and circulation pump with distribution piping. Mechanical filters are installed in the pump discharge piping to remove solid impurities, as well as heat exchanger for cooling (or heating) the functional bath, and measuring and regulation elements. The pump discharge piping leads back in the tank through mixing nozzle frames on its bottom. The suction of pumps is driven from the tank itself and from its overflow pocket (used during operation), which ensures defoaming of the upper layer of the functional bath (the foam is generated when the paint drips from the removed product, falls to the surface and aerates it). The circuit of the paint is fitted with basic temperature and pressure sensors to control the technological and technical parameters of the bath and device. Since the deposition process produces considerable heat, it is necessary to cool the bath during operation. This is usually ensured by a plate heat exchanger whose secondary circuit is connected to a cooling unit or another source of cooling water. Cooling is performed automatically.

### 11.4.5 Circuit of the anolyte

The anolyte circuit stabilizes the pH of the coating bath and it is described functionally in the previous text. In practice, it consists of several anode cells placed in the coating bath, circulation tank with a pump and distribution piping. A conductivity meter is installed in the pump discharge piping and is connected with a magnetic valve for control of make-up DI water supply from the distribution pipeline which allows maintaining the set values of the anolyte conductivity. The cells are designed as circuits hydraulically separated by an anion-selective membrane. The anode that is installed in the centre of the cell is usually made of stainless steel. During operation, the anolyte solution must circulate through cells in an amount corresponding to their size and design type.

## 11.5 Design arrangement of electrophoretic anode cells

In Fig. 11.3, common types of anode cells are shown. At present, tubular cells are mainly used. The new product in the market is a seamless extruded tube, whose advantage is its high mechanical resistance without a joint. The design of the most widely used tubular cells and the actual product are shown in Figs. 11.4 and 11.5.

The most important manufacturers of electrophoretic paints are the American company PPG and also BASF, Du Pont and AKZO Nobel. It is positive that the electrophoresis process has a long-time tradition in the Czech Republic and the first cataphoresis in Eastern Europe was built by the company MEGA a. s. in 1992, which is currently the most important Czech company in this area. It supplies and ensures comprehensive servicing of cataphoretic baths for approximately 80 % of operators in the Czech Republic and Slovakia. In the area of turnkey cataphoretic lines, the company MEGA-TEC s.r.o. is very active, producing planar and tubular electrophoretic

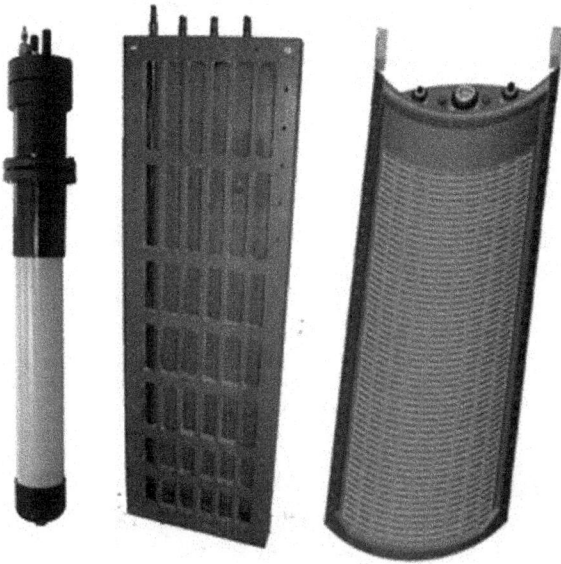

**Fig. 11.3:** Common types of anode cells: (a) tubular, (b) planar, (c) C-type cell (it is a combination of two previous types).

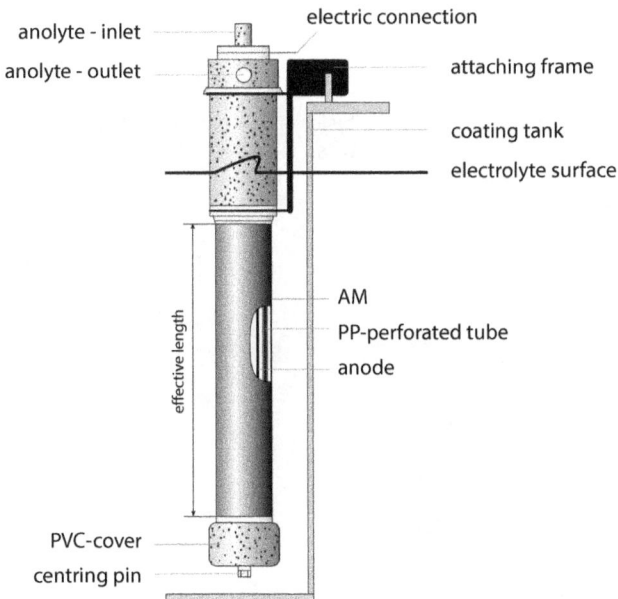

**Fig. 11.4:** Scheme of tubular EFC cell.

**Fig. 11.5:** Tubular anode cell of type EFC-V1 by MEGA a.s. (photo MEGA a.s.).

cells on the basis of RALEX® AMH5E-HD membrane, developed and manufactured by MEGA a.s.

# References

[1]   A. Goldschmidt and H. Streitberger, BASF Handbook on Basics of Coating Technology. Vincentz Network, Hannover, Germany, 2007. ISBN 978-3-86630-903-6.
[2]   Electrocoating Guide, PPG Industrial Coatings, 1999.
[3]   Booklet POWERCRON 6000, PPG, 2009.

David Tvrzník, Aleš Černín, Luboš Novák

# 12 Main industrial applications of electromembrane separation and synthesis processes

## 12.1 Introduction

Since their beginning and after the first commercial success, electromembrane separation processes have played an important role in water treatment and in the modern industry, thus, directly impacting the quality of products and the efficiency of the technological processes. As the only processes with membrane technique, they offer the possibility to separate components based on their charge, allowing the particular demineralization of electrolyte solutions or their effective concentrating up to the solubility limits. With respect to water, they allow, with high recovery and selectivity, the separation of ions, water treatment of special or service water, as well as processing of wastewater, to efficiently resolve many environmental problems. They offer recycling of valuable components or produced solutions and are, therefore, key to non-waste or low-waste technologies.

The principle of EMP separation also offers potential for separating electrolytes from non-electrolytes, which allows the purification of various organic or food-processing solutions by removing unwanted salts. Significant is also the application importance of the electromembrane synthesis processes EDBM and EDM, which are used in the production of inorganic acids and bases from their salts or in the production of organic acids and other special chemical substances.

The processes of electrodialysis (ED) and electrodeionization (EDI) have the largest application potential and wide industrial utilization. In the use of these technologies, a new concept preferred today is their integration into a single technological unit, which allows comprehensive solutions of the defined requirements to increase the product quality, while reducing investment and operating costs. This approach replaces the traditional idea of mutual process confrontation between individual membrane and the so-called conventional separation processes. On the contrary, it utilizes their different principles and separation functions and synergetically combines their different application potentials with respect to the concentration and nature of treated solutions. Therefore, electromembrane separation or synthesis processes tend to be used today with other types of electromembrane and pressure processes or they are combined with conventional separation and synthesis processes (media filtration, coagulation, flocculation, sedimentation, evaporation, bioreactors, catalysis, advanced oxidation processes, etc.). This widens the application potential and apart from the field of water treatment, which affects all modern industries, these processes target the

https://doi.org/10.1515/9783110739466-014

chemical, power generation, food, pharmaceutical processing and other industries, with the application potential in the production of valuable chemicals [1–3].

## 12.2 Application in water and wastewater treatment

### 12.2.1 Introduction

Of all electromembrane separation processes, ED [1–3] still has the highest application potential in this field. It is used as an independent separation process or as part of integrated technologies mainly in the following areas:
- desalination of brackish water, municipal wastewater and surface water to service or special water,
- water treatment by softening and in selective removal of ions,
- treatment and recycling of wastewater and industrial waste solutions,
- recycling of valuable components from industrial wastewater,
- integrated technologies for intentional concentrating of electrolytes.

However, the importance of the EDI process has been growing in recent years, becoming a vital part of the technology to prepare ultrapure water for various purposes.

### 12.2.2 Desalination of brackish water, municipal wastewater and surface water

Water with salt content above the limits for drinking or irrigation water is desalinated by separate ED and RO processes or their combination, depending on the capacity and concentration of the feed. For the desalination of waters with feed TDS up to 8,000 mg dm$^{-3}$, which typically corresponds to some municipal wastewaters from sewage treatment plants or industrial wastewaters, brackish waters or river waters, single-pass ED polarity reversal (EDR) is the most common process. It is intended for 40 to 90 % desalination of the feed water. The biggest commercial successes in this area belong to Ionics (now Suez Water Technologies & Solutions) and MEGA a.s. companies.

The EDR technology can be used independently or in combination with RO and other membrane processes for the following purposes:
- preparation of service or process water from brackish water (EDR or RO, possibly RO + EDR),
- removal of nitrates from drinking water (EDR or RO),

- preparation of service or process water from municipal wastewater, from sewage treatment plants (tertiary treatment) or river water (MBR + EDR),
- preparation of special water for cooling towers and loops (EDR or RO + EDI).

The pre-treatment, before the EDR process itself, which was generally mentioned with this process in Section 7.6, cannot work in these applications without sand filtration and the use of safety filters such as candle filters. In rare cases, it is necessary to dose acid or substances to reduce scaling (deposition of poorly soluble inorganic compounds), so-called antiscalants, or to ensure dechlorination.

In the technology design, salt removal is usually considered between 40 % and 60 % (most often 50 %) in one hydraulic and electric stage, if the content of dissolved salts in the feed of a particular EDR stage is up to, approximately, 2,000–3,000 mg dm$^{-3}$. For higher concentration of salts in the feed, salt removal in a single-pass EDR process decreases. However, the requirement for a higher salt removal can be met by the inclusion of additional hydraulic and electric stages, which is equivalent to increasing the flow path length. Today, the line with EDR stacks can contain up to four hydraulic and electric stages. The common requirement for 75 % reduction of TDS in the processed feed water is usually achieved in two hydraulic and electric stages of the EDR technology. For these purposes, units are created comprising more technological lines operating in parallel with high-flow EDR stacks containing up to 600 cell pairs.

The operating costs of EDR depend on the amount of salt removed from the treated solution in the diluate chambers. In the case of 75 % demineralization of the feed water with TDS of 2,000 mg dm$^{-3}$ and a temperature of 25 °C in a two-stage single-pass line with EDR stacks and at a water recovery of 80 %, the total power consumption of the process is between 0.8 and 1.0 kWh m$^{-3}$ of the produced diluate (TDS 500 mg dm$^{-3}$). For deep demineralization of the feed water with maximum TDS (8,000 mg dm$^{-3}$), the power consumption can be approximately 4 kWh m$^{-3}$. Even though the power consumption of the EDR process decreases for feeds with lower salt content, it will probably never go below 0.3–0.4 kWh m$^{-3}$ because pumps must be used for the feed water and concentrate recirculation.[1] In the field of service water preparation, ED becomes competitive with reverse osmosis in terms of operating costs, if the feed water TDS is lower than approximately 2,000–3,000 mg dm$^{-3}$. Compared to RO, EDR has many other advantages. These are mainly lower requirements for pre-treatment, longer service lifetime of ion-selective membranes (7–10 years), production of low volumes of concentrate with high salts content and, especially, higher water recovery (up to 95 %). The advantage of EDR compared to RO is also the fact that EDR operation does not concentrate dissolved neutral compounds,

---

[1] The numbers are given for RALEX® membranes and spacers of thickness 0.8 mm produced by MEGA a.s.

which, in the case of RO, would cause problems with scaling (e.g. silica). On the other hand, EDR does not allow, due to the separation principle, to add microbial separation to desalination, which is a clear advantage in the use of RO for drinking water production.

EDR operation usually does not require adding chemicals (in exceptional cases, acid or antiscalants are dosed to the concentrate circuit to prevent deposition of insoluble inorganic substances such as $CaCO_3$). As discussed in Section 7.6, the main consumption of chemicals is due to the periodical cleaning of the system with solutions of diluted acids (HCl, $H_2SO_4$ or $HNO_3$) and hydroxides (NaOH), which is done, approximately, once in 4–8 weeks at the place of operation, using the system called **cleaning-in-place** (CIP).

An interesting example is the EDR application for tertiary wastewater treatment after biological treatment by MBR for preparation of irrigation water, implemented by MEGA a.s. in 2010. See Fig. 12.1. It is one of the biggest EDR technologies with a capacity of approximately 2,100 $m^3$ $h^{-1}$ of the processed solution, with water recovery above 86 %. The block diagram of the technology is shown in Fig. 12.2.

**Fig. 12.1:** EDR technology of the company MEGA a.s. for production of irrigation water from municipal wastewater after biological treatment in the Depurbaix area (Spain) (photo: MEGA a.s.).

Power generation is an important consumer of surface (raw) water. This water is treated and added to cooling circuits to remove the unusable energy of working steam in the form of heat. In cooling towers, the circulating cooling water is cooled by air and it is partially released into the atmosphere in the form of steam. Typically, two-thirds of the water added to a cooling circuit evaporates and the remaining one-third is released from the system with proportionally concentrated dissolved substances in the form of cooling tower blowdown (CTBD). Nowadays, cooling tower blowdowns represent more than 90 % of all wastewater from power plants. Treatment of cooling tower make-up water usually includes mechanical filtration or clarification. In many cases, however, it is not necessary to treat water chemically for these purposes at all.

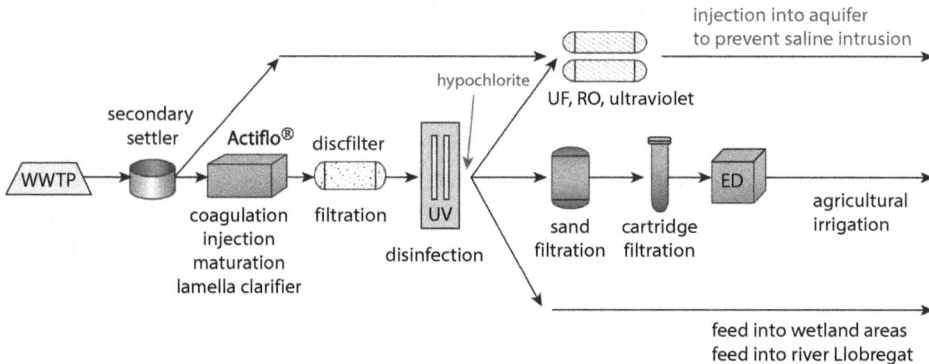

**Fig. 12.2:** Block diagram of the EDR technology for water treatment after biological cleaning for irrigation water.

Cooling water can be recycled mainly due to the fact that this water is not as polluted as boiler blowdowns and sludges. Although the treatment of CTBD is not particularly needed for large power plants, in the Czech Republic, because there is enough surface water available, it is important for small plants for cooling instead of the expensive water from the water mains and also in countries that suffer from lack of surface water. The ED technology seems to be the best for the treatment of cooling tower blowdowns. Unlike alternative technologies (such as RO), it has lower demands for pretreatment of the feed and provides a product quality that is sufficient for cooling circuits. It is also safe to maintain the value of the Langelier saturation index (LSI)[2] outside the corrosion range (LSI > 0). See eq. (7.51). In addition, ED achieves a nominal water recovery that is 10–15 % higher than RO, with comparable power consumption. One example of the actual application is by the company ARAK in Iran, which installed the EDR technology from the company Ionics, around the year 2000. It replaced it with the EDR technology from the company MEGA a.s. in 2010. The feed is a mixture of wastewaters treated at a water treatment plant. After biological purification,

**2** There are other names or other indexes suggested in literature, for example, the Ryznar stability index, more often used to describe the corrosive effects of water. Most frequently, the LSI is used to evaluate the saturation of water with $CaCO_3$.

the treated water with TDS of about 1,250 mg dm$^{-3}$ (conductivity of about 2,200 µS cm$^{-1}$) is desalinated in a three-stage EDR technology, resulting in an amount of 210 m$^3$ h$^{-1}$ of water produced with final conductivity of approximately 200–250 µS cm$^{-1}$. The EDR technology confirms a high nominal water recovery (over 85 %) and power consumption of 0.5 kWh m$^{-3}$.

## 12.2.3 Treatment to produce water for special purposes and ultrapure water

In some industrial applications, it is necessary to use special waters that meet demanding requirements for purity and quality, corresponding to ultrapure water. This applies, for example, to water for steam boilers in power generation, which is one of the most important industrial applications, with respect to water consumption. Standards prescribe small values of conductivity, contents of $CO_2$ in all forms, $SiO_2$ and other components in the boiler water. Due to the expensive preparation of make-up water, economical steam boiler operation requires minimization of blowdown, that is, high quality of make-up water. Usually, water is required with a maximum conductivity of 0.2 µS cm$^{-1}$ and a maximum $SiO_2$ content of 20 µg dm$^{-3}$.

To prepare ultrapure water on an industrial scale, ion-exchange technologies are commonly used, whose disadvantages are discussed in Section 8.9. To ensure continuous operation, the demineralization lines must be doubled. Due to the requirements for the duration of the service cycle, high-volume ion-exchange columns must be selected.

An alternative to conventional water treatment technologies is the use of membrane separation processes. For applications in power generation, it is possible to utilize pressure membrane processes, especially reverse osmosis (RO), microfiltration (MF) and EMP, namely ED, EDR and EDI. Even more advantageous, however, is the complete replacement of the existing demineralization lines by RO + EDI systems, operating continuously without the need for regeneration chemicals.

Most of the early EDI systems operated on pre-treated tap water and gave water quality similar to that produced by separate-bed deionizers [4]. At present, EDI is almost exclusively used in combination with RO to prepare high purity water with resistivity of 5–18 MΩ cm, thus competing with conventional mixed-bed columns. As an alternative to water demineralization in mixed-bed columns, the EDI process in more demanding applications originally required a safety measure in the form of a mixed-bed polisher placed after the EDI system, but today, due to technical progress, it represents the final step of water treatment. In processing permeates from double-pass RO or deionized water, EDI can steadily produce water with resistivity ranging from 16 to 18 MΩ cm.

Since the introduction of the first commercial system in the market, the main application area of the use of EDI systems is the preparation of high-purity or ultrapure water, especially in power generation, pharmaceutical and semiconductor industries. The reason for the success of EDI in the pharmaceutical industry, among others, is the fact that EDI modules did not induce bacterial growth and they even had, in some cases, bactericidal properties [4]. However, the current trend in this field are the HWS modules, that is, modules sanitizable with hot water at temperatures of 80–85 °C, which allow the removal of the biofilm formed normally on the surface of ion-selective membranes and suppress the growth of microorganism colonies. Today, the HWS modules are available from most major producers, who guarantee the preservation of the demineralization performance of the EDI modules even after 80–150 cycles of hot water sanitization. In the power generation industry, EDI is used almost exclusively to prepare make-up water for steam boilers. The success of EDI in the power generation and semiconductor industries was due to the development of reliable high-capacity modules or systems with guaranteed high removal of specific components such as $SiO_2$, $H_3BO_3$ and TOC [5]. EDI is also used in biotechnology and food processing industries and as a source of high-quality rinse water in the microelectronic industry and surface treatment of metals or optical glass, in the automotive industry, etc. EDI systems have also been installed in many institutions like hospitals, universities and dialysis centres. To a lesser extent, EDI has been tested to purify wastewater, urea, fruit juices, steam condensates, antifreeze liquids and sugar solutions.

The diagram of a typical RO + EDI technology is shown in Fig. 12.3a [6]. UV sterilization of water and TOC removal is carried out only in cases where bacteriological safety is required for the produced water, especially in the pharmaceutical and food industries or in the production of beverages.

The company USFilter used the EDI system to produce bottled water [8]. In another case study, the same company describes using EDI to prepare very pure rinse water for a facility manufacturing solar panels. The entire technology includes multimedia filters, softening, activated charcoal filtration, RO, EDI, UV sterilization, mixed bed filter and candle filter (0.2 μm) [9].

An example of the use of EDI technology for the preparation of rinse water for semiconductor components is shown in Fig. 12.3b and described in [7]. Except in cases of increased flow rates, EDI steadily produces water with resistivity above 16 MΩ cm. On an average, feed water contains is 2 ppb silica, conductivity of up to 10 μS cm$^{-1}$ and $CO_2$ content of 12 mg dm$^{-3}$. At peaks, the $CO_2$ content is and 29 mg dm$^{-3}$ and the average silica content is 66 ppb (as $SiO_2$).

In the work [10], the use of EDI is described in three cases as a part of a complex ZLD system. The EDI processes the condensate from the evaporator, which concentrates the cooling tower blowdown. The EDI product is used as make-up water for boilers. In some of them, the EDI system has a peak capacity of up to 113.5 m$^3$ h$^{-1}$. Water produced by the EDI systems has resistivity of 14–16 MΩ cm. The authors point to a higher content of organic carbon and solid particles in the processed condensate, which is manifested by an increase in pressure loss and deterioration of

**(a)**

**(b)**

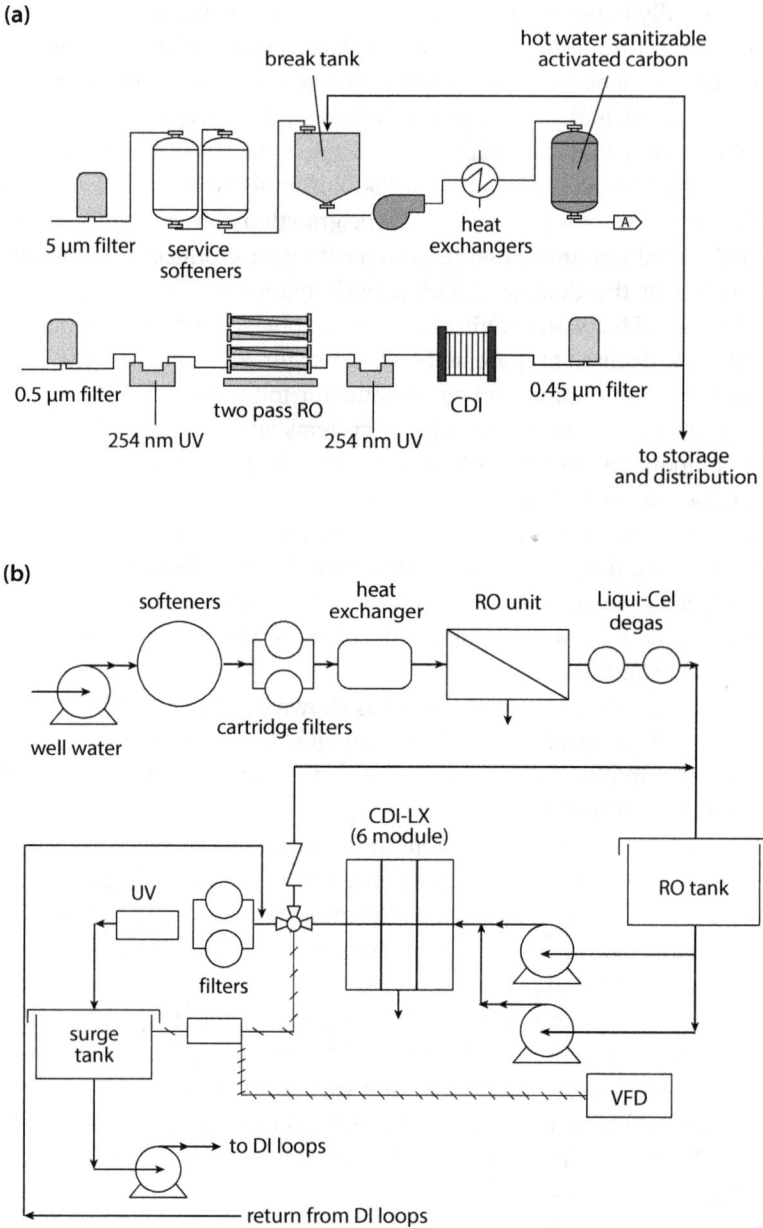

**Fig. 12.3:** Typical example of EDI technology [6] (a) and in production of rinse water for semiconductor components [7] (b).

the demineralization performance of EDI modules. The authors state that the original operating parameters of these systems can be fully recovered by chemical cleaning, in combination with rinsing of the EDI modules.

In literature, particular attention is paid to the possibility of separation of $NH_3$ and $NH_4^+$ ions. In the work [11], the use of the EDI process to remove $NH_4^+$ ions was studied. From a model solution containing approximately 200 mg $dm^{-3}$ $NH_4^+$, its content was decreased to less than 1 mg $dm^{-3}$ $NH_4^+$ (removal above 99.5 %) by using a two-stage EDI system. The goal of the work was to show the possibility of recycling waste (sewage) water containing $NH_4^+$ ions, resulting from the gradual decomposition of urea as water suitable to prepare drinking water. The authors of the work [12] discuss the possibility of EDI process to recycle the blowdown containing $NH_3$ from a steam boiler in a nuclear power plant. During a six-month pilot test, the authors demonstrated the ability of EDI to reduce the $NH_3$ content from 1,000 ppb to 3–4 ppb in the product, while the conductivity of produced water was 0.07 $\mu S\ cm^{-1}$.

The problem of treating water containing ammonium ions and ammonia, similar to waters containing $CO_2$ forms ($CO_2$, $HCO_3^-$ and $CO_3^{2-}$), lies in the back diffusion of neutral $NH_3$ through anion-selective membrane from the concentrate to the diluate, causing the product contamination.

Many authors discuss the possibility of EDI process to remove ions of heavy metals, for example, Ni, Cu and Pb from water [13, 14]. However, similar works are only at the level of laboratory or pilot studies and no industrial application of the EDI process is known in this area. For example, a three-chamber cell was used with a central diluate chamber filled with a cation-exchange resin and separated from the anode chamber by an anion-selective membrane, and from the cathode chamber by a cation-selective membrane [15–17]. The regeneration of the cation-exchange resin bed in the diluate chamber was performed by water splitting at the bipolar interface between an anion-selective membrane and cation-exchange resin, and by the migration of $H^+$ ions through the cation-exchange resin bed, that is, in the same way as in EDI systems with layered or separated ion-exchange resin beds; see Section 8.2. Nickel ions were removed from the solution circulating through this chamber.

# 12.3 Concentrating of solutions

Concentrating the electrolyte solutions is a frequent requirement of the industry. Further technological applicability of the electrolyte solutions may be conditioned by requirements for their high concentration. This is the case, for example, in the production of fertilizers or the recovery of salt from sea water or thermal water. Another purpose can be the recovery of valuable components from waste solutions or the effort to minimize their volumes. In the chemical industry, reaction heat is often used to evaporate water, for example, heat generated by neutralization reactions is used in the production of $NH_4NO_3$- or $Ca(NO_3)_2$-type fertilizers. In other cases, it is necessary to supply heat and an evaporator is usually used for this purpose. The heat of vaporization of water is 2,257 kJ kg$^{-1}$ at the temperature of 100 °C. However, despite their modern design, in combination with heat recovery, current evaporators still have high power consumption, typically 10–300 kWh per 1 m$^3$ of evaporated water.

Section 7.3.3.2 discusses the transport of the solvent by ion-selective membranes in the ED process. Unlike evaporation, the specific power consumption in the ED process does not depend on the amount of evaporated water but on the amount of electrolyte transferred by the ion-selective membranes and on the electric current load of the ED stacks. By reducing the electric current load, the specific power consumption decreases, but at the same time, a larger area of ion-selective membranes is necessary – it means the investment costs grow. In most cases, the ED process is not capable of concentrating electrolyte solutions to an industrially applicable level, directly. However, the treatment of diluted solutions significantly decreases the volume of water, which must be subsequently evaporated from the solution, and also maintains the energy-efficient conditions. Therefore, the ED process, combined with the evaporator, is an interesting alternative especially for concentrating solutions with electrolytes contents of up to tens of kg m$^{-3}$.

The basic limits for efficient concentrating of electrolyte solutions by the ED process can be divided into two areas. The first limitation is related to the composition of the treated solution, which generally does not allow the ED process to exceed the solubility products of components contained in the concentrate under given operating conditions. Logically, technological designs focus on the pre-treatment of the solutions by softening, that is, by removal of poorly soluble components or by adjustment of the pH value of the produced ED concentrate. To prevent the precipitation of poorly-soluble components in the concentrate chambers, polarity reversal in the EDR process is also used.

The second limitation of the ED process is related to the ED stack itself. It is connected with the insulation resistance against the destructive effects of the so-called current leakage, which are particularly significant in the area of a high concentration gradient between the diluate and the concentrate (Section 7.3.4). We may not forget the properties of the ion-selective membranes used, especially the unwanted transport of the solvent through the membrane from the diluate to the concentrate and the permselectivity decreasing with the growing concentration of both the diluate and concentrate. The issue of effective concentrating was studied thoroughly in Japan where the ED process is used widely to produce salt from sea water. The result was an ED concentrator stack (EDC) of a special design, with a very low linear velocity of electrolyte inside the concentrate chamber and a complicated structure of a flow distributor, shown in Fig. 7.32, or with a non-flow-through concentrate chambers with individual outlet of the concentrate from each chamber, Fig. 12.4 [18]. In this EDC design, the concentrate was formed almost exclusively by electroosmotic transport of water through ion-selective membranes along with NaCl and flowed from the upper part of the vertically-oriented ED stack. The mass fraction of the salt in the concentrate solutions ranged from 18 to 20 mass %.

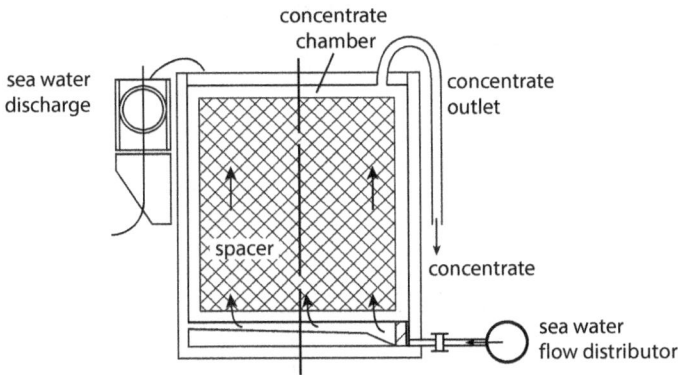

**Fig. 12.4:** Concentrate chamber of the electrodialysis stack to extract salt from sea water [18].

This solution was further improved by Zabolotsky [19], who placed the concentrate outlet in the bottom part of the non-flow-through concentrate chambers of the vertical ED stack. He demonstrated that natural convection of the solution contributes to increasing the concentration of the electrolyte inside the concentrate chambers. However, the problem of this solution lies primarily in keeping the concentrate chambers completely filled with this electrolyte and in the method of filling these chambers with electrolyte when the ED stack is put into operation. Thus, attention is focused on finding a suitable filling for concentrate chambers and various types of hydrophilic non-woven textiles are considered. An example of the results achieved with this type of EDC in treating $NH_4NO_3$ solution is shown in Fig. 12.5 [20].

As we have discussed in the previous text, a general problem of concentrating solutions by the ED process is the composition of the solutions treated in connection with the unwanted formation of precipitates of insoluble substances in the concentrate chambers; see also Section 7.3.3.6. For example, sea water contains, besides $Na^+$ and $Cl^-$ ions, significant amounts of $Ca^{2+}$ and $SO_4^{2-}$ ions also. Therefore, $CaSO_4$ can precipitate in concentrate chambers. The prevention of this precipitation and chemical cleaning of ED stacks with non-flow-through concentrate chambers without the need to use CIP seems to be unresolved at present.

A practical example of the use of the EDM process can be the treatment of water, saturated with $CaSO_4$ (Fig. 7.6). This water usually cannot be treated by most conventional methods and membrane separation processes without oversaturation of the solution and the subsequent precipitation of $CaSO_4 \cdot 2H_2O$ directly in the device. Other processes that do not lead to precipitation of $CaSO_4 \cdot 2H_2O$ (e.g. ion exchange) are unsuitable for the treatment of these solutions due to economic reasons. However, the

EDM process implemented as a four-circuit ED stack, according to Fig. 7.6, allows conversion of $CaSO_4$ to well-soluble $Na_2SO_4$ and $CaCl_2$. Therefore, high content of these substances can be achieved in both concentrates, which leads to high recovery of demineralized diluate that can be subsequently recycled in the customer's water management system. Only by mixing the two concentrates outside the ED stack, $CaSO_4 \cdot 2H_2O$ gets precipitated and can be stored in the solid state. The mother liquor, after precipitation, is saturated with $CaSO_4$ and is mixed with the treated solution and the entire cycle is repeated [21].

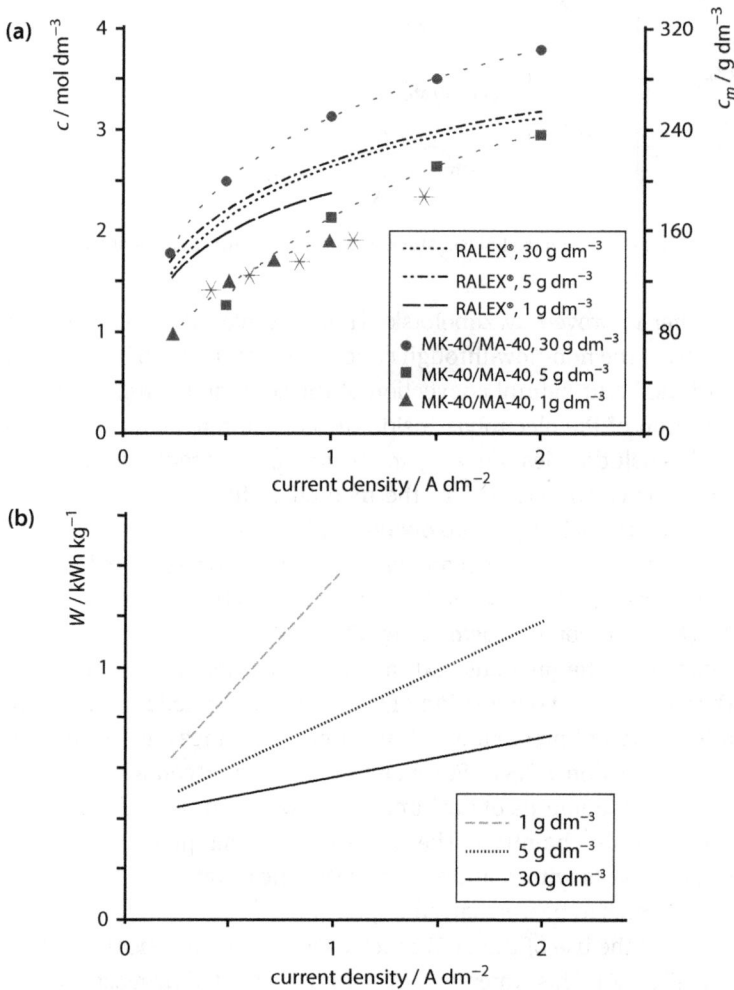

**Fig. 12.5:** (a) Dependence of the $NH_4NO_3$ molar concentration $c$ and mass concentration $c_m$ in the concentrate on the current density, electrolyte concentration in the diluate and type of ion-selective membrane. (b) Specific power consumption $W$ per kilogram of transferred $NH_4NO_3$ [20].

This principle is used in complex technologies, called ZDD (zero discharge desalination) or ZLD (zero liquid discharge), whose aim is to minimize liquid waste and to maximize the use and saving of water resources, at least in some geographic areas. The ZDD technology combines RO, silica removal systems (dosing of NaOH and $MgSO_4$, filtration by ceramic membranes) and EDM. The goal of EDM in these technologies is to demineralize the retentate from RO, saturated with $CaSO_4$, and to return it back to the feed of RO, thus increasing the total water recovery. The company Veolia Water demonstrated a pilot system of ZDD and achieved water recovery of 95 to 98 % [22, 23].

The maximum content of the electrolyte in the feed water and retentate, and the water recovery in the RO process are limited by some fundamental factors (concentration polarization, osmotic pressure, scaling). Compared to RO, ED allows treating more concentrated solutions and achieving higher content of electrolytes in the concentrate and tolerates higher levels of oversaturation of poorly soluble substances in the concentrate. Therefore, the combination of the RO and ED processes is currently considered as an answer to problems with large volumes of waste from RO.

## 12.4 Treatment of waste industrial solutions

In general, the use of electromembrane processes in the treatment of industrial wastewaters and solutions brings these options:
- recovery of valuable components of wastewaters,
- reduction of wastewater volumes by their concentrating,
- reuse of liquid waste using the ZLD technology with the possibility of recycling demineralized water and produced concentrates.

In the area of recovery of valuable substances from wastewaters, ED is applied to recover metals (Cr, Ni or Zn) from rinse waters after electroplating. New possibilities to use ED are brought to the chemical and processing industries in the separation of phosphorus, lithium and boron or in the recovery of specific organic substances (e.g. waste glycerol).

For many chemical production and other industries, it is also interesting to use ED to concentrate electrolyte solutions without the need for phase conversions. As stated in Section 12.3, the conventional method of solution-concentrating is a vacuum evaporator whose operating costs are often too high (power consumption of around 10–300 kWh per 1 $m^3$ of evaporated water). While the operating costs of the evaporator are related to the amount of evaporated water, the operating costs of ED are linked to the amount of electrolyte transferred by ion-selective membranes. Concentrating the solution by ED can substantially reduce the volume of the solution entering the vacuum evaporator, which reduces the energy demands of the entire process. To reduce the volume of wastewater, it is convenient to use the ED or EDR process wherever the composition of treated solution does not represent a concentration limit.

Since a common requirement to the process is to gain in the same operation also diluted product which can be reused as process water, ED, or its combination with other separation processes, is considered a very useful alternative or a supplement to the ZLD technology application.

As an example of the ZLD technology, an integrated technological solution is presented that offers the unique treatment of waste sulphate waters from uranium mining in the DIAMO s. p., plant GEAM in Dolní Rožínka, Czech Republic. This unique and zero-discharge technology, not used anywhere in the world yet, integrates conventional evaporators (annual capacity of 210,000 $m^3$ of clean waters and approximately 8,000 t of $Na_2SO_4$) and membrane processes of ED and RO, installed by the company MEGA a.s. and DIAMO s.p., respectively, in 2007. To protect the pipes against incrustation, membranes from poisoning by heavy metals (Mo, U, Cu) and formation of deposits, the system is complemented with the treatment and pretreatment of the feed solution. The treated solution can be described as a multicomponent waste solution with TDS of approximately 35 g $dm^{-3}$. The only output of this technology is the RO permeate, which is released into the waterline after partial recycling or treatment and crystalline sodium sulphate, used for the production of washing powders. The basic block diagram and examples of the technology are shown in Figs. 12.6 and 12.7.

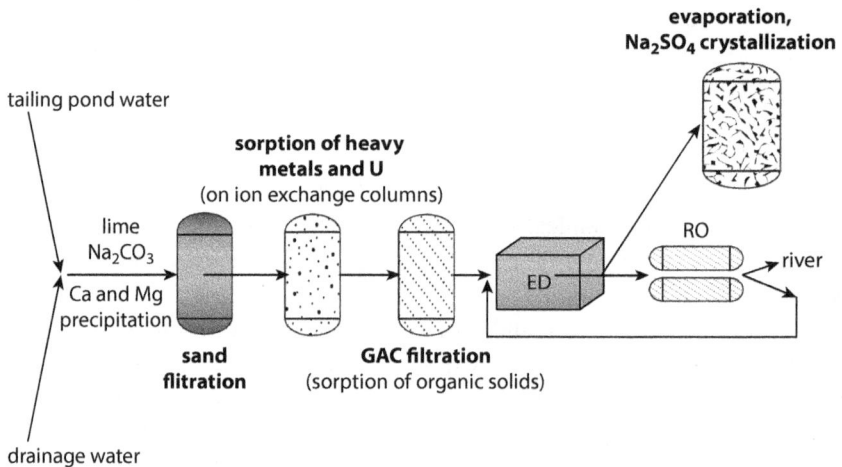

**Fig. 12.6:** Block diagram of integrated zero-discharge technology of the company MEGA a.s. for processing sludge sulphate waters from uranium mining in the Czech Republic.

**Fig. 12.7:** Industrial implementation of the ED process in integrated zero-discharge technology of the company MEGA a.s. for processing sludge sulphate waters from uranium mining in the Czech Republic (photo: MEGA a.s.).

Another example of the use of ED to process industrial wastewaters is the production of ammonium nitrate ($NH_4NO_3$), which produces waste vapours containing, after condensation, $NH_4NO_3$ in grams per $dm^{-3}$ in some countries. This condensate cannot be economically reused and it must be disposed off in ion-exchange columns, with high operating costs. The process that allows the processing of this solution would have to be able to produce, on one hand, a diluate with residual salt content in the order of mg $dm^{-3}$, which can be recycled as technological water, and simultaneously a concentrate with the salt content of 100–200 g $dm^{-3}$, which can be reused to produce $NH_4NO_3$. Under certain conditions, ED allows meeting both requirements. However, parallel production of highly diluted diluate and highly concentrated concentrate cannot be achieved simply in ED. The reason lies in the current leakage, back diffusion and internal leakage between the diluate and the concentrate, with a simultaneous high concentration gradient of ions. Therefore, it is much more advantageous to separate both processes and propose suitable conditions for demineralization in an independent multi-stage EDR line supplemented with an independent ED line for concentrating. An alternative solution in the development can be the application of an integrated technology based on the combination of EDC and ED.

# 12.5 Use in food industry

The introduction of electromembrane technologies in the food industry significantly contributes to lowering prices and improving the food quality. They are mainly used in the following application areas and industries [24]:
- Dairy – ED, EDR:
    - demineralization of milk whey;
    - demineralization of UF permeate from the production of protein concentrates;
    - demineralization of mother liquor in the production of lactose;
    - demineralization of skimmed milk.

- Viticulture and fruit growing:
    - tartrate stabilization of wine, that is, reduction of bitartarate concentration in wine by ED;
    - adjustment of pH of wine by EDBM;
    - deacidification of fruit juices by ED.

- Sugar industry:
    - purification of beet or cane juice where the UF process can increase the purity of the sugar solution by removing non-sugar substances and the MF, UF, NF and ED processes can remove colouring and other impurities;
    - demineralization of molasses by the ED process.

## 12.5.1 Use in dairy production

In the dairy industry, ED (or EDR) is currently used mainly in demineralization of milk whey (sweet, acid, casein, salty). The approaches of various suppliers of this technology differ and there are batch designs at relatively low temperatures (10–15 °C) as well as feed-and-bleed multi-stage lines working at temperatures of around 35 °C. The most successful companies in the application of whey demineralization are Novasep, Eurodia, GE Water & Process Technologies (now Suez Water Technologies & Solutions) and MEGA a.s. Its design of the ED technology for whey demineralization is shown in Fig. 12.8.

The entire problem of demineralization includes, besides the controlled transport of ion components, the handling of the microbial stability of whey during its processing and the periodic removal of unwanted surface layers of proteins and mineral deposits from membranes. Many industrial projects of whey demineralization have been suspended due to unclear logistics or possible utilization of the concentrate, generated by the membrane process. It is very difficult to process this mineralized solution with a low content of proteins and lactose with any commercial advantage, and since the ratio is 0.8–1.0 ton of the concentrate per ton of

**Fig. 12.8:** Electrodialysis units for whey demineralization by the company MEGA a.s. in Russia (left) and in the Czech Republic (right) (photo: MEGA).

processed raw material, the method of handling the concentrate is crucial for the final implementation of the technological plan. It is usually released into the operator's wastewater system and processed in a water treatment plant or is discharged directly to the sea in seaside regions. Contemporary technologies are capable of combining several processes to reduce the volumes of the concentrate and all other liquid wastes or convert them to an utilizable product. Unfortunately, their investment and operating costs are currently commercially unattractive for many operators.

For demineralization of whey, ion exchange and nanofiltration (NF) are considered, in addition to EDR. Each of these technologies has its specifics and scope of usability. However, EDR, as the only alternative technology, allows high demineralization of all types of whey in differently concentrated forms. It can demineralize natural whey with the dry matter content of max. 6 %, concentrated and partially demineralized whey using NF with the dry matter content of max. 20 % and concentrated whey using RO or evaporation process with dry matter content of max. 20–24 %. The degree of concentrating substantially increases the efficiency of demineralization by EDR. On the other hand, there is the disadvantage of decreasing the desalination efficiency with the increasing degree of demineralization. Protein and lactose losses for EDR are about 2 mass %. Nanofiltration can only be used for partial demineralization of whey (max. 30–40 %) and it primarily removes monovalent ions, whereas multivalent ions are concentrated in the product. The advantage is that the raw material is concentrated to the dry matter content of 18–20 % at zero loss of proteins and lactose. To achieve higher degrees of demineralization (70 %, over 90 %), it is possible to use ED and ion exchange. The main disadvantage of the latter technology is the production of waste solutions during the regeneration in this process.

At present, the trend of integrated technologies is on the rise also in this application area, combining all described demineralization methods. A clear map is shown in Fig. 12.9.

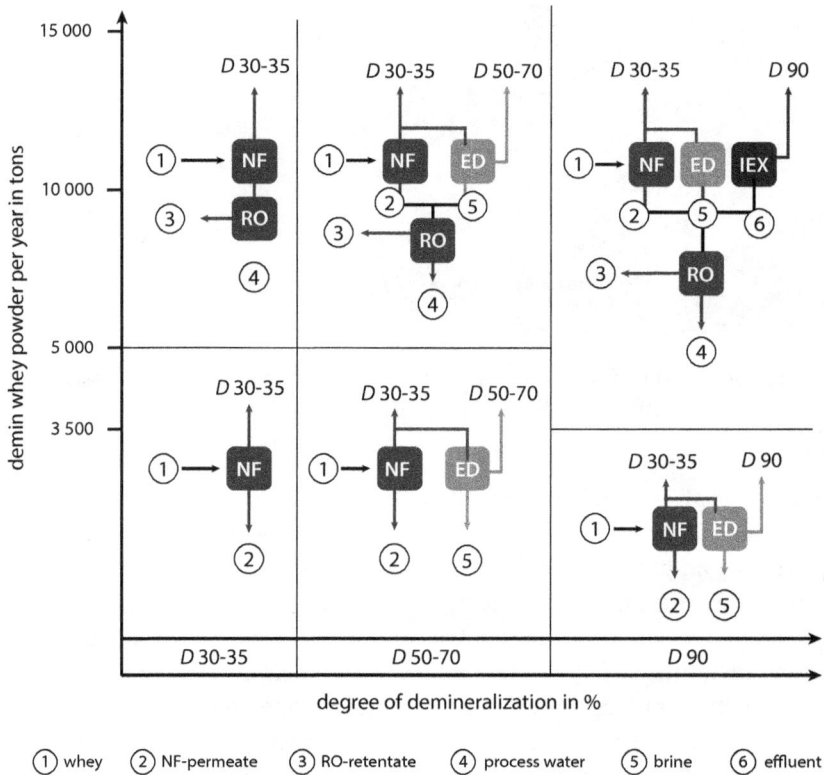

**Fig. 12.9:** Technological map of efficient combinations of integrated demineralization technologies for processing of milk whey (*D* is degree of demineralization in %) [27].

Removal of residual salts from concentrated whey using EDBM is also studied by some research teams. It is a modern technology, described, for example, by Shee et al. This research group managed to reduce the content of fats from 0.76 % to 0.21 % using EDBM [25]. Another technology by J. Balster et al. [26] described milk acidification and whey desalination using EDBM in one production process. This technology is shown in Fig. 12.10.

It is evident that the feed milk is acidified with HCl and the treated milk enters the system with bipolar membranes (BM), where desalination occurs in the space between CM and AM (Fig. 12.11). Separated ions pass through the appropriate ion-selective membrane and they are used to produce an acid or base. The produced base is then used to neutralize whey and the acid is used for initial acidification of the feed milk [26]. Sometimes, it is also possible to see that the feed milk is acidified directly by EDBM, as shown in Fig. 12.12.

Bipolar membranes and their acidification ability are also used to separate proteins from milk, soya or other sources. Bazinet et al. [28, 29] used electro-acidification with bipolar membranes to separate β-lactoglobulin from α-lactoalbumin from whey, and they achieved 98 % purity of the product and 44 % yield of β-lactoglobulin. The same authors also dealt with the separation of protein fractions from soya [29]. Other authors describe the production of highly pure casein from milk using EDBM [30].

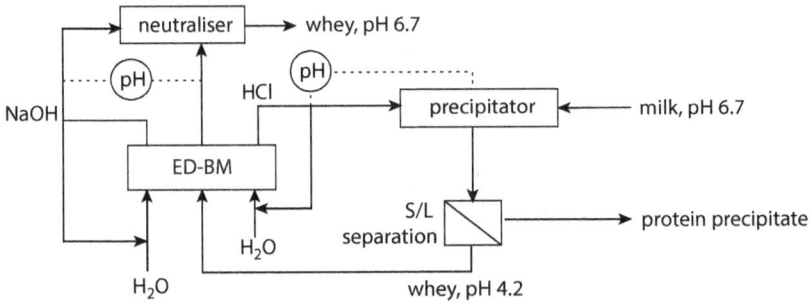

**Fig. 12.10:** Milk acidification and whey desalination using EDBM [26].

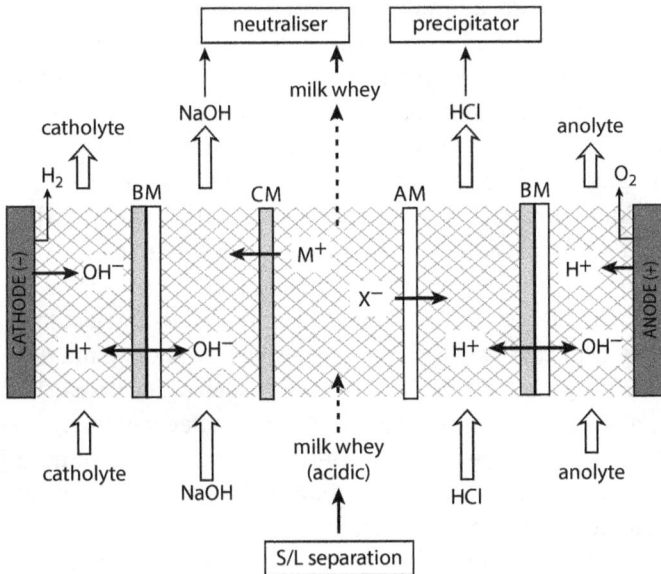

**Fig. 12.11:** Function of bipolar membranes for whey processing [26].

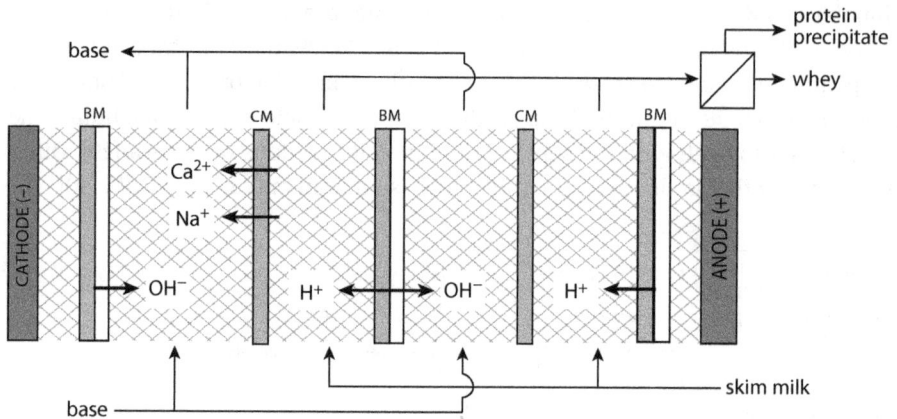

**Fig. 12.12:** Acidification of milk using EDBM [26].

## 12.5.2 Use in viticulture and fruit growing

Wine stabilization is a general term for a number of technological interventions to ensure that bottled wine remains clear and its changes, in terms of colour and taste, caused by ageing are slow. The stability of wine is impaired mainly by protein and microbial turbidity, caused by excess iron or precipitation of cream of tartar (KHT, chemically potassium bitartrate). Wine stabilization is a phase of treating wine before bottling. Recently, the process of tartrate stabilization has expanded and is applied to most wines today. Traditional techniques of wine stabilization are [31, 32]:

–  **Cold stabilization** (cooling just above the freezing point, −4 °C). In this state, the wine is left in isothermal tanks, usually for one week, and crystals of potassium bitartrate are formed, which are filtered, and the wine is then heated to ambient temperature. Treating of wine with cold stabilization requires much handling and low temperatures can affect the organoleptic properties of wine, such as its aroma. The process is very lengthy, requires filtration and heating to ambient temperature.

–  **Use of chemical additives**. For example, metatartaric acid (monoester of tartaric acid) is added to the wine to inhibit potassium bitartrate precipitation. Its effect is limited by time and temperature, however.

–  **Ion exchange**. By using a strong acid-cation exchange resin, $K^+$ and $Ca^{2+}$ ions in the wine are replaced with $H^+$ or $Na^+$ ions. Replacement of $K^+$ and $Ca^{2+}$ ions with $H^+$ ions reduces the pH of wine. Replacement of $K^+$ ions with $Na^+$ ions is based on a higher solubility of sodium bitartrate in wine than potassium bitartrate. Both techniques may negatively affect organoleptic properties of wine. Therefore, usually only a small part of the wine is treated in this way, followed by blending with untreated wine.

An alternative to these conventional techniques uses electromembrane separation processes. The French National Institute for Agricultural Research INRA, in cooperation with the company Eurodia Industrie, has developed a method of tartrate stabilization using ED [33]. Their method has been recognized by the International Organization of Vine and Wine (OIV) and approved for use in the European Union (AOC certificate), USA (TTB/FDA certificate) and other countries for all types of wine. Examples are shown in Fig. 12.13.

**Fig. 12.13:** ED technology for wine stabilization by the company Eurodia Industrie [34]. Left: capacity 30 hL h$^{-1}$, right: capacity 90 hL h$^{-1}$.

To achieve tartrate stability, the ED process uses electromembrane separation, in which the electric field action removes ions from the wine. It is a very fast process, much easier and cost-effective than the conventional method. The problem with respect to the applicability of ED for tartrate stabilization is in the low selectivity of the process because anions of the other weak organic acids present in the wine are transported through anion-selective membranes together with (bi)tartrates, and equivalent amounts of $K^+$ or $Ca^{2+}$ ions are transported through cation-selective membranes. The result is that, in order to achieve tartrate stability in a wide temperature range (i.e. low value of the saturation temperature of KHT), wine must often be desalted from 15 % to 30 % or more.

In practice, the ED process is not applicable for all types of wine because, at higher degrees of desalination, it negatively affects the organoleptic properties of the wine. According to experimental results, wine has acceptable organoleptic properties at the degree of desalination up to 20 %. Information from the company Eurodia [34, 35] suggests that 21 % of its ED installations are used for tartrate stabilization and it has been used, especially, in the last ten years. ED saves additives (chemicals) and it is an automated process which can also be mobile and save energy compared to cold stabilization. In particular, Eurodia states that operating costs are lower by 30–40 % in comparison with cold stabilization and represent no more than 0.76 EUR per bottle of wine. At the end of 2010, the company had more than 80 systems with a total of 200 ED stacks in operation in France, Italy, Spain, Portugal, Germany, Australia, New Zealand, South Africa, Argentina, Chile, Canada and the USA.

Another interesting application of the electromembrane synthesis process is the use of EDBM to adjust wine acidity. By using EDBM, pH of wine, wine juice or must and various fruit juices can be adjusted without changing their taste and colour. Two-circuit arrangement of EDBM is the most suitable for processing wine products. See Fig. 7.8. Thanks to EDBM, it is possible to adjust the pH of the treated solutions with great accuracy. A range of pH changes of 0.1–0.3 for wine and must and of 0.7–0.9 for wine juice is used. To adjust the pH of fruit juices, the two-circuit arrangement of EDBM is theoretically the most suitable, but the three-circuit arrangement is necessary in practice; see Fig. 7.7. More information about these applications can be found in [35].

# 12.6 Use in purification and production of chemicals

Electromembrane separation and synthesis processes or integrated membrane processes using ED are widely used in the production of special substances and pharmaceutical solutions, particularly in the purification of organic media, on the basis of selective separation electrolyte-nonelectrolyte by ED or for conversion of substances directly in the membrane stack of electromembrane reactors (multi-chamber EDM or EDBM). These can be applied in the production of inorganic and organic acids from their salts or directly for the production of chemical substances. As well as in other areas presented before, separation and synthesis processes in the production of valuable chemical substances (downstream bioprocessing, cell cultivation) involve more and more types of membrane techniques. The resulting products are subsequently used in the food, pharmaceutical and chemical industries.

## 12.6.1 Use in purification of organic solutions

An excellent example of this application field is the ED technology by the company MEGA a.s., applied to purify organic solutions for subsequent use in the pharmaceutical industry. The feed is a mixture of organic substances, unwanted mineral impurities (salts, acids or bases – NaCl, NaOH) and organic additives (secondary products and raw materials, that is, ethanol, 4-hydroxycrotonic acid, etc.). Projects were carried out for tfor purification technologies by Czech and foreign companies to process butyrobetaine (15 t per day), L-carnitine (3 t per day), thymidine (10 t per day) and dihydroxyacetone. The technological line contains from one to four ED stacks connected in parallel. The vertical arrangement of the hydraulic system is advantageous not only because of the small footprint area but also for the possibility of cleaning in place (CIP). The typical operating mode of these technologies is the batch process arrangement.

## 12.6.2 Use in production of organic and inorganic acids

In this area, the EDBM process has the best application possibilities in the production of acids and bases (mineral and mainly organic) from their salts. Organic acids are widely used especially in the food industry, production of beverages, pharmaceutical and cosmetic industries, production of detergents and other biochemical or chemical products. In industrial practice, EDBM is mainly used in the production of the inorganic compounds, NaOH, HCl, HF, $HNO_3$, $H_2SO_4$ and $H_3PO_4$.

Organic acids are produced either by fermentation or by chemical synthesis. The resulting salts of a particular acid are subsequently regenerated using EDBM, which prevents generation of by-products, reduces the use of additional chemicals and totally contributes to the simplification of the production process. The EDBM technology is used, for example, to produce lactic, succinic, salicylic, formic, acetic, propionic, itaconic, citric and many other acids [36]; more information about some of them can be found further in the text.

**Lactic acid** (2-hydroxypropanoic acid) is a very important organic acid used mainly in the food industry. It is produced either by chemical synthesis or by the fermentation process, which is cost-effective. Since the fermentation broth contains many impurities, ED is performed first to concentrate and purify the lactic acid salt. This concentrated salt is split by EDBM into lactic acid and NaOH, which is used to control the pH in the fermenter [37]. Figure 12.14 shows the production of lactic acid, including the technical evaluation of the process, and Tab. 12.1 shows the economic evaluation of the production of the ED technology for a capacity of 5,000 t per year [38].

**Succinic acid** (butanedioic acid) is produced in the fermentation production of 1,3-propanediol, one of the basic monomers for the production of polyester. During the microbial conversion of glycerol to 1,3-propanediol, a mixture containing 1,3-propanediol, alcohols and salts is produced. Most often, these are salts of succinic, acetic and sulphuric acids, which must be removed from the process. EDBM [39–41] is used to remove succinic and acetic acids from the mother liquor of 1,3-propanediol; see Fig. 12.15.

**Citric acid** (2-hydroxypropane-1,2,3-tricarboxylic acid) is used in the food, pharmaceutical and biochemical industries. At present, it is produced mainly by fermentation, which produces a low-concentrated salt that must be converted to acid and concentrated. The usual methods are precipitation, acidification, etc. However, the problem, in comparison with the conventional procedure, is the production of waste $Na_2SO_4$. An alternative is the use of the EDBM technology, producing very pure citric acid but not the aforementioned waste $Na_2SO_4$. The resulting by-product, NaOH, can be reused in the fermentation process. Xu et al. state that the optimum input concentration of sodium citrate is 0.5–1.0 mol $dm^{-3}$. It was also ascertained that the maximum achievable concentration of citric acid is approximately 30 g $dm^{-3}$ [42].

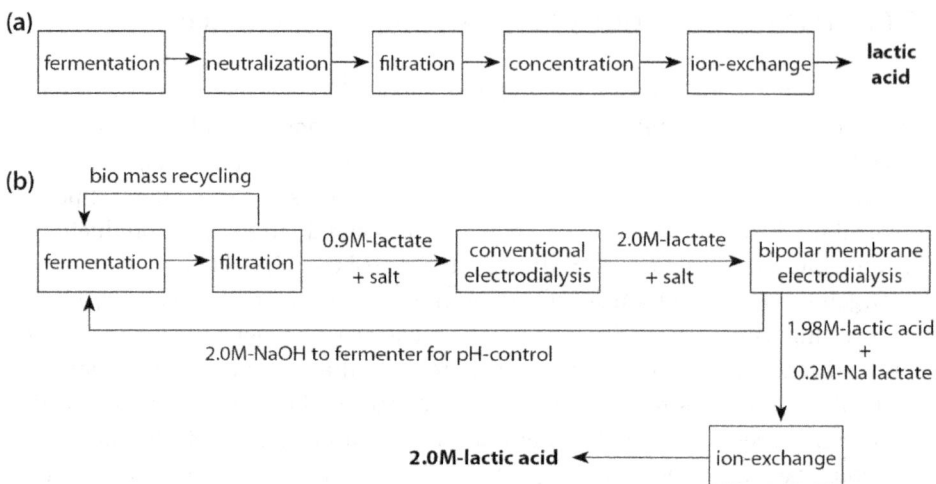

**Fig. 12.14:** Diagram of the production of lactic acid [38]: (a) conventional production process and (b) EDBM application.

**Tab. 12.1:** Economic evaluation of the production of lactic acid with the capacity of 5,000 t per year.

| | |
|---|---|
| Technology design | System with two ED units, 900 m$^2$ of membranes in a conventional ED unit, 280 m$^2$ of membranes in an EDBM unit |
| Process characteristics | Current efficiency 60 %, yield 96 % |
| Pre-treatment | Cross-flow microfiltration (500 m$^2$ of membranes), conventional ED (900 m$^2$ of membranes) |
| Final treatment | Ion-selective technology (9,000 L of ion-selective resins) |
| Economy | Life cycle and price of membranes: Ion-selective membranes 7,000 h, 0.05 USD per 1 kg of lactic acid Bipolar membranes 10,000 h, 0.11 USD per 1 kg of lactic acid Power consumption 1.43 kWh per 1 kg of lactic acid |

Other compounds that can be regenerated or concentrated by the EDBM process are amino acids, for example, L-lysine, L-histidine and L-arginine [43, 44].

The number of the installed and operated EDBM devices worldwide is relatively small and this technology, unfortunately, is not developing sufficiently fast. This is influenced by a number of factors, for example, the imperfect properties of bipolar membranes and of the system bipolar membrane/anion-selective membrane in EDBM, higher investment and operating costs, limited application range, etc. The economic demands of the process are crucial for its industrial application. With respect to the membrane properties, the potential of heterogeneous bipolar membranes, representing

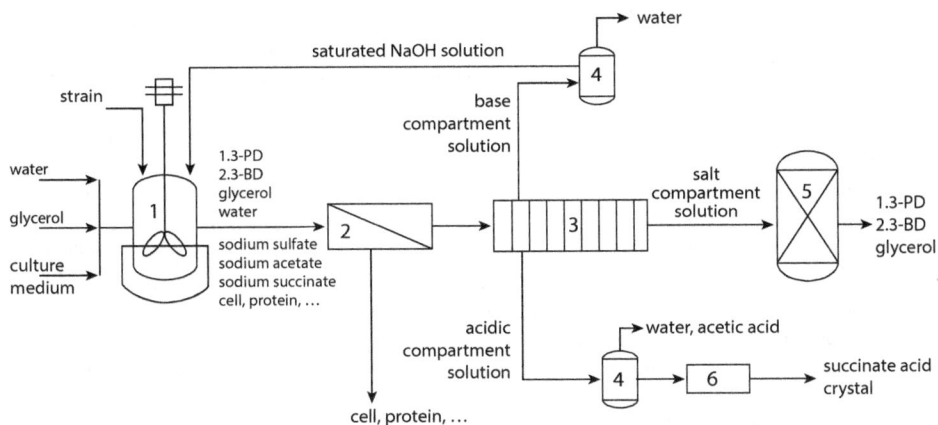

Fig. 12.15: Diagram of the production of succinic acid.

lower costs from the economic viewpoint, is primarily limited, in comparison with homogeneous types, by their worse characteristic properties (permselectivity, thickness, kinetics of water splitting – water-splitting potential, limited number of repeating units in the EDBM stack, combination of heterogeneous BM/AM in EDBM, etc.).

Despite the above facts, demand has been growing in the last decade for bipolar membranes or the EDBM technology among technology suppliers and some end users (producers of organic acids). This is motivated by environmental requirements and cost optimization of the production of bipolar membranes and EDBM. The production of organic acids is promising, thanks to the growing importance of the production of biodegradable polymers, in which organic acids play an important role, especially lactic and succinic acids. An interesting application of EDBM with heterogeneous bipolar membrane is the pH control of solutions without adding chemicals. However, these are rather small-capacity special applications, studied currently by a number of research centres.

# References

[1]    A. B. Koltuniewicz and E. Drioli, Membranes in Clean Technologies. Wiley-VCH Verlag, Weinheim, 2008. ISBN 3527320075.
[2]    R. W. Baker, Membrane Technology and Applications. 2nd Edition. John Wiley, Chichester, 2004. ISBN 978-0-470-85445-7.
[3]    H. Strathmann and H. Chmiel, Die Elektrodialyse, Ein Membranverfahren mit vielen Anwendungsmöglichkeiten. Chem. Ing. Tech. 56(3): 214–220, 1984.
[4]    F. DiMascio, J. Wood, and J. M. Fenton, Continuous deionization – production of high-purity water without regeneration chemicals. Electrochem. Soc. Interface 7(3): 26–29, 1998.

[5]   A. D. Jha, L. Liang, J. Gifford, Advances in CEDI Module Construction and Performance. Corporate literature of Ionpure.
[6]   D. Beattie, Using RO/CEDI to Meet USP 24 on Chlorinated Feed Water. Ultrapure Water Expo, 2001.
[7]   D. Schwarz, J. Gifford and J. Telepciak, Case study of an EDI system at the Honeywell solid state electronics center. Ultrapure Water 19(5): 22–26, 2002.
[8]   Leading Bottled Water Company Relies on USFilter CDI® System to Meet Water Quality, Environmental and Safety Requirements. Case study of the company USFilter, 2001.
[9]   BP Solar Meets Expansion Needs, Reduces Operating and Maintenance Costs with USFilter CDI-LX™ High-Purity Water System. Case study of the company USFilter, 2002.
[10]  T. Prato and C. Gallagher, New EDI Application: Feed from Brine Concentrator Distillate. International Water Conference, 2001.
[11]  E. F. Spiegel, P. M. Thompson, D. J. Helden, H. V. Doan, D. J. Gaspar, and H. Zanapalidou, Investigation of an electrodeionization system for the removal of low concentrations of ammonium ions. Desalination 123: 85–92, 1999.
[12]  C. Goffin and J. C. Calay, Use of continuous electrodeionization to reduce ammonia concentration in steam generators blow-down of PWR nuclear power plants. Proceedings of the Conference on Membranes in Drinking and Industrial Water Production, 2, 505–509, 2000. ISBN 0-86689-060-2.
[13]  O. Souilah, D. E. Akretche, and M. Amara, Water reuse of an industrial effluent by means of electrodeionization. Desalination 167: 49–54, 2004.
[14]  A. Mahmoud, L. Muhr, S. Vasiluk, A. Aleynikoff, and F. Lapicque, Investigation of transport phenomena in a hybrid ion exchange-electrodialysis system for the removal of copper ions. J. Appl. Electrochem. 33: 875–884, 2003.
[15]  P. B. Spoor, W. R. Veen, and L. J. J. Janssen, Electrodeionization 1: Migration of nickel ions absorbed in a rigid, macroporous cation-exchange resin. J. Appl. Electrochem. 31: 523–530, 2001.
[16]  P. B. Spoor, W. R. Veen, and L. J. J. Janssen, Electrodeionization 2: Migration of nickel ions absorbed in a rigid ion-exchange resin. J. Appl. Electrochem. 31: 1071–1077, 2001.
[17]  P. B. Spoor, L. Koene, W. R. Veen, and L. J. J. Janssen, Electrodeionization 3: The removal nickel from dilute solution. J. Appl. Electrochem. 32: 1–10, 2002.
[18]  R. E. Lacey and S. Loeb, Industrial Processing with Membranes. Wiley-Interscience, Toronto, 1972.
[19]  RU2008137905.
[20]  V. I. Zabolotsky, Technical report of the Kuban State University in Krasnodar. Russian Federation, 2009.
[21]  US Patent 7,459,088.
[22]  E. Gilbert, B. Biagini, B. Mack, M. Capelle, and T. Davis, ZDD – Achieving Maximum Water Recovery. Veolia Water Solutions & Technologies.
[23]  ZDD – a Zero Discharge Desalination Solution for Brackish Water Treatment. Technology Data Sheet, Veolia Water Solutions & Technologies.
[24]  K. J. Valentas, E. Rotstein, and R. P. Singh, Eds., Handbook of Food Engineering Practice. CRC Press LLC, 1997. ISBN 0-8493-8694-2.
[25]  F. L. T. Shee, P. Angers and L. Bazinet, Delipidation of a whey protein concentrate by electroacidification with bipolar membranes. J. Agric. Food Chem. 55: 3985–3989, 2007.
[26]  I. Balster, D. F. Pünt, H. Stamatialis, A. B. Lammers, and W. M. Verver, Electrochemical acidification of milk by whey desalination. J. Membr. Sci. 303: 213–220, 2007.
[27]  F. Rouset, J. Kincl, and M. Bobak, Optimization of Whey and Whey Derivatives Demineralization Process. DAIRY Ingredients symposium, California, 2012.

[28] L. Bazinet, D. Ippersiel, and B. Mahdavi, Fractionation of whey proteins by bipolar membrane electroacidification. Innov. Food Sci. Emerg. Technol. 5: 17–25, 2004.

[29] L. Bazinet, D. Ippersiel, R. Labrecque, and F. Lamarche, Effect of temperature on the separation of soybean 11 S and 7 S protein fractions during bipolar membrane electroacidification. Biotechnol. Progr. 16(2): 292–295, 2000.

[30] M. P. Mier, R. Ibanez, and I. Ortiz, Influence of process variables on the production of bovine milk casein by electrodialysis with bipolar membranes. Biochem. Eng. J. 40: 304–311, 2008.

[31] C. Fernandes, P. Cameira Dos Santos, and M. Norberta De Pinho, Wine Tartaric Stabilization by Electrodialysis. CHISA, Prague, 2006.

[32] P. Ribéreau-Gayon, Y. Glories, A. Maujean, and D. Dubourdieu, Handbook of Enology Volume 2: The Chemistry of Wine – Stabilization and Treatments. John Wiley & Sons, Ltd, 2006. ISBN: 0470010371.

[33] Patent WO 95/06110.

[34] F. Lutin and D. Bar, Reducing the Water Consumption in Electrodialysis Processes. NAMS, Chicago, May 12–15, 2006 2006.

[35] www.eurodia.com.

[36] H. Strathmann, Bipolar Membranes and Membrane Processes. In Encyclopedia of Separation Science, Level II, Membrane Separations. Academic Press Twente, 2000, 1667–1676. ISBN 978-0-12-226770-3.

[37] T. Franken, Bipolar membrane technology and its applications. Membrane Tech. 2000(125): 8–11.

[38] P. Boyaval and G. Corre, Production of propionic acid. Lait 75(4–5): 453–461, 1995.

[39] D. A. Glassner and R. Datta, Process for the Production and Purification of Succinic Acid. US Patent 5143834. 1. 9. 1992.

[40] T. Xu and Y. Weihua, Citric acid production by electrodialysis with bipolar membranes. Chem. Eng. Proc. 41: 519–524, 2002.

[41] T. Xu and Y. Weihua, Effect of cell configurations on the performance of citric acid production by a bipolar membrane electrodialysis. J. Membr. Sci. 203: 145–153, 2002.

[42] P. Pinacci and M. Radaelli, Recovery of citric acid from fermentation broths by electrodialysis with bipolar membranes. Desalination 148: 177–179, 2002.

[43] H. Wang and L. Yu, Separation of proline and serine by bipolar membrane electrodialysis. Membr. Sci. Technol. 25(5): 21–25, 2005.

[44] H. Wang and L. Yu, Separation of mixed amino acids by bipolar membrane electrodialysis. J. Tsinghua Sci. Technol. 44(12): 1588–1591, 2004.

Part III: **Electromembrane conversion processes**

Typical electromembrane processes for energy conversion may be divided into two main groups. The first traditional group is represented by electrolysis used to convert electrical energy into the chemical energy of the products. These processes represent a base, among others, for the largest electrochemical technology, that is, chlorine and caustic production by brine electrolysis. Nowadays, attention is focussed on water electrolysis as a source of so-called green hydrogen. These topics are treated in Chapter 13.

The second group of processes is related mainly for the conversion of chemical energy stored in a fuel or in a concentration gradient to electrical energy. This includes fuel cells or less-known flow batteries with ion-selective membrane, reverse electrodialysis (RED) and membrane supercapacitors that convert chemical energy into electrical energy. These issues are discussed in Chapter 14.

https://doi.org/10.1515/9783110739466-015

Martin Paidar, Karel Bouzek

# 13 Membrane electrolysis

Nowadays, a wide range of electrochemical processes are used in the industry, including production of both organic and inorganic materials, galvanic plating, surface treatment and separation processes. As far as the production capacity is concerned, the most important is the technology of production of chlorine and sodium or potassium hydroxide by electrolysis of brine (aqueous solution of NaCl or KCl). Practically, all new production facilities installed in the last two decades are based on membrane electrolysis. This clearly documents the importance of membrane electrolysis as an industrial technology. The goals are primarily to reduce the energy demands compared to traditional technologies, decrease the environmental impacts and ensure sufficient product purity for majority of applications. The same can be said for most membrane electrolyses. This state can be achieved only by rapid development in the area of material research of solids, mostly polymer electrolytes. In particular, the availability of new ion-conductive or ion-selective materials with sufficient stability has opened this perspective field with a great potential for further growth.

One of the main problems that had to be resolved since the first efforts to utilize electrochemical processes industrially was the need to separate anolyte and catholyte, especially due to the necessity to isolate the product or to prevent its degradation. The products of large-scale processes were originally deposited onto the electrode surface or taken near the working electrode. Therefore, they could not be transported to the counter-electrode, which prevented their degradation or contamination. Porous separators were also studied and applied to reduce unwanted mixing of electrode solutions. However, only the invention of the solid electrolyte allowed a completely new approach based on the utilization of the ion-selective membrane.

The first intensively developed technology dealing with the solution to this problem was the electrolytic decomposition of water. In particular, this was primarily the separation of the generated hydrogen and oxygen to prevent the formation of an explosive mixture. In 1902, more than 400 electrolysers were in operation [1]. However, this process is discussed in more detail in Chapter 14, which is devoted to membrane processes in energy conversion. The second intensively developed process, motivating the development of separation partitions, was the electrolytic production of chlorine. As early as 1851, a patent was filed, which focused on this process using inorganic porous diaphragms. Similar to water electrolysis, this separator was usually based on asbestos. Only the successful development of ion-selective polymer materials on the basis of perfluorinated sulphonated polymers provided materials that are sufficiently chemically stable and mechanically resistant to replace diaphragms in both mentioned processes as separation partitions and to allow the development of new technology. Now, let us pay closer attention to the basic construction materials and design of electrolytic cells, with particular regard to membrane cells.

https://doi.org/10.1515/9783110739466-016

# 13.1 Design of electrochemical membrane reactors

The design of the first built electrochemical reactor was substantially limited by available materials (glass, carbon, steel, wood). The shape of the individual parts was determined by the possibilities of machining these materials. Only the intensive development of materials, especially organic polymers, enabled the significant advancement and variability of electrochemical reactors. Electrochemical reactors generally contain more cells that form separate units. Today, ion-selective membranes represent the highest level of current development of materials for the design of electrochemical cells. In general, each electrochemical reactor consists of the outer body (tank), electrodes and electrolyte. The electrolyte compartment can be divided by separator (membrane or diaphragm) to the anolyte and catholyte. The diaphragm is generally an inert porous barrier[1] [2]. The presence of the separator brings many advantages and it is often crucial for the technology implementation. On the other hand, the use of the separator is connected with higher investments and demands for process management (risk of the separator clogging, etc.). The decision, whether to use a separator and which one, depends on the requirements of a particular process. The overall design is based on the effort to minimize the energy demands and to achieve the highest possible current efficiency. Also considering the investment cost, the CAPEX and OPEX ratio must be balanced. With respect to the wide range of electrochemical processes, only large-capacity operations like brine and water electrolyses have cells optimized specifically for a particular process. Other reactors are usually based on modifications of the existing designs [3]. However, with the development of mathematical modelling (Chapter 6), it is possible to optimize electrolytic cells much more efficiently, which is the current trend in the design of electrolysers.

## 13.1.1 Design of membrane electrolysers

From the design point of view, the easiest arrangement is the immersion of two plate electrodes in a rectangular tank separated by a membrane into the cathode and anode spaces. This arrangement is mainly used in processes where it is necessary to remove the electrodes from the electrolyser regularly (for example in galvanic processes). If the electrodes need not be replaced or otherwise mechanically treated, the system is much more common with frames bound together as shown in Fig. 13.1, the so-called **frame and plate** or **filter-press design** [4, 5]. The cell body itself then consists of the frames. The thickness of one compartment is often designed to the necessary minimum and it is typically determined by the thickness of

---

1 In literature, a thin diaphragm is sometimes incorrectly referred to as a "membrane". This can be confusing because it is not a separator with ion-selective properties.

the seal. In general, it is advantageous to work with rectangular compartments due to design reasons and electric current distribution. Higher pressure is also often used in electromembrane processes, especially in water electrolysis. In this case, a cylindrical shape of the device is more suitable because it is easier to seal the cell. A special case represents pressure-asymmetrical cells with a significant pressure difference between the anode and the cathode compartments (asymmetrical electrolysis of water, electrochemical compressors). Then, it is necessary to ensure sufficient mechanical support for the membrane to prevent its deformation or even rupture.

For industrial production using sufficiently high concentrations of electroactive substances, the filter-press arrangement is usually adequate. In processes requiring low concentration (e.g. wastewater treatment), the use of plate electrodes is inappropriate. The transport of ions to the electrode surface becomes the main rate-determining step of the process (see Chapters 2 and 3). In this case, it is convenient to use a working electrode with the largest possible surface or to intensify the electrolyte mixing near the electrode surface. Most often, porous 3D electrodes are used, consisting of packed particles, conductive foams, felts, etc. [6]. Alternatively, in purification of gases, the 3D electrode also functions as the absorption column; Fig. 13.2 [7]. In this case, it is convenient to use a tubular arrangement of the cell.

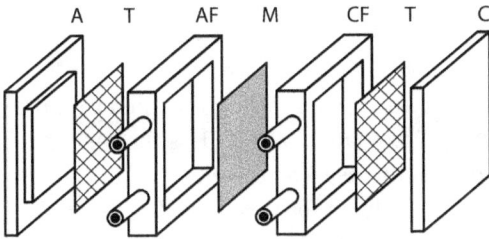

**Fig. 13.1: Scheme of filter-press arrangement [4, 5].** A, anode; T, turbulizer; AF, anode frame; M, membrane; CF, cathode frame; C, cathode.

These designs maintain the traditional arrangement with the anode and cathode in contact with the liquid electrolyte separated by the membrane. In the standard concept of electrolysis, the presence of the supporting (base) electrolyte is necessary and it is subsequently separated from the electrolysis products. The development of ion-selective membrane materials enabled to carry out completely new processes without the necessity of an ionically conductive liquid solution of the electrolyte. In the so-called **zero-gap** (zero distance) arrangement, the electrodes are pressed directly to the membrane surface. It is not needed to maintain sufficient conductivity of the circulating medium. The membrane itself serves as the electrolyte. In this arrangement, the membrane is referred to as the **solid electrolyte** (typically, but not exclusively, *solid polymer electrolyte*, SPE), Fig. 13.3. The arrangement is most often used in energy conversion processes (see Chapter 14) but also, for example,

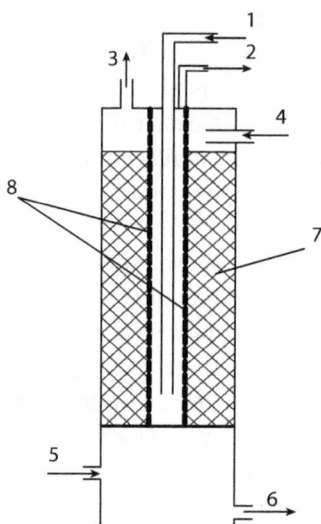

**Fig. 13.2: Scheme of electrochemical absorption column/cell with membrane and porous 3D cathode [7].** 1, anolyte inlet/anode; 2, anolyte outlet; 3, gas outlet; 4, catholyte inlet; 5, gas inlet; 6, catholyte outlet; 7, porous cathode (absorption column); 8, membrane.

in electro-organic syntheses [8]. Pressing the electrodes to the membrane surface has the advantage of decreasing the distance between the electrodes to the membrane thickness, that is, below 1 mm. The electrolyser itself is distinctly more compact than in the traditional concept. The material and design of the electrodes represent a separate topic and they are discussed in Section 13.1.3.

A fundamentally different approach must be applied for high-temperature processes with operating temperatures above approximately 500 °C. It is the case when melts or membranes consisting of solid oxides are used as the electrolyte. In these cases, the shape of the cell must be based primarily on the design and material possibilities. The main criterion is to secure the power supply and tightness of the cell. High temperatures do not allow using flexible seals of organic polymers and most structural elements are of high-temperature ceramics that are difficult to work with. Since high-temperature electromembrane processes are mainly used for energy conversion, this issue will be discussed in more detail in Chapter 14.

## 13.1.2 Electric connections of electromembrane cells

From the point of view of supplying the electric current to electrolysers, it is possible and advantageous to interconnect them in larger units, both to increase the capacity and for easier transformation of the electricity from the power grid. The operating voltage on one electrolytic cell ranges in the order of volts. On the other hand, current densities in industrial applications achieve up to tens of kiloamperes per square metre [4]. Therefore, it is obvious that a high current and relatively low voltage per one cell must be supplied. This influences the choice of the electrode connection in

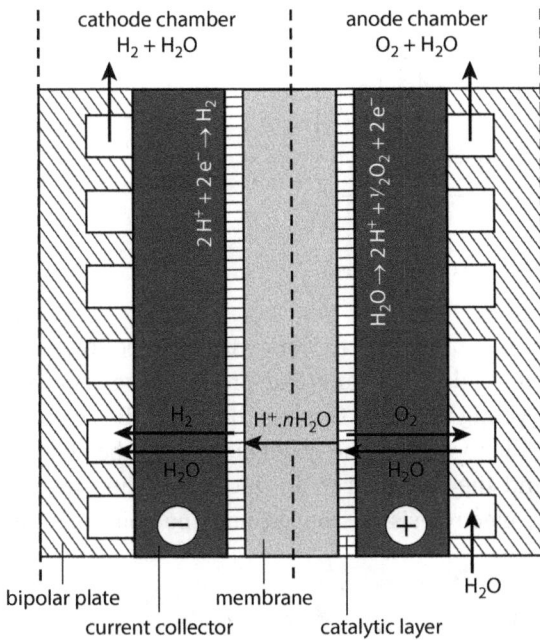

cathode chamber $H_2 + H_2O$

anode chamber $O_2 + H_2O$

$2H^+ + 2e^- \rightarrow H_2$

$H_2O \rightarrow 2H^+ + \frac{1}{2}O_2 + 2e^-$

$H_2$   $H^+ \cdot nH_2O$   $O_2$

$H_2O$   $H_2O$

$-$   $+$

$H_2O$

bipolar plate   membrane

current collector   catalytic layer

**Fig. 13.3:** Scheme of an electrochemical cell with the solid electrolyte [9].

the electrolyser. In the connection of electrodes and electrolysers, Kirchhoff's laws apply. Therefore, the electrolysers are often connected in series to increase the required output voltage on the rectifier, while maintaining constant electric current load.

In the electrolyser itself, there are two ways to connect the electrodes. If each electrode has its own current supply and, therefore, the polarity determined, we speak about the **monopolar** connection (see Fig. 13.4a). If current is supplied only to the terminal electrodes between which other electrodes are inserted, one side of an inserted electrode must work as the cathode and the other side as the anode when the current passes through. This arrangement is called **bipolar** (see Fig. 13.4b). Bipolar electrodes are often made of different materials on the cathode and anode sides. In the case of the bipolar connection, the total voltage on the electrolyser is determined by the sum of the voltages of individual cells, while in the monopolar arrangement, the voltage of the electrolyser equals the voltage on one cell. From the point of view of the electrolyser design, there is a considerable advantage in the possibility of arranging cells in a stack in which cells repeat regularly, and the entire stack can contain up to hundreds of cells (filter-press design and solid electrolyte design).

In terms of capacity, the electrolyser output is determined by the surface of electrodes and the applied current density. If the capacity needs to be increased, it is necessary to connect another electrolyser in the operation. Electrolysers work only exceptionally in the single-pass mode of the electrolyte. Even from the point of view

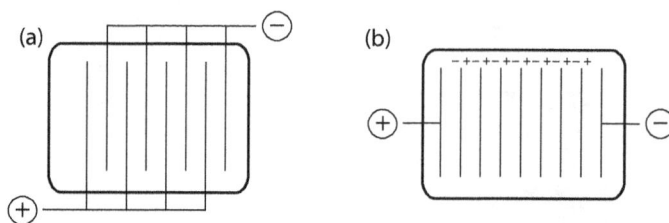

**Fig. 13.4:** Monopolar (a) and bipolar (b) connection of electrodes.

of hydrodynamics and mass transfer, it is preferable to work with the electrolyte re-circulation when the reactants are refilled in the tank and products are removed from it (similar to the continuous recycling mode for ED, see Chapter 7). Besides the continuous operation, it is also possible to work in the batch mode with the electro-lyte circulating between the tank and the electrolyser. In this case, when the required concentrations are achieved, electrolysis is stopped and the solutions are replaced with new ones.

## 13.1.3 Materials and design of electrodes

The electrode is the basic working tool in electrolysis on which the reaction takes place. The development of new electrochemical processes simultaneously with the growing pressure on economical operation brings the necessity to develop new electrodes based on new materials and with new construction designs. From the point of view of their shape stability, electrodes can be divided into shape changing and dimensionally stable.

### 13.1.3.1 Electrode materials

In general, electrode material must meet many requirements. First, it must be an electron conductor with sufficient electrical conductivity, necessary for uniform distribution of the current. Another requirement is the mechanical stability of electrodes. Besides a stable shape, good machining properties are desirable, since difficult machining properties are negatively reflected in their price. The electrode material should be sufficiently chemically stable in the applied electrolyte also in the currentless state (to avoid spontaneous dissolution when the power supply is switched off). Important is also the **overvoltage** of the desired electrode reactions on the material. Here, it must be noted that the overvoltage depends on a number of factors, especially the temperature, electrolyte pH, real surface of the electrode and mixing intensity. In industry, the high overvoltage of the electrode material is considered a negative property (due to the higher energy demands of the process), with the exception of supressing unwanted reactions. In aqueous solutions, the range of usable potentials of the electrode material is determined by water decomposition, that is, evolution of hydrogen on cathode and oxygen on anode (or by anodic dissolution). This range is referred to as the **potential window of the electrode**. If other reactions are occurring on the electrode than water decomposition, the working potential of the electrode must be within the potential window; otherwise, water decomposition is the dominant reaction. Therefore, much attention is paid to anhydrous systems, for example, ionic liquids [10].

Electrodes changing their shape with the electrolysis time are most often connected with the galvanic industry, where metal is deposited on the cathode or dissolved on the anode. These are electrodes of the first kind (metal immersed in a solution containing ions of a particular metal). In addition, other systems can be mentioned, for example, lead accumulators or production of $MnO_2$. Although ion-selective membranes are slowly being used even in the galvanic industry, their utilization is very limited. They are used as separators in selected types of batteries or in removal of metals from wastewater.

Much greater has been the advancement of electromembrane processes with dimensionally stable electrodes. Dimensionally stable electrodes serve as the inlet/outlet of the electric charge from/to the electrolyte and the electrode material does not change during the electrode reaction. From the point of view of physical chemistry, we speak about **oxidation–reduction (redox) electrodes**. In the electrochemical industry, dimensionally stable electrodes have a wide range of utilization. They are used in both inorganic and organic productions (production of $O_2$, $Cl_2$, $H_2O_2$, adiponitrile, etc.), wastewater treatment of organic impurities, electrometallurgy, electroplating practice, fuel cells, etc. In the case of cathode, the choice of materials is relatively wide because the reduction potential prevents dissolution. As the cathode material, we can use common materials like steel, stainless steel, copper, carbon, platinum-group metals, etc. The choice of the anode material is very limited due to the oxidative environment. At the laboratory scale, the choice of electrode materials is not limited by the price; so, even very expensive materials (platinum sheet, etc.) are often used. In industrial settings, the situation is quite different and the investment costs of electrodes can present a very substantial problem for technology implementation.

For dimensionally stable anodes, it is necessary to understand that there is no ideal stable anode – every anode is subject to degradation processes during its life and both producer and operator want to minimize these processes. The stability of dimensionally stable anodes is determined by the electrolysis conditions, mainly by the composition, pH of the electrolyte and applied current densities.

As already mentioned, the anode material is a critical part of electrochemical reactors. Ordinary materials, nickel and stainless steel are used in alkaline environments. At low current densities, it is also possible to use inexpensive carbon materials, especially graphite. Electrodes made of $SnO_2$ or $PbO_2$ also have acceptable stability in acid solutions. A unique electrode material of special properties is boron-doped diamond, the so-called BDD [6].

Thanks to their stability, platinum-group metals appear as the best anode materials but their use is limited by the high price. In practice, therefore, composite electrodes prove to be the most suitable, with a small amount of platinum-group metal applied on a cheaper carrier and acting as the electrode/electrolyte interface. However, the contact between the carrier and electrolyte and the risk of anode dissolution of the carrier material may not be ruled out. Therefore, titanium and graphite are most often used as carriers. Titanium, as a valve metal, is coated with a passive layer of highly resistant $TiO_2$, by which further dissolution is limited. A layer of platinum or oxide of platinum-group metals (Ir or Ru) is deposited on its surface. These anodes are called MMO (*mixed metal oxides*) [11], ATA (*activated titanium anodes*) [12] or DSA (*dimensionally stable anodes*) [13] and their introduction into industrial practice meant significant prolongation of electrode working life and lower risk of impurities formation from the dissolved anode [14]. Somewhat different is the situation when using carbon support. Carbon is often used in processes in which the main emphasis is on the surface area of the platinum catalyst. In this case, it is necessary to work only in a range of potentials in which the carbon oxidation is negligible and the anode is not significantly deteriorated. The carbon carrier, with the applied catalyst, is mainly used in gas diffusion electrodes in which gas acts as the reactant.

### 13.1.3.2 Electrode design

Besides the electrode material, the shape and position of the electrode in the reactor are also important for the proper operation of electrochemical processes. A very common product of the electrode reaction is gas ($H_2$, $O_2$, $Cl_2$) and its presence in the space between electrodes should be minimized. If the electrode, in the form of a compact desk, is positioned horizontally and gas is produced, the electrode surface easily gets covered with non-conductive bubbles and the electrolysis process stops or the electrolyser is even damaged. However, desk-shaped electrodes are very convenient from the point of view of handling and assembly, for example, in the case of a filter-press electrolyser (Fig. 13.1). Therefore, the necessity of electrode modification depends on the specific process. Also for the zero-gap arrangement, it is necessary to ensure the supply of reactants and removal of products. For this reason, it is preferable to use electrodes in the form of grates and grids. The openings secure flow of the electrolyte or carrier medium for solid electrolyte electrolysers. In the industry, the so-called **expanded metal (or expanded mesh)** proved to be very good shape; made by cutting out openings in a metal sheet and stretching it. The resulting structure has diamond- or hexagon-shaped holes (Fig. 13.5). These help to deflect the produced gas behind the electrode, which prevents the interelectrode space resistance from increasing. If the electrode is to be pressed to the membrane surface, it is necessary to ensure that it does not cut or otherwise mechanically damage the membrane. Expanded metal electrodes are also used flat in a compressed form.

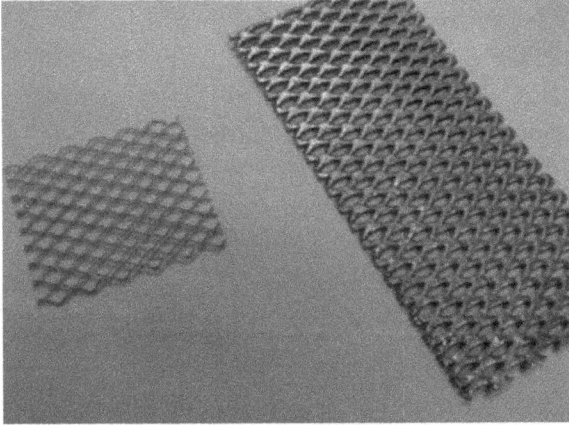

**Fig. 13.5: Expanded metal** (photo: author).

In recent years, the electrodes are increasingly common in systems with gaseous substances as reactants. The combination of gas with the electrolyte and electron-conductive material requires a special electrode design. The electrode must contain a sufficient number of places with **three-phase contact** (all mentioned phases are in contact). These electrodes are referred to as **gas diffusion electrodes** (GDE). Major advancements in gas diffusion electrodes have been achieved in the development of fuel cells [15] and production of chlorine and hydroxide [16]. Based on the process type, these electrodes can be divided into three types:
- The electrolyte is liquid and penetrates into the electrode pores (Fig. 13.6a).
- The electrolyte is the membrane itself (Fig. 13.6b).
- The electrode material demonstrates concurrently ion and electron conductivity (Fig. 13.6c).

If the gas-diffusion electrode is in contact with the electrolyte on one side in the form of a solution or membrane and on the other side, a gas is led into the electrode, the penetration of the liquid through the electrode into the gas compartment (like in the HCl solution electrolysis) must be avoided. In the case of the solid electrolyte reactor with zero gap configuration, it is necessary to incorporate an ion-conductive polymer in the structure of the catalytic layer (membrane electrolysis of water). The simplest case occurs when the electrode material is both ion conductive and electron conductive; this simplifies the entire system to two-phase contact (high-temperature electrolysis of water).

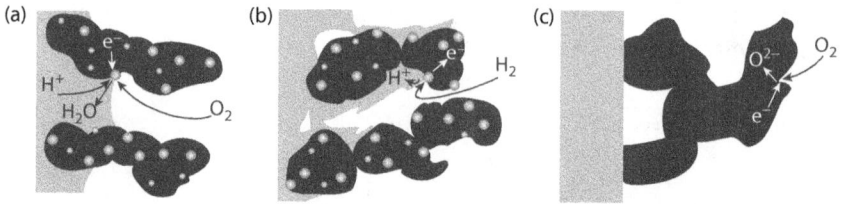

**Fig. 13.6: Three-phase contact in gas diffusion electrodes.** (a) Liquid electrolyte/gas/electron-conductive phase, (b) solid polymer electrolyte/gas/electron-conductive phase and (c) solid electrolyte/gas/electron-ion conductive phase.

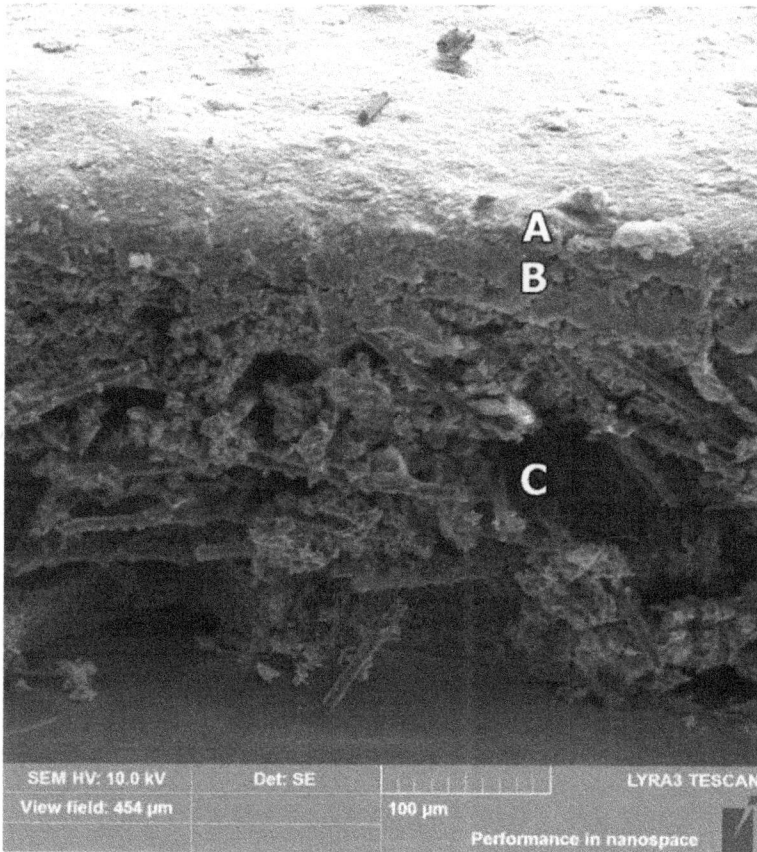

**Fig. 13.7: Section of gas diffusion electrode.** A, catalytic layer; B, microporous layer; C, porous carrier (photo: author).

In general, the gas diffusion electrode contains three layers, as shown in Fig. 13.7. A microporous layer of electron-conductive particles, bound, for example, by PTFE, is applied to the porous carrier consisting of fibres of a conductive material (carbon, titanium). The hydrophobic nature of the microporous layer prevents flooding of the reaction zone and simultaneously allows contact between the gas and catalytic layer deposited on the surface of the microporous layer in the catalytic layer. In the case of the solid electrolyte electrolyser, with both products and reactants in gaseous state, it is also possible to use hydrophilic polymers as the binder of the microporous layer, for example, cation-exchange perfluorinated polymers.

## 13.1.4 Perfluorinated membranes

Unlike separation processes, the high intensity of production is emphasized in membrane electrolysis, while the operating environment is often extreme (e.g. 5 kA m$^{-2}$, 33 % NaOH at 90 °C). This means high demands on the chemical and mechanical stability of membranes. From the point of view of the use, cation-selective perfluorinated membranes are completely prevailing in electrolytic processes. As already mentioned in Chapter 4, perfluorinated polymers with ion-selective groups on a polytetrafluoroethylene skeleton (tetrafluoroethylene copolymer with perfluorvinylethersulphate) are highly chemical-resistant membranes of unique properties. On the other hand, the preparation of fluorinated polymers is very difficult and, therefore, the number of their producers in the world is limited. The basic structural formula is the same for all producers (Fig. 13.8). Individual polymers mainly differ by the length of the side chains carrying functional groups and their concentration (see Tab. 13.1). The discovery of these membranes allowed developing the membrane process of hydroxide and chlorine production. They are also the key material for low-temperature fuel cells and membrane electrolysis of water.

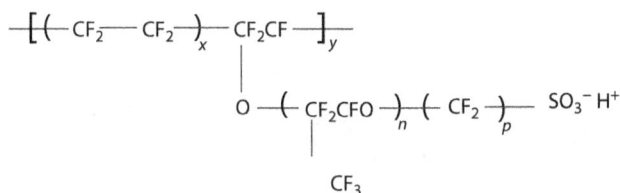

**Fig. 13.8:** Structural formula of perfluorinated sulphonated polymer ($x = 3.6$–$13.5$; $n = 0$–$2$; $p = 1$–$5$) [17].

The uniqueness of these membranes lies in the formation of hydrophilic and hydrophobic domains where ion-selective groups are oriented inwards into hydrophilic domains. Thus, the concentration of functional groups is much higher locally (Fig. 13.9). Thanks to this property, perfluorinated membranes can selectively separate strongly concentrated solutions, for example, saturated solution of NaCl from concentrated NaOH solution.

**Tab. 13.1:** Structure and properties of selected perfluorinated sulphonated membranes [17].

| Structural parameters (see Fig. 13.8) | Producer/mark | Ion-exchange capacity (mmol g$^{-1}$) | Thickness (µm) |
|---|---|---|---|
| $n = 1, x = 5–13.5, p = 2$ | **Dupont** | | |
| | Nafion$^®$ 120 | 0.83 | 260 |
| | Nafion$^®$ 117 | 0.91 | 175 |
| | Nafion$^®$ 115 | 0.91 | 125 |
| | Nafion$^®$ 112 | 0.91 | 80 |
| $n = 0–1, p = 1–5$ | **Asahi Glass** | | |
| | Flemion$^®$ T | 1.00 | 120 |
| | Flemion$^®$ S | 1.00 | 80 |
| | Flemion$^®$ R | 1.00 | 50 |
| $n = 0, p = 2–5, x = 1.5–14$ | **Asahi Chemicals** | | |
| | Aciplex$^®$ S | 0.83–1.00 | 25–100 |
| $n = 0, p = 2, x = 3,6–10$ | **Dow Chemical** | | |
| | Dow$^®$ | 1.25 | |
| | **Solvay** | | |
| | Hyflon$^®$ Ion | 1.11 | |

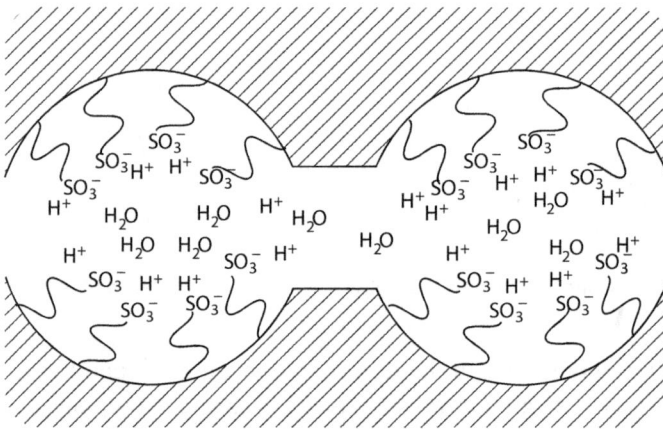

**Fig. 13.9:** Structural model of arrangement of functional groups within perfluorinated sulphonated membrane.

## 13.2 Industrial applications of membrane electrolysis

Membrane electrolysis represents an effective tool for electrochemical synthesis of various compounds. Its utilization is varied, from large-scale processes to specialties production in small volumes. From an energy consumption point of view, electrochemical processes are very demanding and the expenses for electricity are the most important part of the operating costs. On the other hand, easy automation, simple design, relatively low investment costs and, often, the only possible method of production make membrane electrolysis an irreplaceable industrial process.

Membrane electrolysis is mainly applied in inorganic technology but its importance is also growing in the production of organic substances. Due to the versatility of membrane electrolysers, it is not possible to describe all these processes; so, only selected technologies with significant production volume are mentioned, adequately illustrating the possibilities of membrane reactors.

### 13.2.1 Membrane electrolysis of brine

Clearly, the most important industrial electrolysis process is the production of hydroxides of alkali metals (especially sodium hydroxide) and chlorine. The worldwide production of chlorine exceeds 60 Mt per year, which ranks both commodities among the most produced chemicals in the world.

The electrolysis of brine was originally carried out in the diaphragm electrolyser and then in the mercury cathode electrolyser. Both these production procedures are gradually replaced, due to economic and environmental reasons, by the third most recent method using the cation-selective membrane. In the last 10 years, the share of the membrane method in the production of hydroxide and chlorine has increased from 25 % to 58 % of the volume of production [18]. Besides legislative pressures, this change is also due to the lower electricity consumption while achieving high-quality products. The essence of the membrane electrolysis of brine is described by the overall electrode reaction:

$$2\,NaCl + 2\,H_2O = Cl_2 + 2\,NaOH + H_2 \tag{13.1}$$

In this process, hydrogen is produced as a by-product on the cathode. It is often perceived only as an unwanted product but this hydrogen produced by electrolysis represents approximately 4 % of the worldwide hydrogen production. As shown in Fig. 13.10, chloride is oxidized on the anode by electrolysis to chlorine, according to the reaction:

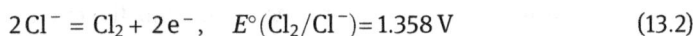

$$2\,Cl^- = Cl_2 + 2\,e^-, \quad E^\circ(Cl_2/Cl^-) = 1.358\,V \tag{13.2}$$

As already mentioned, hydrogen is developed on the cathode according to the reaction:

$$2\,H_2O + 2\,e^- = H_2 + 2\,OH^-, \quad E^\circ(H_2O/H_2, OH^-) = -0.828 \text{ V} \tag{13.3}$$

From the anode compartment, surplus sodium ions pass through the ion-selective membrane into the cathode compartment where hydroxide ions, produced according to the reaction (13.3), form the final product. The result is sodium hydroxide solution with concentration of 33 mass %. The anode consists of a titanium electrode activated by ruthenium oxides. From the thermodynamic point of view, oxygen should develop on the anode due to the lower potential of the reaction. However, due to the higher overvoltage of the oxygen evolution reaction on the ruthenium oxide-based electrode, the chlorine development reaction is preferred (13.2). The cathode is nickel or steel. In modern operations, electrodes are pressed directly to the membrane surface. The total voltage on the cell is approximately 3.0–3.6 V at a current load of 3–5 kA m$^{-2}$. The operating temperature of electrolysis is 80–90 °C [19].

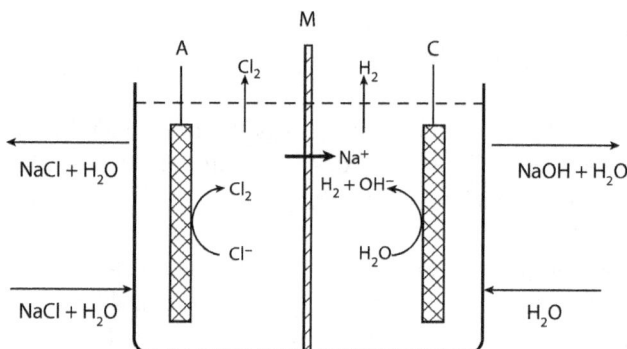

**Fig. 13.10: Scheme of membrane electrolyser of brine.** A, anode; M, membrane; C, cathode.

In this technology, the membrane must meet many requirements, especially the high ion conductivity and chemical resistance. Thanks to the separation capabilities of the membrane, the unwanted penetration of chloride ions from the anolyte to the cathode space allows to keep NaCl content in a final product below 50 ppm (i.e. mg of chloride in 1 kg of 50 % NaOH). These parameters cannot be achieved only with a sulphonated perfluorinated membrane. In general, sulphonated membranes have very good ion conductivity but their ion selectivity for hydroxide ions is much lower, compared to membranes containing carboxyl groups. Therefore, the membrane electrolysis of brine utilizes specially developed perfluorinated membranes with a thin layer of polymer carrying carboxyl groups on the cathode side, preventing hydroxide ions to penetrate the anode compartment. The membrane then represents the main component of the entire process but also the most sensitive part of the technology. The membrane is in contact with the weakly-acidic brine solution from the anode side and with the concentrated hydroxide from the cathode side. This means a significant leap of the pH value on the membrane. Any presence of impurities, which can precipitate by alkalization, then means a serious risk of irreversible blockage of the membrane. For the smooth operation of the technology, it is necessary to work with a highly pure solution of brine, purified using ion exchangers.

An important innovation is the introduction of the gas diffusion cathode (see Fig. 13.6) into the brine electrolysis process. In this case, hydrogen is not produced on the cathode but, on the contrary, oxygen is consumed according to the reaction:

$$H_2O + \tfrac{1}{2}\,O_2 + 2\,e^- = 2\,OH^-, \; E^\circ(H_2O, O_2/OH^-) = 0.401 \text{ V} \tag{13.4}$$

This cathode is referred to as the **oxygen depolarized cathode** (ODC). From the comparison of the standard electrochemical potentials of reactions (13.3) and (13.4), it is obvious that this step represents a decrease in the equilibrium voltage on the cell by more than 1 V. Overall, it means approximately 30 % savings in electric energy, which is the main operating cost of the technology. Due to the technological difficulties with the ODC operation, their introduction into practice is slow. Therefore, operators of the hydroxide and chlorine productions more often try to use the produced hydrogen, for example, in membrane fuel cells (Chapter 14).

## 13.2.2 Membrane production of ozone

Ozone is used as an oxidation agent with disinfection effect. For this reason, it is a possible substitute for the currently predominantly-used chlorine. Its main advantage is the elimination of the risk of the formation of chlorinated organic substances. On the other hand, the impossibility of storage and difficult production are causes for the small extension of ozonization. One of the possibilities of ozone production is the electrolytic decomposition of water in the electrolyser with solid polymer electrolyte. Water is dosed to the anode as the starting raw material. The cathode consists of a gas diffusion electrode with a platinum catalyst on which hydrogen is produced. By the choice of the anode material, the desired ozone can be produced instead of oxygen, according to the reaction (13.5). Various anode materials are used, but most often it is $PbO_2$ [20]:

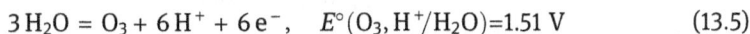

$$3\,H_2O = O_3 + 6\,H^+ + 6\,e^-, \quad E^\circ(O_3, H^+/H_2O) = 1.51 \text{ V} \tag{13.5}$$

Since the standard potential of the ozone production reaction (from water) is by approximately 280 mV higher than of the oxygen production reaction, it is obvious that oxygen will represent a substantial portion of the product (gas generated on the anode). In addition, the energy demands of electrolysis will be higher than in the case of simple electrolysis of water (see Chapter 14). The voltage on the cell typically ranges from 3 to 5 V [21] and the composition of the output gas depends on the method of the electrolysis implementation. Current electrolytic generators provide ozone with content up to 30 mol. %. It is clear that the demands put on the membrane are very high and its working life represents a key element of the entire generator. Only functionalized fluoropolymers are able to meet these conditions.

### 13.2.3 Membrane production of hydrogen peroxide

Besides conventional procedures, hydrogen peroxide can also be produced by membrane-electrolysis. The main advantage is the generation of the peroxide solution directly at the spot; so it is not necessary to transport and store it. Hydrogen peroxide is used for the disposal of organic pollution and in the so-called advanced oxidation processes. By the reaction of hydrogen peroxide with ferrous ions, highly reactive $OH^\bullet$ radicals are produced. This reaction is known as the Fenton's reaction [22].

Hydrogen peroxide is produced on the cathode in the electrolyser with the cation-selective membrane, according to reaction (13.6). Protons and oxygen are produced on the anode by electrolysis of water. The protons pass through the membrane. Therefore, the only input raw material for the hydrogen peroxide production is demineralized water and oxygen, originating typically from air:

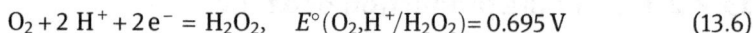

$$O_2 + 2\,H^+ + 2\,e^- = H_2O_2, \quad E^\circ(O_2, H^+/H_2O_2) = 0.695\,V \tag{13.6}$$

Traditionally, the cathode was made of porous material but gas-diffusion electrodes are mainly used nowadays. As the cathode materials, reticulated vitreous carbon, carbon black and, more recently, carbon nanotubes and nanofibers can be used [23, 24]. Due to the considerable aggressiveness of the produced peroxide, it is necessary to use a highly resistant membrane. At present, only perfluorinated sulphonated membranes are used, for example, Nafion®. The anode is of titanium activated by a layer of $IrO_2$ or platinum. Technical literature also mentions an alternative procedure of hydrogen peroxide production using a modified fuel cell; the fuel cell for hydrogen- and oxygen-generated hydrogen peroxide on the cathode instead of water [25]. However, the advantage of the simultaneous production of electric energy and chemically valuable product also brings many operation complications, so the electrolytic method of the $H_2O_2$ production is preferred in practice.

### 13.2.4 Membrane electrolysis in electro-organic production

Generally, electromembrane electrolysis is mainly used in inorganic technologies. The electrochemical production of organic substances is problematic primarily due to the low conductivity of water-free systems. This is the main reason why electrochemical methods are not applied in large-scale productions. The breakthrough came with the introduction of an electrochemical process in the production of adiponitrile (intermediate production for the production of Nylon) from acrylonitrile by the company Monsanto in 1965 [26]. Originally, this production was only carried out in a membrane-separated electrolyser but today there are also productions in undivided cells [5]. The main step of the production is the electrochemical reduction connected with dimerization on the cadmium cathode:

$$2\,CH_2 = CH-CN + 2\,H^+ + 2\,e^- = NC-(CH_2)_4-CN \qquad (13.7)$$

In acid environment, oxygen is produced on the anode. The main function of the cation-selective membrane is to prevent the oxidation of organic substances on the anode. At the same time, $H^+$ ions pass through the membrane, participating in the reaction in the cathode compartment (Fig. 13.11).

A significantly large area of the application of electromembrane electrolysis is the production of specialties with high added value. At present, there are many productions using ion-selective membranes and others are in the pilot plant phase [3, 27]. Electrolytic production of organic substances differs from inorganic productions in some respects. These differences must be taken into account when designing and controlling the process. Non-polar organic substances usually penetrate ion-selective membranes. Unwanted oxidation reactions on the anode or even polymerization represent other risks. Therefore, much attention is paid to the selection of electrode materials in practice [21].

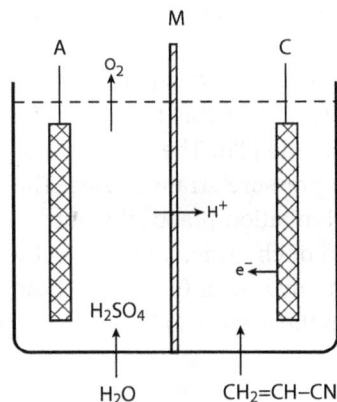

**Fig. 13.11:** Scheme of membrane electrolyser for production of adiponitrile [21].

In addition to electrolysis with a supporting electrolyte, it is convenient to use for organic synthesis a solid polymer electrolyte electrolyser. In this case, it is not necessary to ensure the conductivity of the processed solutions and the step of separation of the supporting electrolyte is avoided. The wide possibilities of using a solid electrolyte electrolyser (primarily with the Nafion® membrane) have been described in detail by Ogumi [28–32]. Among other processes, membrane electrolysis can be used for hydrogenation of olefins, reduction of nitrobenzenes to anilines or decarboxylation. A particular use of membrane electrolysis in the production of specialties is usually subject to secrecy so the scope of the utilization of electromembrane synthesis of organic substances cannot be determined as it is for inorganic productions.

### 13.2.5 Membrane compressors/concentrators

Electromembrane electrolyser with differential pressure on the membrane can replace traditional compression of gases using mechanical compressors. The basic advantage of this method of compression represents the absence of moving parts. There is also no risk of contamination by lubricants, noise, etc. [33]. In practice, asymmetrical pressure is mainly used for membrane electrolysis of water (see Chapter 14), in which hydrogen is generated and already compressed to a pressure that is suitable for storage or further compression.

In a membrane compressor, oxidation of hydrogen to $H^+$ occurs on the anode. It subsequently passes through the membrane and is reduced back on the cathode to produce hydrogen [34]. Since the reactions on the anode and cathode are identical, although in opposite directions, the equilibrium voltage on the electrolyser is determined only by the difference in hydrogen activities on the electrodes (see Chapter 3). Thus, compressors work at operating voltages lower than 0.3 V [35]. The entire arrangement is very similar to the membrane fuel cell (see Chapter 14) with the main difference consisting in using a membrane with sufficient thickness to endure the required pressure difference.

A concentrator represents a similar process when hydrogen is separated from gases with its low content using membrane electrolysis. Other gases remain in the anode flow and pure hydrogen is generated on the cathode [35]. The concentrator can be operated in both symmetrical and asymmetrical pressure arrangements. The main disadvantage of membrane compressors is their limitation practically only to hydrogen (theoretically, they could also process oxygen or chlorine) and sensitivity of the device to catalytic poisons that often accompany hydrogen (e.g. CO). These are major reasons why they have not been recognized in the market so far.

# References

[1]  E. Zoulias, et al. A review on water electrolysis 2006 cited 12. 10. 2013; Available from: http://www.cres.gr/kape/publications/papers/dimosieyseis/ydrogen/A%20REVIEW%20ON%20WATER%20ELECTROLYSIS.pdf

[2]  V. S. Bagotsky, Phase Boundaries (Interfaces) between Miscible Electrolytes. In Fundamentals of Electrochemistry. John Wiley & Sons, New York, 2005, 69–77. ISBN 0471700584.

[3] A. M. Couper, D. Pletcher, and F. C. Walsh, Electrode materials for electrosynthesis. Chem. Rev. 90(5): 837–865, 1990.

[4] D. Pletcher and F. C. Walsh, Industrial Electrochemistry. 2nd Edition. Chapman & Hall, London, 1990. P. 654. ISBN 0412304104.

[5] E. Steckhan, Electrochemistry 3. Organic Electrochemistry. In Ullmann's Encyclopedia of Industrial Chemistry. Wiley-VCH, Verlag, 2000. ISBN 978-3-527-30673-2.

[6] G. Chen, Electrochemical technologies in wastewater treatment. Sep. Purif. Technol. 38(1): 11–41, 2004.

[7] K. Jüttner, U. Galla, and H. Schmieder, Electrochemical approaches to environmental problems in the process industry. Electrochim. Acta. 45(15–16): 2575–2594, 2000.

[8] J. Jörissen, Electro-organic synthesis without supporting electrolyte: Possibilities of solid polymer electrolyte technology. J. Appl. Electrochem. 33(10): 969–977, 2003.

[9] H. Vogt and G. Kreysa, Electrochemical Reactors. In Ullmann's Encyclopedia of Industrial Chemistry. Wiley-VCH, Verlag, 2000. ISBN 978-3-527-30673-2.

[10] T. Torimoto, et al.. New frontiers in materials science opened by ionic liquids. Adv. Mater. 22 (11): 1196–1221, 2010.

[11] D. H. Kroon and L. M. Ernes, Case histories: MMO-coated titanium anodes for cathodic protection; Part 2. Mater. Performance. 46(6): 24–210, 2007.

[12] V. Panić, et al. The influence of the aging time of $RuO_2$ and $TiO_2$ sols on the electrochemical properties and behavior for the chlorine evolution reaction of activated titanium anodes obtained by the sol-gel procedure. Electrochim. Acta. 46(2–3): 415–421, 2000.

[13] C. Comninellis and G. P. Vercesi, Characterization of DSA®-type oxygen evolving electrodes: Choice of a coating. J. Appl. Electrochem. 21(4): 335–345, 1991.

[14] S. Trasatti, Electrocatalysis: Understanding the success of DSA®. Electrochim. Acta 45 (15–16): 2377–2385, 2000.

[15] S. Litster and G. McLean, PEM fuel cell electrodes. J. Power Sources 130(1–2): 61–76, 2004.

[16] I. Moussallem, et al.. Chlor-alkali electrolysis with oxygen depolarized cathodes: History, present status and future prospects. J. Appl. Electrochem. 38(9): 1177–1194, 2008.

[17] R. Souzy, et al.. Functional fluoropolymers for fuel cell membranes. Solid State Ionics 176 (39–40): 2839–2841, 2005.

[18] Eurochlor. Chlorine Industry Review 2012–2013, 2013 cited 10. 12. 2013; Available from: http://www.eurochlor.org/media/70861/2013-annualreview-final.pdf.

[19] T. Brinkmann, G. G. Santonja, F. Schorcht, S. Roudier, and L. D. Sancho, Best available techniques (BAT) reference document for the production of chlor-alkali. Eur. IPPC J. 2014, doi:10.2791/13138.

[20] Y.-H. Wang and Q.-Y. Chen, Anodic materials for electrocatalytic ozone generation. Int. J. Electrochem. 7, 2013. https://www.hindawi.com/journals/ijelc/2013/128248/.

[21] K. Scott, Membranes for Electrochemical Cells. In Handbook of Industrial Membranes. K. Scott Ed.. 2nd Edition. Elsevier Science, Oxford, 1995, 773–790. Section 18. ISBN 1856172333.

[22] E. Brillas, I. Sirés, and M. A. Oturan, Electro-Fenton process and related electrochemical technologies based on Fenton's reaction chemistry. Chem. Rev. 109(12): 6570–6631, 2009.

[23] A. Da Pozzo, et al. An experimental comparison of a graphite electrode and a gas diffusion electrode for the cathodic production of hydrogen peroxide. J. Appl. Electrochem. 35(4): 413–419, 2005.

[24] A. Alvarez-Gallegos and D. Pletcher, The removal of low level organics via hydrogen peroxide formed in a reticulated vitreous carbon cathode cell, Part 1. The electrosynthesis of hydrogen peroxide in aqueous acidic solutions. Electrochim. Acta 44(5): 853–861, 1998.

[25] F. Alcaide, P. L. Cabot, and E. Brillas, Fuel cells for chemicals and energy cogeneration. J. Power Sources 153(1): 47–60, 2006.

[26] M. T. Musser, Adipic Acid. In Ullmann's Encyclopedia of Industrial Chemistry. Wiley-VCH Verlag, Weinheim 1–11, 2000. ISBN 978-3-527-30673-2.

[27] C. A. C. Sequeira and D. M. F. Santos, Electrochemical routes for industrial synthesis. J. Brazilian Chem. Soc. 20(3): 387–406, 2009.

[28] Z. Ogumi, et al.. Application of the SPE method to organic electrochemistry. VII. The reduction of nitrobenzene on a modified Pt-nafion. Electrochim. Acta. 33(3): 365–369, 1988.

[29] Z. Ogumi, et al.. Application of the SPE method to organic electrochemistry. XIII. Oxidation of geraniol on Mn, Pt-Nafion. Electrochim. Acta. 37(7): 1295–1299, 1992.

[30] Z. Ogumi, K. Nishio, and S. Yoshizawa, Application of the SPE method to organic electrochemistry. II. Electrochemical hydrogenation of olefinic double bonds. Electrochim. Acta. 26(12): 1779–1782, 1981.

[31] Z. Ogumi, S. Ohashi, and Z. Takehara, Application of the SPE method to organic electrochemistry. VI. Oxidation of cyclohexanol to cyclohexanone on Pt-SPE in the presence of iodine and iodide. Electrochim. Acta. 30(1): 121–124, 1985.

[32] Z. Ogumi, et al.. Application of the solid polymer electrolyte (SPE) method to organic electrochemistry. III. Kolbe type reactions on Pt-SPE. Electrochim. Acta. 28(11): 1687–1693, 1983.

[33] S. A. Grigoriev, Electrochemical systems with a solid polymer electrolyte. Part II. Water electrolyzers, bifunctional elements, and hydrogen concentrators. Chem. Petrol. Eng. 48 (9–10): 535–539, 2013.

[34] S. A. Grigoriev, Electrochemical systems with a solid polymer electrolyte. Part I. General information about electrochemical systems with an SPE. Chem. Petrol. Eng. 48(7–8): 478–483, 2012.

[35] S. A. Grigoriev, et al.. Description and characterization of an electrochemical hydrogen compressor/concentrator based on solid polymer electrolyte technology. Int. J. Hydrogen Energy 36(6): 4148–4155, 2011.

Martin Paidar, Karel Bouzek, Petr Mazúr, Aleš Černín

# 14 Electromembrane processes for energy conversion

Electromembrane processes play an increasingly important role in the area of energy conversion, too. It was primarily perceived exclusively as a highly efficient conversion process of the energy stored in chemical bonds into electrical energy. Gradually, this view extended to other areas, especially to processes of accumulating energy and recovering it. The research and development of unorthodox solutions in the area of energy acquisition also gained importance. At present, most attention is paid to studying processes within the so-called hydrogen economy.

## 14.1 Hydrogen economy

The term "hydrogen economy" [1] covers a number of technological steps that allow converting electricity into the energy of chemical bonds of hydrogen, its storage and subsequent conversion back to electrical energy, when needed. The main objective is the highest possible efficiency of the entire cycle shown in Fig. 14.1. There are many variations of this cycle, reflecting specific local conditions and needs. For example, a possible variation can be seen in the use of surplus hydrogen as a chemical raw material or fuel for mobile and other applications.

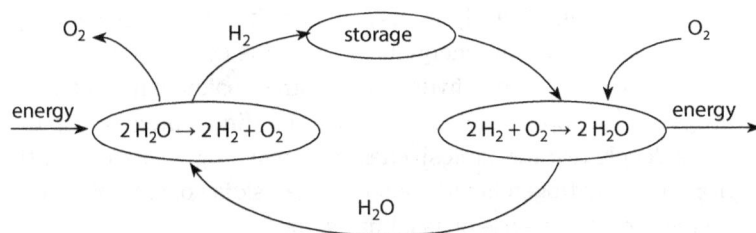

**Fig. 14.1:** Scheme of basic cycle of hydrogen economy.

The cycle is named after hydrogen, that is, the element at its centre. The reason for selecting hydrogen as an energy carrier lies in its properties as well as in the amount and form in which it is present on Earth.

Let's deal with the first of these two reasons. It is the energy contained in the H–O bond in the water molecule. As indicated in the chemical reaction described by eq. (14.1), this energy reaches the value of 286.0 kJ mol$^{-1}$ for one mole of water formation:

https://doi.org/10.1515/9783110739466-017

$$H_2(g) + \tfrac{1}{2}\,O_2(g) = H_2O(l), \quad \Delta H_r^\circ = -286.0\,\text{kJ mol}^{-1}, \quad \Delta G_r^\circ = -237.2\,\text{kJ mol}^{-1} \quad (14.1)$$

Of the value of $\Delta H_r$, 237.2 kJ mol$^{-1}$ represents the useful work, while 48.8 kJ mol$^{-1}$ is irreversibly released during the reaction in the form of thermal energy. This fact affects, to a certain extent, the design of technological units. The value of useful energy shows that the mass density of hydrogen energy is very high and reaches more than 140 MJ kg$^{-1}$. On the other hand, for practical applications it is also necessary to take into account the volume energy density of this substance [1] – this is due especially to the fact that hydrogen is a gas under normal conditions, and the volume energy density is crucial for storage. This has led to the research and development of several storage methods, based on the amount of hydrogen to be stored and possible need of mobility of the storage system. The comparison of the energy densities of selected major fuels with hydrogen is shown in Fig. 14.2. An important property of hydrogen is also its high reactivity, which allows the reactions to occur to the necessary extent with a minimum amount of catalyst.

The second aspect is the occurrence of hydrogen. It is known that although hydrogen is the most widespread element in the universe, it does not occur in its molecular form in a significant amount on Earth. This is particularly due to its reactivity at appropriate initiation. It reacts with all elements except noble gases. For these reasons, hydrogen only occurs on the Earth in the form of compounds, especially in water. Therefore, its reserves are distributed relatively evenly, specifically from the geographic point of view, and this raw material is sufficiently accessible for most countries.

Besides these positives, the use of hydrogen as an energy carrier also has some problems. Probably the most serious is the already mentioned gaseous state of hydrogen under usual conditions. It significantly limits the installation storage capacity and has a negative impact on the overall energy balance of the cycle due to the energy necessary to compress stored gaseous hydrogen. Another major complication is the reactivity of hydrogen at appropriate initiation. Since the Hindenburg airship disaster in 1937 [3], people deeply fear the explosiveness of the hydrogen mixtures with air. Although this property of hydrogen is indisputable, the risk is comparable if not lower than the risk connected with the use of gasoline as fuel.

As shown in Fig. 14.1, the hydrogen economy contains two basic electromembrane processes:
- Fuel cell converting the energy of chemical bond between hydrogen and oxygen to electrical energy.
- Electrolysis of water as a means of conversion in the opposite direction, that is, storing electrical energy in the energy of chemical bond.

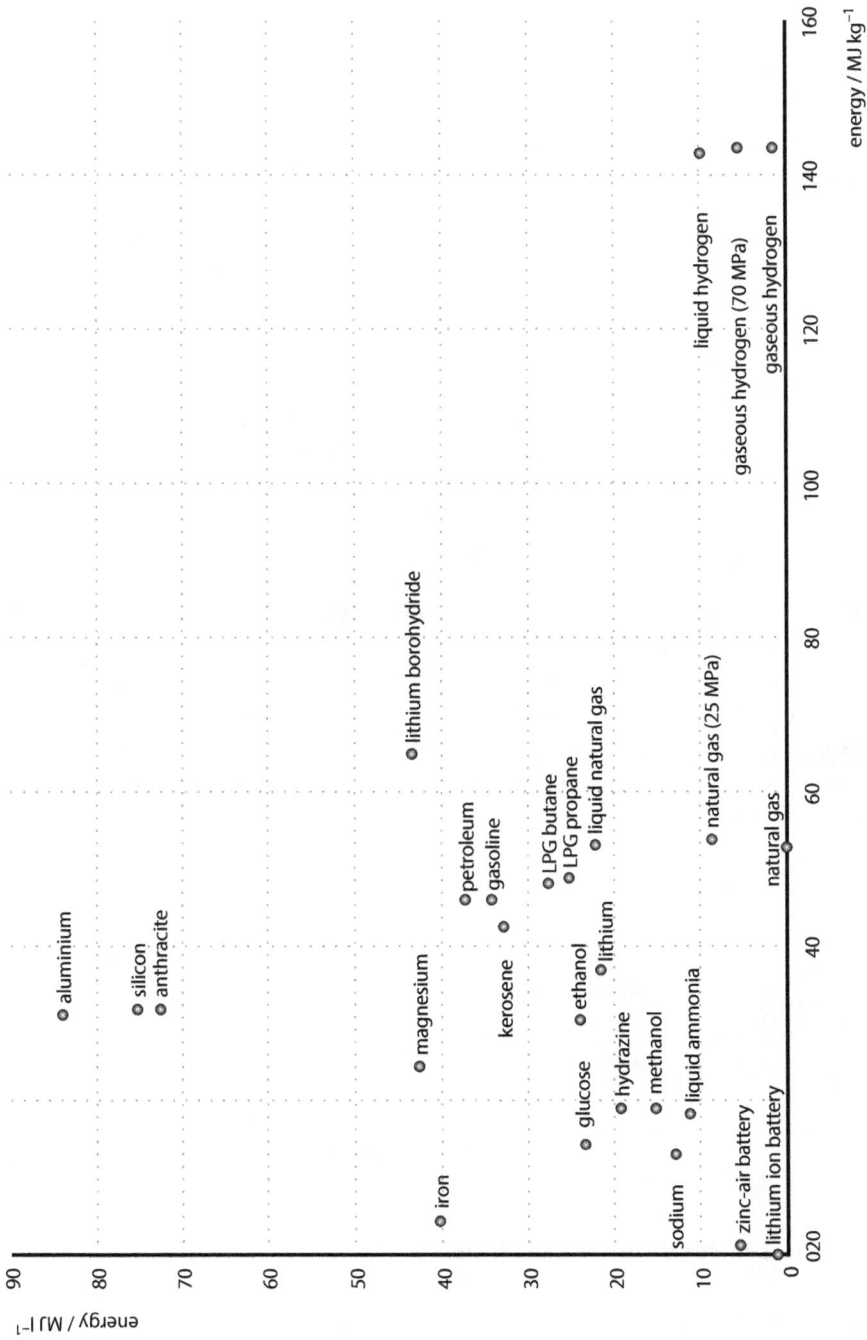

**Fig. 14.2:** Comparison of energy densities of selected fuels and substances [2].

## 14.1.1 Fuel cells

The fuel cell is generally an electrochemical reactor that allows the continuous process of a redox reaction with spacial separation of the reduction and oxidation step. While in the case of a purely chemical reaction, the reaction described by eq. (14.1) produces water and releases, exclusively, the thermal energy corresponding to $\Delta H_r$ of the reaction, the electrochemical reaction in its ideal form releases the thermal energy $T \Delta S$ and performs the useful work corresponding to $\Delta G_r$, see eqs. (3.73) and (3.74).

The reason for this difference is the already mentioned separation of the reduction and oxidation steps. Thus, the summary reaction described by eq. (14.1) can be divided into two partial reactions. In this model case, we will use the most frequently cited case when protons are used as ions carrying the electric charge. In this case, the partial reactions can be written in the form of equations as follows:

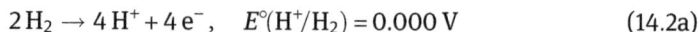

$$2\,H_2 \rightarrow 4\,H^+ + 4\,e^-, \quad E^\circ(H^+/H_2) = 0.000\ V \tag{14.2a}$$

$$O_2 + 4\,H^+ + 4\,e^- \rightarrow 2\,H_2O, \quad E^\circ(O_2, H^+/H_2O) = 1.229\ V \tag{14.2b}$$

It is clear that in the case of non-electron-conductive separation of these reactions, equilibria potentials are achieved on electrodes shortly after reactants are led to them. Thus, between the electrodes where these reactions occur, a potential difference is generated, ideally corresponding to $\Delta G$ of the overall reaction according to eq. (3.74). In a real system, however, the occurrence of electrode reactions is connected with irreversible processes, which reduce the potential difference – that is, the voltage on the cell.

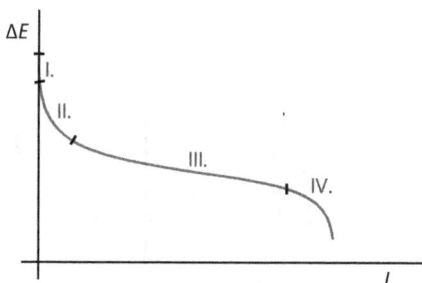

Fig. 14.3: Scheme of the load curve of a fuel cell with four basic areas (explanation in the text).

### 14.1.1.1 Factors limiting the fuel cell efficiency

The factor limiting the efficiency and use of the fuel cell can be best documented on its **load curve**. It is one of the basic characteristics describing the efficiency of a device and range of convenient operating parameters, in this particular case, of the current load. The scheme of the load curve is shown in Fig. 14.3.

The figure illustrates the basic non-idealities of the behaviour of a real fuel cell. They can be divided into the following groups:
- penetration of reactants through the electrolyte,
- activation (polarization) overvoltage,
- ohmic loss,
- mass transfer.

The **penetration of reactants** through the electrolyte to the opposite electrode is the only factor influencing the efficiency of the fuel cell in the entire current load extent, including the open circuit potential, that is, cell not loaded with the current. In this context, the penetration of hydrogen as the fuel to the oxygen electrode (cathode) is usually mentioned [4]. The loss of the faradaic efficiency of the cell is then listed as a primary effect. Faradaic efficiency and its definition are discussed in Section 14.1.1.2. The decrease of faradaic efficiency corresponds to the fact that a part of the fuel is not used for the desired electrode reaction, but is lost by reaction with oxygen flowing through this space. A side effect is the change of the electrode potential of the oxygen electrode. The presence of hydrogen at the cathode surface leads to formation of the mixed electrode potential of this electrode, which, in this case, is lower that the equilibrium potential. This effect represents a second mechanism causing the loss of the energy conversion efficiency in the fuel cell. Unlike faradaic efficiency, this aspect is also important for the possible penetration of oxygen to the hydrogen electrode. In this case, only the increase of the hydrogen electrode potential is considered as loss. This non-ideality of the electrolyte (in the present case, membrane) function is usually indicated by the voltage decrease on the cell when the electrical circuit is disconnected; see area I in Fig. 14.3.

The **activation overvoltage** determines the efficiency of the fuel cell in the area of low current loads. This area (area II in Fig. 14.3) is characterized by a rapid decrease of the voltage on the cell, connected with the current load increase. The rate-determining step in this area is the charge transfer kinetics, described by the **polarization curve**; see Section 3.5.4. The voltage decrease on the cell then represents the driving force necessary to ensure the proper rate of electrode reactions. The cathodic reduction of oxygen is responsible for a significantly higher share of this driving force.

The **ohmic loss** is characterized by a linear decrease of the cell voltage with the increasing current load; see area III in Fig. 14.3. In this area, the electrode reaction is already so rapid that the passed electric charge or the current density on the electrodes is determined by the ohmic resistance of the system, that is, conductors, endplates and bipolar plates, electrodes and, especially, electrolyte. In the description of the behaviour of solid electrolytes, the use of the linear ohmic resistance represents a substantial simplification, albeit without any physical basis. This simplification does not represent a significant deviation from the reality within the range of the current densities and conditions corresponding to the concerned part of the load curve. The ohmic resistance is, therefore, often used in practice to evaluate the quality of these components.

**Mass transfer** begins to apply as a rate-determining step in the area of high current loads. The rate of the reactants transfer to the active surface of the electrode reaches values comparable to the rate of the electrode reactions and begins to limit it. As shown in area IV in Fig. 14.3, this phenomenon is characterized by a sharp decrease of the voltage with the increasing current density.

### 14.1.1.2 Fuel cell efficiency

The efficiency of the fuel cell and method of its expression were discussed, in general, as the efficiency of the galvanic cell in Section 3.5.1. The highest thermodynamically obtainable cell efficiency can be expressed in the form comparing the reaction Gibbs energy and reaction enthalpy; see eq. (3.82). Using eq. (3.74), this relation can be expressed as:

$$\eta_{T,GC} = - \frac{nFE_{eq}}{\Delta H_r} \tag{14.3}$$

where $E_{eq}$ corresponds to the equilibrium voltage of the fuel cell.

However, this description does not take into account the energy losses caused by the kinetics of the electrode reactions or ohmic losses of the voltage due to the electric current passage through individual components of the fuel cell and the entire system. These can be expressed by the following equation:

$$\eta_{kin,GC} = - \frac{nF\Delta E_r}{\Delta H_r} \tag{14.4}$$

where $\eta_{kin,GC}$ corresponds to the kinetically limited voltage efficiency of the fuel cell and $\Delta E_r$ to the actual voltage on the current-loaded fuel cell.

It is clear that both of these efficiencies assume zero faradaic losses of the process in the fuel cell and count in only the voltage losses. Faradaic efficiency $\Phi_I$ is defined by relations (3.85) and (3.86) as the ratio between the electric charge used to conduct the required electrode reaction and the total charge passed through the system. In the case of the fuel cell, it is more convenient to express this efficiency as the ratio between the passing electric current and the theoretical electric current calculated from the amount of the fuel consumed using Faraday's law, as follows:

$$\Phi_{I,GC} = \frac{I}{\dot{n}_f z_f F} \tag{14.5}$$

where $\dot{n}_f$ corresponds to the amount of substance of the fuel consumed in the fuel cell per unit of time (unit: mol s$^{-1}$), $z_f$ is the number of electrons transferred in the fuel oxidation reaction and $I$ is the electric current passing through the fuel cell.

Since the instantaneous value of the electric current and consumed [5] fuel is used to measure the efficiency, it can be described as the instantaneous faradaic efficiency.

The total efficiency of energy conversion in the fuel cell can be expressed by the relation analogic to eq. (3.86) as follows:

$$\eta_{tot,GC} = \eta_{kin,GC} \, \Phi_{I,GC} \qquad (14.6)$$

It is clear that the fuel cell efficiency can be further influenced by operating parameters such as the temperature or pressure of individual gases entering the system or current load. When the pressure of the entering gases is increased, in addition to the fuel voltage increase according to eq. (3.76), we can expect an increase in mass transfer intensity due to the increase of the volume concentration of reactants, extending the practically usable current loads of the fuel cell.

Then, it is necessary to include in the expression of the total system efficiency the so-called energy overhead of the system, in particular the energy consumption connected with pumping or compression of circulating media, system and temperature control.

### 14.1.1.3 Fuel cell design

The details of the fuel cell design vary according to the particular type and certain aspects according to the particular version. However, their basic principles are the same for all types. They are based on the effort to minimize the negative effects decreasing the efficiency of the resulting unit, while maintaining the design sufficiently simple to ensure the lowest possible production costs and longest possible lifetime. In the following text, we will document the basics of the design on a PEM fuel cell. The abbreviation PEM can be explained in two ways, either **proton exchange membrane**, or **polymer electrolyte membrane**. However, most of the facts mentioned above have general validity, and differences between particular types of devices types primarily result from the nature of the materials used or requirements of their operating conditions.

The basic building blocks of any electrochemical system, including fuel cells, are two electrodes and the electrolyte separating them. This is schematically shown in Fig. 14.4. In this device type, maximum effort is made to use solid electrolytes or at least electrolytes filling a relatively precisely defined space. The specific composition of the electrolyte will be dealt with in the discussion about individual types of fuel cells, emphasizing their limitations and advantages.

The requirements for both electrodes, oxygen and hydrogen, or cathode and anode, are very similar and result from the fact that the considered reactants, with some exceptions, are in the gaseous state in technically important fuel cell applications at their operating temperature. These facts lead to the requirement of using electrodes with a large active surface, that is, electrodes that would allow working with a high

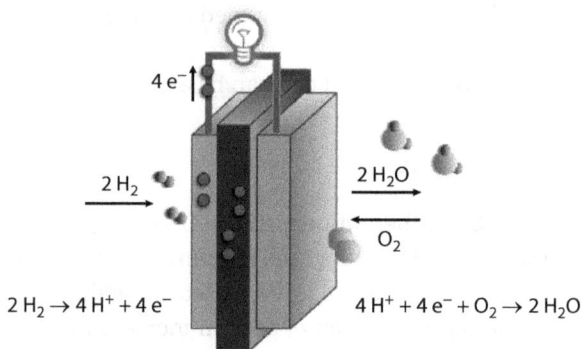

**Fig. 14.4:** Scheme of the basic design of one cell consisting of anode/ion selective membrane/cathode composite called membrane electrode assembly (MEA).

current density relative to the geometric surface, even if the reactant is very "diluted" in volume, from the electrochemical point of view. The second aspect is the fact that the gaseous phase represents an electron and ion insulator. Thus, besides the transport of the electric charge in the form of electrons, the electrode must ensure effective transport of ions from the active area, or their supply to the active area where the electrode reaction occurs. An effective solution, according to the present state of knowledge, is only the use of a three-dimensional porous electrode providing a large active surface as also sufficient ion and electron conductivity. There are three basic approaches, which allow meeting all requirements, namely electron-conductive electrodes:
-   with pores partially filled with liquid electrolyte,
-   with pores partially filled with solid electrolyte,
-   of material exhibiting both electron and ion conductivity.

The principle of the function of these electrode types is discussed in more detail in Section 13.1.3.2.

This system must be supplemented with a component securing the effective distribution of reactants and removal of products and electric charge. This role is usually played by distribution plates between which the MEA is bound. Distribution plate active area is usually equipped with distribution channels whose geometry must meet many requirements. Let us list at least the basic ones:
-   uniform surface coverage of the electrode with reactants,
-   limited pressure gradient associated with the reactant passage through the entire channel length,
-   sufficient electrical contact with the electrode, limiting the ohmic loss of voltage,
-   low production costs.

Typically, three basic types of the geometric arrangement of distribution channels are used, as shown in Fig. 14.5.

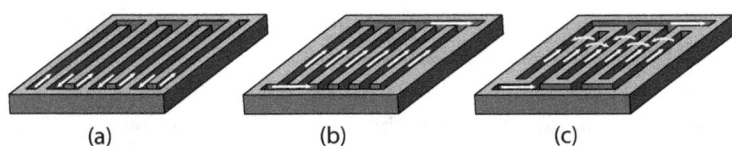

**Fig. 14.5:** Scheme of the basic types of the distribution channel geometry: (a) serpentine, (b) parallel channel and (c) interdigitated.

Of these types, the serpentine arrangement is the most common (Fig. 14.5a). It is marked out by the fact that the reactant is, in an ideal case, forced to flow along the entire active surface of the electrode, and if the amount of the reactant is sufficient, there cannot be any area on the electrode without adequate reactant supply. However, the serpentine arrangement is convenient mainly for small electrode surfaces; otherwise, the reactant trajectory inside the channel would be too long, leading to a significant pressure loss. It could negatively influence the mechanical stability of the cell or the reactant penetration through the membrane.

The arrangement in parallel channels (Fig. 14.5b) avoids the problem of excessive pressure losses of the reactant on larger active surface of the electrode. There are no significant changes of the reactant composition on the electrode surface but it is more difficult to ensure a uniform distribution of the reactant flow along the electrode surface. The risk that areas where the reactant flow dynamics is substantially slower may occur is much higher in this case. In addition, the linear velocity of the reactant flow is slower, which negatively influences the mass transfer to the catalytic electrode layer. For larger electrode surface, a combination in which several parallel channels run in the serpentine arrangement is used.

The interdigitated arrangement of distribution channels (Fig. 14.5c) is a system of feed and collection channels, which are not directly interconnected, and reactants are thus forced to pass through the electrode structure. To a certain extent, this arrangement combines the advantages of the previous two. It secures a relatively uniform distribution of reactants along the electrode surface, while the length of the channel passed by the reactant is relatively short. The fact that the reactant is forced to pass directly through the structure of the gas diffusion electrode significantly increases the mass transfer between the flowing reactant and active part of the electrode, that is, its catalytic layer. It also leads, however, to a substantial pressure gradient. In addition, the system becomes significantly dependent on the constant permeability of the electrode structure for the reactant over the life of the cell as a whole. However, in the course of the operation, the carbon parts of the gas diffusion electrode may erode (e.g. due to the volume changes of the polymer electrolyte or pressure gradient in the gaseous phase). The released particles may be drifted by the reactant flow and subsequently block another part of the electrode. Thus, the distribution of the reactant along the electrode surface becomes uneven.

A similar effect will occur if the cell is not bound completely uniformly, which can cause local collapses of the porous structure of the electrode.

The material of the distribution plates is also very important. As shown in Fig. 14.5, the channel structure is rather complex, and from the economic point of view, it is preferable to produce them by injection moulding, hot pressing or stamping, that is, technologies that allow producing a large number of parts simply with a single mould or stamp. An economically more demanding alternative is machining of metals or other materials. However, it is necessary to bear in mind the requirements on the distribution plate material. The primary requirements are a good electrical conductivity and high corrosion stability. The second requirement has become crucial for several reasons. Let us cite the following two as the main ones. Even though reactants are gases in most cases and a properly working system should not contain any major amount of aggressive liquids, the environment can be aggressive and cause the corrosion. Also, the solid electrolyte may contain radicals or other aggressive substances that can contribute to the corrosion of distribution plates. Any corrosive degradation leading to the release of metal cations can have fatal consequences for the lifetime of the system. A shorter lifetime can be primarily caused by the degradation of the polymer electrolyte whose functional group will be blocked by the ions released from the corroding plates. Some ions can accumulate on the hydrogen electrode surface and disrupt its catalytic activity by deposited layers. In the first stages of the research, the natural choice of the material for this purpose was carbon in low-temperature applications, primarily in the form of graphite, and high-alloyed steel in high-temperature applications [5, 6]. However, in low-temperature applications, the use of this material has many problems, especially high production costs and considerable weight. Much effort has been invested in the possibility of using polymer or composite materials. So far, the most common problem is their insufficient electrical and thermal conductivity. The possibility of using metal plates has been also studied, whose resistance to corrosion is sufficient and allows resolution of the remaining problems. In high-temperature variants, alloyed steels or other metal alternatives remain irreplaceable for these purposes.

### 14.1.1.4 Types of fuel cells

There are many types of fuel cells, using different construction materials, operating under different conditions and suitable for different applications. There are many approaches to classifying fuel cells. The most important is the classification based on the operating temperature, while further categorization is usually done on the basis of the ion mediating the charge transfer through the electrolyte [7].

Depending on the operating temperature, we distinguish between three basic groups of fuel cells:

(i)  low temperature
(ii)  medium temperature and
(iii) high temperature.

As already stated, these individual types usually allow a finer division, primarily according to the electrolyte used.

**(i) Low-temperature fuel cells** include two process variants, differing, as has been already mentioned, in the type of the ion transporting the charge through the electrolyte. Historically, in the first type of process described by Grove [8], the charge is transported by protons – it is the so-called **acid path**. Nowadays, the liquid electrolyte is replaced with a cation-selective membrane based on perfluorinated sulphonated polymer (see Section 13.1.4). This type of fuel cell, known as PEM, was used in Section 14.1.1.3 as a model example to describe the fuel cell design, so it is not necessary to discuss its design in detail at this place. The basic advantages of this type of fuel cells include its high flexibility and high intensity of the process. In extreme cases, it can reach up to 10 kW m$^{-2}$ (related to the membrane surface), although the performance is usually about half of this value. On the other hand, this type of fuel cell is completely dependent on relatively expensive construction materials, represented, in particular, by the catalyst based on platinum nanoparticles and perfluorinated polymer electrolyte. The operating temperature of this device typically ranges between 50 and 80 °C, depending on the producer and application type. This operating temperature does not provide much space to utilize waste heat, so this type of fuel cell is mainly intended for use in areas where high flexibility and specific performance prevail over the mentioned negatives. These primarily include mobile applications, especially in the automotive industry. However, the possibility of using this device for small cogeneration units is still being pursued.

A variant of this process is represented by **direct methanol fuel cells** [9]. Unlike the conventional fuel cells of the PEM type, they use methanol as the fuel instead of hydrogen. This substitution is motivated by the high specific volume energy of methanol, which is in the liquid state, under normal conditions. The reaction of methanol on the platinum anode is usually described as a sequence of electrode reactions, as shown in Fig. 14.6.

The main problem is the fact that the rate determining the last step of this mechanism is extremely slow. The reason is the strong CO bonding as an intermediate product of oxidation onto the platinum catalyst surface. Therefore, the reaction is substantially slowed down by catalyst poisoning. In practice, this is resolved by using higher catalyst load and/or Pt–Ru alloy instead of pure platinum. Thanks to the bifunctional reaction mechanism, adding Ru decreases the platinum catalyst poisoning. Another problem is the penetration of electroneutral methanol through the membrane into the cathode surface. This leads to the occurrence of mixed potential on the cathode, as was already discussed in connection with the penetration of hydrogen through the membrane, but

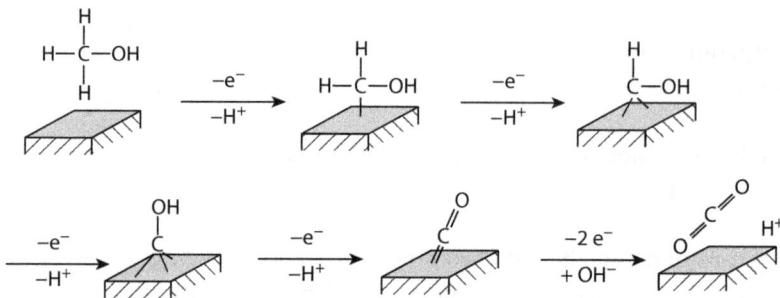

**Fig. 14.6:** Scheme of reaction mechanism of methanol oxidation on platinum electrode.

mainly causes the cathode catalyst poisoning. For these reasons, the attention paid to this type of device has gradually decreased, and its use is primarily expected in low-performance mobile devices preferring maximum independence from fuel supplies.

Although the second variant of low-temperature fuel cells, utilizing the OH$^-$ ion as the charge carrier in the electrolyte and called the **alkaline path,** was described as the second, it was the first to see any significant technical development and practical application [10]. The electrode reactions occurring in this type of fuel cell are summarized by the following equations:

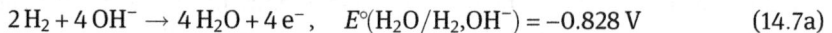

$$2\,H_2 + 4\,OH^- \rightarrow 4\,H_2O + 4\,e^-, \quad E^\circ(H_2O/H_2,OH^-) = -0.828\ \text{V} \tag{14.7a}$$

$$O_2 + 4\,H_2O + 4\,e^- \rightarrow 4\,OH^-, \quad E^\circ(O_2,H_2O/OH^-) = 0.401\ \text{V} \tag{14.7b}$$

The reason for the priority development of alkaline fuel cells consisted, and to a great extent still consists, in the material issues. The alkaline environment allows utilizing a substantially wider range of construction materials and avoiding problems associated with the acid variant. In terms of stability, nickel or mixed oxide of nickel and other metal (or just mixture of individual oxides) meet the requirements. With these, it was possible to take the first step and to start designing gas diffusion electrodes, which led to practically applicable values of the power output of fuel cells. Several demonstration units were built in the mid-twentieth century, documenting the practical applicability of fuel cells. These demonstrations culminated in the Apollo space programme [7]. A significant disadvantage of the alkaline path is the absence of a suitable solid anion-selective electrolyte, which would exhibit properties, at least, similar to commercially available cation-selective perfluorinated sulphonated materials.

**(ii) Medium-temperature fuel cells** basically include only two variants of one fuel cell type. In this context, the medium temperature typically means the range between 150 and 200 °C. The advantages of the increased operating temperature consist in the higher rate of electrode reaction and lower sensitivity of the platinum catalyst to some types of catalytic poisons, especially to carbon monoxide. The second motivation for the development of this technology is the higher temperature at which heat energy is released. This is connected with the possibility of its efficient use and increase in the total effectiveness of the process. From this point of view, medium-temperature fuel cells can be considered a technology suitable primarily for stationary cogeneration units and applications requiring lower outputs and not completely stable production. In terms of design, this type of fuel cells is very close to acidic low-temperature units, and the main difference is in the electrolyte used. At the operating temperatures of a medium-temperature fuel cell and atmospheric pressure, perfluorinated sulphonated polymers dry out and lose their ion conductivity. Their chemical stability can be limited, too. Thus, phosphoric acid fixed in a porous matrix was used as the electrolyte, in the past. At temperatures up to 200 °C, this acid exhibits negligible vapour pressure, and in the presence of a small amount of water, also sufficient proton conductivity. However, the problem with losses of the acid from the porous matrix led to the search for a more stable alternative. The option of fixing the acid in a suitable polymer matrix was selected. At present, derivatives of polybenz-imidazole are the most commonly used for these purposes [11]. This also influences the structure of the catalytic layers of the electrodes and the differences area, mainly in the area of the binders used. There is an unpleasant side effect of the relatively significant corrosion aggressiveness of phosphoric acid at the operating temperature of the cell. For this reason, possibilities are being studied about extending the life of this type of fuel cells, which have not reached the values typical for low-temperature units, so far.

(iii) In **high-temperature fuel cells**, the operating temperatures reach even higher values, up to 600–1,000 °C. However, the upper temperature limit is used relatively rarely. In oposite, and there is tendency to work at temperatures below 800 °C. There are two basic variants of this type of fuel cells. In the past, the more intensively developed molten carbonate fuel cell was the cell using a eutectic mixture melt of $Li_2CO_3$ with $Na_2CO_3$ or $Li_2CO_3$ with $K_2CO_3$ as the electrolyte, fixed in a porous matrix of $y$-$LiAlO_4$. The operating temperature is typically around 650 °C, that is, above the melting point of the carbonate mixture [7]. The resulting melt provides an electrolyte with sufficient conductivity. In this case, the ion ensuring the charge transfer through the electrolyte is carbonate. This leads to some modification of the electrode reactions, and they are described by the following equations:

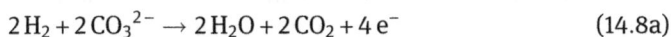

$$2\,H_2 + 2\,CO_3^{2-} \rightarrow 2\,H_2O + 2\,CO_2 + 4\,e^- \tag{14.8a}$$

$$CO + H_2O \rightarrow H_2 + CO_2 \tag{14.8b}$$

$$O_2 + 2\,CO_2 + 4\,e^- \rightarrow 2\,CO_3^{2-} \tag{14.8c}$$

Naturally, eq. (14.8b) is not an electrode reaction but a reaction of the CO present in the fuel with water vapour under operating conditions of the fuel cell.

In this type of cell, it is also not necessary to use electrocatalysts based on platinum-group metals. This is primarily due the rapid kinetics of electrode reactions caused by the high operating temperature. However, it means substantially higher demands on the construction materials in term of corrosion resistance. In this case, the use of carbon-based material is no longer possible, and the oxygen electrode is typically made of porous nickel. It undergoes many significant changes during the cell operation, the first of which is the oxidation of metallic nickel to NiO. The oxide structure is penetrated by the lithium ions, changing the material porosity. The oxygen electrode achieves the highest performance when its pores are partly filled with the electrolyte melt. Nickel is also used in the hydrogen electrode design, in the form of a sintered powder. This electrode requires a different pore structure than the oxygen one. To prevent sintering of the electrode during operation, the electrode is made of nickel powder containing 2 to 10 mass % of chrome. The distribution plates are usually made of stainless steel and nickel-coated on the hydrogen electrode side.

The second type of high-temperature fuel cells are fuel cells with the ceramic (oxidic) electrolyte. The development of this cell type has become predominant especially in the last years. It is due to the fact that the originally used ceramic electrolyte required high operating temperatures up to 1,000 °C, which meant inadequately high demands on the construction materials used. The continuing research in this field allowed decreasing the operating temperature to the range of approximately 600–800 °C, which is more acceptable from the material point of view. Besides, the electrolyte used is solid and does not contain any corrosive liquid components, which makes the design easier. For these reasons, this variant has been attracting more and more attention. The traditional solid oxide electrolyte consists of $ZrO_2$ stabilized by $Y_2O_3$. This material is an $O^{2-}$ ion conductor due to the oxygen vacancies in its structure. It leads to a change of the electrode reactions, which can be expressed in the form of the equations:

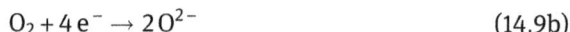

$$2\,H_2 + 2\,O^{2-} \rightarrow 2\,H_2O + 4\,e^- \qquad (14.9a)$$

$$O_2 + 4\,e^- \rightarrow 2\,O^{2-} \qquad (14.9b)$$

The electrolyte composition has undergone some changes over the years. $ZrO_2$ doped with $Y_2O_3$ (YSZ) is currently considered to be primarily an electrolyte for high-temperature electrolysis of water vapour, in which, unlike high-temperature fuel cells, the operating temperature is generally maintained higher, that is, at about 800 °C, to decrease the equilibrium potential of the reaction. For high-temperature fuel cells, where the preferred operating temperature is around 600 °C, this material is no longer suitable, and materials with higher ion conductivity at optimum operating conditions are selected. Today, the most intensively explored materials include especially $Ce_{1-x}Gd_xO_{2-\delta}$ (GDC) [12].

Besides electrolytes transporting oxygen ions, there are also proton-conductive ceramic materials. The most common material of this type is $BaCeO_3$, doped at the cerium position with an element with more electrons, for example, terbium or samarium [13]. However, the research on fuel cells with proton-conductive electrolyte is currently less intense than the research of systems with GDC-based electrolytes.

Like the electrolyte, the electrodes of high-temperature fuel cells are of ceramic materials that best resist the operating conditions. As in low-temperature PEM systems, it is necessary to ensure effective transport of both electrons and ions in the 3D structure of the electrode. With the cathode, the situation is quite simple: materials with the perovskite structure and general formula $ABO_3$ are used, and these can be doped at A and B positions to achieve both ion and electron conductivity. In high-temperature fuel cells, the most frequently used materials are of type $La_{1-x}Sr_xMnO_3$ (LSM), $La_{1-x}Sr_xCoO_3$ (LSC) doped at position A, and $La_{1-x}Sr_xFe_yCo_{1-y}O_3$ (LSCF) doped at both positions [12]. The simultaneous ion and electron conductivity of the anode is achieved by mixing two materials, most commonly, nickel and ion-conductive electrolyte material. The electrode made of such composite must have a convenient microstructure to meet other requirements, for example, low resistance to diffusion transport of gases, sufficient share of the electron-conductive phase allowing its percolation (mutual interconnection of its forming particles) and probability of the three-phase contact occurrence (see Section 13.1.3.2).

The motivation for the development of high-temperature fuel cells is clear and includes several factors. The high operating temperature eliminates platinum-group metals as catalysts completely, while maintaining, unlike the alkaline low-temperature system, the relatively high process intensity. The produced waste heat can be used effectively, so this type of units is ideal in the field of cogeneration of heat and electrical energy. The high operating temperature also significantly changes the requirements for the fuel. The hydrogen electrode is not sensitive to the presence of CO, but uses it as a fuel. In an ideal case, these cells can even directly burn hydrocarbons, preferably natural gas, fed into the anode space. This is allowed by its steam conversion, occurring, thanks to the high operating temperature, directly in the anode chamber of the fuel cell. However, in the long run, these systems usually do not exhibit sufficient time stability due to the gradually developing carbon deposits blocking the electrode. Therefore, steam reforming unit is placed directly before the fuel entry to the cell. Even this configuration represents a significant simplification compared to the low-temperature PEM cell, thanks to the absence of purification of the gaseous mixture resulting from hydrocarbon conversion by water vapour. At the same time, the heat necessary for the reaction of hydrocarbon with water vapour is taken directly from the fuel cell. Such a unit does not depend on hydrogen supplies and can work directly with natural gas whose infrastructure is already available in developed countries.

### 14.1.1.5 Operating arrangement of fuel cells

Fuel cells in the design arrangements presented for their individual types in the previous text are characterized by operating voltage typically lower than 1 V and relatively high electric current generated. This has many negative consequences. The first is that even a small ohmic loss of voltage in the electric circuit at high electric currents leads to relatively high losses of the overall efficiency of the system. Also, further processing of the generated electrical energy is connected with efficiency decrease. Therefore, it is clear that, for practical applications, it is necessary to make modifications in the system to reduce such impacts. The most efficient arrangement of individual cells is their connection into a stack, which is schematically shown in Fig. 14.7.

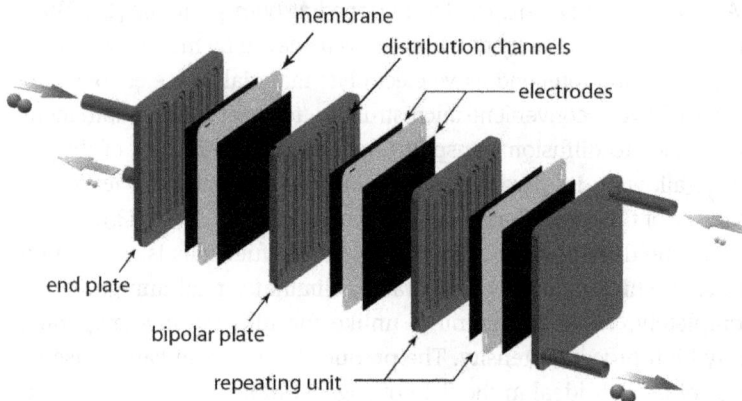

**Fig. 14.7:** Scheme of the connection of individual fuel cells into a stack.

From the point of view of electrical circuit, this arrangement can be considered as a variant of serial connection of individual cells. The electric current flows through individual cells, and the electric potential of the passing charge increases by a value corresponding to the voltage on the cell. The voltage on the stack then corresponds to the sum of the voltages of individual cells, while the electric current remains constant, that is, identical to the current passing through one cell.

Although this arrangement is advantageous from the point of view of the generated electric current, its practical implementation brings many problems and potential risks. It is crucial to ensure a uniform distribution of the fuel and oxidant both between cells in the stack and on the surface of each cell. Depletion of one of the components at any point in the cell could have fatal consequences for its lifetime, as the current will continue flowing through this area. But if there are not enough reactants, inversion will occur, and one cell in the stack might start to work as an electrolyser and its electrodes will degrade particularly rapidly. Another important aspect is the temperature. Concentration of more fuel cells in the stack limits heat exchange with the environment and creates a maximum in temperature field typically at the centre of the stack. This problem is usually resolved by building in a cooling circuit, but it complicates the design substantially. The risk of parasitic currents needs to be mentioned, too. However, it is only present in a system with an ion-conductive liquid electrolyte circulating in a common hydraulic circuit. This risk increases with the number of cells in the stack and with the decreasing ohmic resistance of the collection and distribution channels of the electrolyte, along the stack.

All these aspects usually clearly limit the maximum dimensions of the cell and stack. However, there are considerable differences in where the maximum upper limit is found. It depends particularly on the fuel cell type, materials used, reactants and other aspects. Therefore, it is not possible to determine any definite, generally valid limit values.

## 14.1.2 Water electrolysis

Water electrolysis can be considered to be the second basic technology of the hydrogen economy. It represents a process inverse to the fuel cell. There are many similarities between these technologies, especially in terms of the basic types of design and unit arrangement. However, such simplification can be too concise and very misleading, if it does not contain sufficiently deep understanding of the nature of both processes. This has several reasons. The first is the fact that in water electrolysis, the electrode potentials reach different extreme values than in fuel cells. The second reason is that, in low-temperature water electrolysis units, water in the system is largely in its liquid phase. In this case, it is always necessary to assume a two-phase flow inside the cell. On the other hand, the risk of limiting the rate of electrode reactions by mass transfer is reduced in this configuration. However, it only applies under assumption of a properly designed cell, which decreases the risk of blocking the surface of electrodes by the gas developed. This issue will be discussed separately for individual types of water electrolysis.

The direction in which the water electrolysis process runs corresponds to the summary reaction described by the equation:

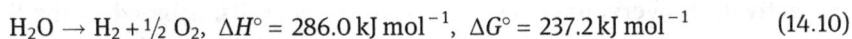

$$H_2O \rightarrow H_2 + \tfrac{1}{2} O_2, \ \Delta H° = 286.0 \, \text{kJ mol}^{-1}, \ \Delta G° = 237.2 \, \text{kJ mol}^{-1} \qquad (14.10)$$

It results from these values of $\Delta H°$ and $\Delta G°$ that the electrolytic decomposition of water is endothermic and can occur with less energy (supplied through electrical work) than corresponds to $\Delta H°$. Under standard conditions, this situation corresponds to cell voltage reduction from 1.48 V down to 1.23 V (25 °C). In this range, therefore, thermal energy must be supplied to the cell in addition to electrical energy to cover the thermodynamic energy demands of the process. However, electrolysis is usually not operated at these voltage levels on the industrial scale. It is due to the fact that the electrode reactions are very slow in this voltage range, primarily due to the activation overvoltage of the anode. At the value corresponding to the upper limit of the interval, that is, 1.48 V, the energy entering the system through the supplied electrical energy reaches the value corresponding to the total energy needed to decompose the water molecule. This voltage is usually referred to as thermoneutral; see Section 3.6, eq. (3.84), that is, as the voltage at which no heat must be supplied to the system or removed from it. This fact somewhat complicates the unique expression of the energy efficiency of the water electrolysis process, as will be explained in the following text.

### 14.1.2.1 Factors limiting the water electrolysis efficiency

Similar to the function of the fuel cell, four basic phenomena limiting the efficiency of the process can be defined for water electrolysis. In this case, it is probably appropriate to slightly modify the order in which they will be discussed as follows:
- activation (polarization) overvoltage,
- ohmic loss,
- mass transfer,
- cross-over of products.

Again, for easy understanding, it is preferable to discuss individual actions using the scheme of the load curve of water electrolysis, as shown in Fig. 14.8.

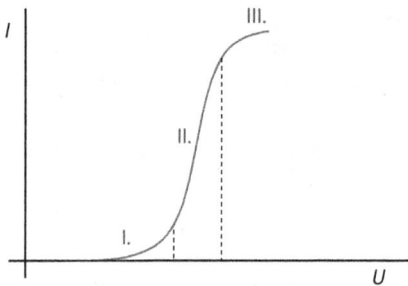

**Fig. 14.8:** Scheme of the load curve of the water electrolysis process.

The **activation overvoltage**, as in the case of fuel cells, is based on the theory of a finite rate of electrode reactions. Unlike fuel cells working as the galvanic cell, the water electrolysis cell represents an electrolyser, so when the current passes, there is no voltage decrease due to irreversible actions, but an increase takes place, see area I in Fig. 14.8. This is essentially the only difference, but it has important consequences for the system. The basic one is the increase of the electrode potential of the anode. Since the decisive part of the activation overvoltage is applied on the anode, that is, on the electrode where oxygen is evolved, there is a relatively high anodic potential established. In addition to the high electrode oxidation potential, the anode material is also subject to the exposure of atomic and molecular oxygen. This substantially limits the selection not only of a suitable electrode material but also other cell materials on the anode side.

The **ohmic loss**, as in the case of fuel cells, becomes more prominent at higher current loads; see area II in Fig. 14.8. Its characteristic shape is due to the linear increase of the cell voltage with increasing current load. Due to the limitted selection of material exhibiting adequate stability, the impact of this effect is more demanding than in most types of fuel cells.

**Mass transfer**, or its negative influence, significantly applies only in some types of electrolysers, particularly in those where water as an input reactant is present in its vapour phase. Therefore, this phenomenon is usually not very important for low-temperature electrolysers of modern design, while for high-temperature units, the action indicated as area III in Fig. 14.8 can occur at high current loads. For water electrolysis, this phenomenon is caused mainly by the decrease of the partial pressure of water vapour on the active surface of the anode to values near to zero. The mass transfer rate from the volume of distribution channels to the active surface of the electrode is, in this case, comparable to the electrode reaction rate, or it is lower. Unlike fuel cells, the voltage characteristically increases fast with the growing current load.

The **cross-over of products** between electrode compartments can be evaluated from two viewpoints in the case of water electrolysis. The critical one usually considered is the danger associated with the formation of an explosive mixture in the anode compartment. This is important primarily in the area of low current loads of the cell, where the amount of oxygen produced on the anode is not enough to dilute the penetrating hydrogen sufficiently. From the point of view of faradaic efficiency, this effect is important, especially in the area of high current loads, where the partial pressures of hydrogen and oxygen grow in the corresponding electrode spaces.

While the first three phenomena influence the voltage efficiency of the process, the last one has impact on faradaic efficiency.

### 14.1.2.2 Water electrolysis efficiency

The efficiency of the electrolytic decomposition of water is expressed in a manner similar to the description of the efficiency of fuel cells (Section 14.1.1.2). In this case, we also build on the definitions introduced in Section 3.6. The thermal efficiency of electrolysis, $\eta_{T,EL}$, is defined by eq. (3.83). It is clear that in the case of reversibly performed water electrolysis under standard conditions, the thermal efficiency reaches the value of 121 %; see eq. (14.10). However, the excess of 21 % by which the value exceeds 100 % does not mean any violation of the law of conservation of energy. This share of energy can only be supplied to the system in the form of thermal energy. In the case of low-temperature and medium-temperature processes, this possibility is usually not used, due to the low rate of electrode reactions at such low overvoltages. The actual voltage in these cases reaches at least the thermoneutral voltage, but usually exceeds it. All energy needed to decompose water molecules is supplied in the form of electrical energy, with the exception of high-temperature processes that cover a considerable part of the energy demands of the reaction using thermal energy, thanks to rapid electrode reactions.

From the practical point of view, it is necessary to evaluate the actual efficiency of electrolysis. Equation (3.83) is usually modified for this purpose into the form:

$$\eta_{ac,EL} = U_{TN}/U_{ac} \qquad (14.11)$$

where $U_{ac}$ is the actual voltage on a current-loaded cell.

Even in the case of water electrolysis, it is necessary to consider the faradaic losses, that is, the efficiency associated with the use of the electric charge passed through the system. In this case, it is defined by eq. (3.87), where the electrolysis product is hydrogen. The overall electrolysis efficiency is then determined by the equation:

$$\eta_{tot,EL} = \eta_{ac,EL}\,\Phi_{I,EL} \qquad (14.12)$$

To evaluate the overall efficiency of the system, it is necessary to take into account, the energy demands of other parts of the process, such as control systems or pumping of circulating media. A separate issue in this process is the compression of gaseous products, primarily hydrogen. In the conventional process arrangement, the produced hydrogen is mechanically compressed after drying to the required pressure. At present, the alternate electrochemical compression process is being discussed intensively. This process is based on the use of different pressures in the cathode and anode compartments. In the case of hydrogen compression, the cathode compartment is maintained at the required pressure, while the anode compartment is maintained usually at the atmospheric pressure. The energy needed to compress hydrogen is supplied, in this case, in the form of the equilibrium voltage increase on the cell according to the Nernst equation (3.69). In practice, however, this energy is significantly lower than the energy needed for mechanical compression, because in the case of electrochemical compression, the irreversible losses are lower. On the contrary, there are increased demands on the mechanical stability of the membrane used and its low permeability for hydrogen, which results from the increase of the driving force of the process with the increase of the hydrogen pressure in the cathode compartment.

### 14.1.2.3 Design of the cell for electrolytic decomposition of water

The principles of the design of the electrolytic cell are based on the principles that apply to fuel cells. The goal of their modifications is primarily to take into account the differences caused by the two-phase flow inside the device and the material selection corresponding to the process needs. There is a tendency to keep the basic arrangement identical to the arrangement described above and documented for the PEM type fuel cell as shown in Fig. 14.4. The same principles apply for the design of electrodes and distribution channels. For water electrolysis, this arrangement has several important advantages. The electrodes adhere directly to the solid electrolyte surface, which minimizes the danger of accumulation of the gaseous phase in the inter-electrode space. The geometry of distribution channel is also selected with the goal of minimizing the danger of accumulation of the gaseous phase in the system. Out of the geometries shown in Fig. 14.5, parallel channels are preferred, in this case. The use of the so-called "zero gap" design (the term is described in more detail in Section 13.1.1) enables circulation of demineralized water through the system as a reactant, in the case of the PEM-type cell. This decreases the demands on construction materials and eliminates the risk of contamination of the environment by aggressive substances. An important difference from this principle is the alkaline electrolysis of water.

The difference of the alkaline electrolysis process results from the absence of a solid alkaline electrolyte that would meet requirements of this process, mainly, sufficient ion conductivity and long-term stability. Besides the membrane, this absence is even more significant with respect to the binder used in the electrode design. The use of the ion-conductive electrolyte is important, especially due to the occurrence of the so-called three-phase contact in the electrode structure, as explained in Section 14.1.1.3. For this reason, a different type of the design of industrial electrolytic cell is still being applied, based on the use of an inorganic electrochemically inactive separator – diaphragm. The scheme of the cell is shown in Fig. 14.9. At present, however, intensive research is underway, aimed at developing a suitable anion-selective polymer electrolyte that would allow implementing this technology in an arrangement similar to the PEM system [14, 15].

### 14.1.2.4 Types of processes for electrolytic water decomposition

Processes for electrolytic water decomposition are usually divided in the same manner as fuel cells, so the operating temperature is applied as the main criterion. A more detailed division can be based on ions carrying the electric charge between the electrodes. According to these rules, we divide processes for electrolytic water decomposition again into three basic groups: low temperature, medium temperature and high temperature.

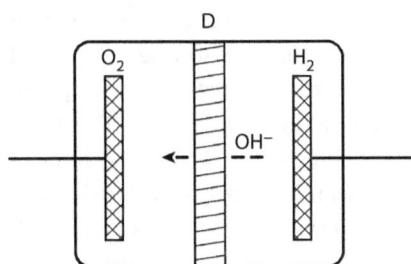

Fig. 14.9: Scheme of diaphragm electrolyser for alkaline water electrolysis; D - diaphragm.

Due to the mentioned fact that the design of the structural arrangement of these technologies is based especially on experience collected in the area of fuel cells, attention in the detailed descriptions of individual types of fuel cells will be paid especially to the differences from fuel cells of the same type, and the facts previously stated will not be repeated.

**Low-temperature electrolysis** of water can be divided according to the ions transferring the charge to the **acidic variant**, called PEM, and **alkaline variant**. In the case of the PEM variant, the main differences are in material demands. This applies mainly to the anode part of the cell, which is exposed to the effects of higher oxidation potentials. For the anode itself, the current state of art is titanium felt or sintered titanium modified with a suitable catalyst, typically $IrO_4$. As far as the material of distribution plates is concerned, the conventional option is titanium protected with an adequate surface layer, decreasing the value of the contact resistance. An alternative is high-grade stainless steel, but only on the assumption of high quality of the protective layer. Its damage would lead to releasing metal ions, which would subsequently degrade the polymer electrolyte and inactivate the catalyst on the cathode side of the cell. This type of cell is mainly characterized by the high intensity and flexibility of the process. However, its main disadvantages include the material demands having significant impact on the investment costs of the process.

The current configuration of the industrial alkaline variant of the process was discussed above. At present, the main focus is on the solution for the absence of a suitable solid, preferably polymer anion-selective polymer. The availability of this material would allow implementing a system with the design arrangement similar to the PEM type cells, but based on less expensive electrode and construction materials. Much attention is currently paid to nickel-based material, activated with a layer of a suitable catalyst [16]. Today, the standard variant of the anode side is a spinel-structure mixed oxide of nickel and cobalt. However, other variants are investigated as well, enabling efficient use, also on the cathode side of the cell. From the point of view of the design of bipolar plates, nickel is considered as the standard material at present, but other alternatives are being studied with the objective of reducing the costs of structural materials.

**Medium-temperature electrolysis** of water represents, unlike the technology of fuel cells, a process largely in the basic research stage. This is also due to material issues. The polymer electrolyte based on polybenzimidazole impregnated with phosphoric acid does not exhibit sufficient chemical stability in the environment. The existing perfluorinated sulphonated materials do not have sufficient long-term mechanical stability. At present, probably the most advanced research in this field is using the melt of potassium hydroxide or sodium hydroxide fixed in an inert porous matrix [17] as the electrolyte. Despite some progress, this process has not yet reached the stage of transferring research findings to industrial practice.

Even though **high-temperature electrolysis** of water has not reached industrial implementation either, there are some advantages connected with its use that are so motivating, that the process attracts continuous attention, leading to substantial progress in the development of this technology. This is a technology exclusively based on the use of solid ceramic electrolyte. Due to the high rate of electrode reactions, problems connected with the high operating temperature can be considered essential from the material point of view, while the increased anode potential is only secondary issue. The materials used do not require major deviations from the fuel cell technology. The research of high-temperature water electrolysis technology is also conducted for capturing and valourizing carbon dioxide. If a mixture of water vapour with carbon dioxide in the proper ratio is led to the cathode space, a synthesis gas [18] is produced on the cathode according to the following summary reaction:

$$H_2O + CO_2 + 4\,e^- \rightarrow H_2 + CO + 2\,O^{2-} \tag{14.13}$$

On the anode, oxygen is developed according to the following reaction:

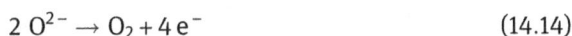

$$2\,O^{2-} \rightarrow O_2 + 4\,e^- \tag{14.14}$$

In this case, the selection of the electrode materials plays an important role. The reason is that it is necessary to achieve sufficient selectivity of cathodic reactions with respect to the required reduction of carbon dioxide. This example shows one of the advantages of high-temperature electrolysis of water. In addition to the conversion of electrical energy to the energy of chemical bonds, it allows processing waste substances to raw materials for new production, or to systems of substance with higher energy content.

### 14.1.2.5 Design arrangement of electrolyser for water decomposition

The design of cells for water electrolysis is based on the same principles applied in fuel cells. By arranging simple cells in a stack, we achieve increase in the voltage on the terminal electrodes, while maintaining the current load value. Production capacity of the stack corresponds to the single- cell capacity multiplied by the number of cells in a stack. Even in the case of an electrolyser, the basic condition remains the necessity of uniform distribution of reactants between individual cells in the stack and over the electrode surface in each cell. Again, there is the exception of the electrolyser for low-temperature alkaline water decomposition. Since a relatively highly concentrated solution of potassium hydroxide circulates through the system, there is the danger of parasitic (shunt) currents and loss of the system efficiency, if cells are connected in series with a common hydraulic circuit. Therefore, it is always necessary to design the cell geometry with respect to this risk.

## 14.2 Redox flow batteries

Redox flow battery (RFB) is a specific type of rechargeable batteries where the energy is stored in electrolytes containing dissolved electrochemically active species (most often, metal ions). The electrolytes are stored in external tanks, and they are pumped into the stack of electrochemical cells where the energy conversion is realized via corresponding electrochemical reactions on the inert electrodes. The basic RFB set-up is schematically presented in Fig. 14.10.

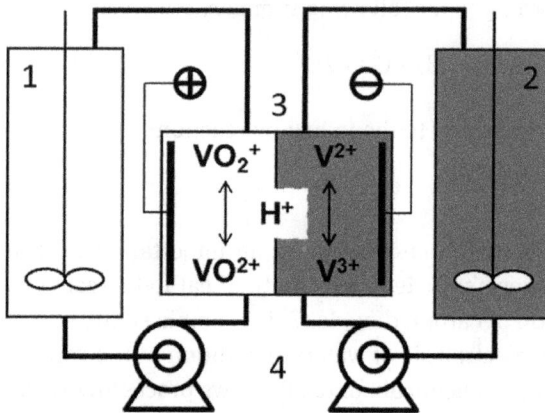

**Fig. 14.10: Scheme of vanadium redox flow battery:** 1, catholyte tank; 2, anolyte tank; 3, electrochemical cell; 4, pumps.

Similar to fuel cell and electrolysers, which have been introduced in previous sections, this arrangement allows independent scaling of power and capacity of the storage, which is the main advantage of RFB when compared to standard accumulators. The battery power depends on dimensions of electrodes and number of cells in the stack, whereas capacity is determined by volume and concentration of electrolytes in tanks. Another benefit originating from the flow arrangement is the simplified removal of waste heat from the battery stack by circulating electrolytes and minimized self-discharge of the battery. The usage of aqueous electrolytes provides the system with safety features of non-flammability and non-explosiveness.

The construction of RFB cell is, in principle, very similar to low temperature fuel cell or electrolysers; see Section 14.1. However it can significantly vary according to the specific requirements of various RFB chemistries. Typically, MEA consists of two electrodes separated by an ion-selective membrane or porous separator. The role of the membrane is to provide ionic connections between both half-cells, while preventing undesired mixing of electrolytes and shorting of the electrodes. The energy conversion is realized via reduction or oxidation reactions of electroactive species on inert electrodes. The absence of phase changes associated with electrode reactions (e.g. metal plating, gas evolution or ion intercalation) is a precondition of efficient and durable battery operation [19]. The individual cells are stacked typically in bipolar arrangement (i.e. serially with respect to the current flow) to increase the battery voltage. The circulation of viscous and ionically conductive electrolytes through the stack (flow-through or flow-by arrangement) requires the optimization of stack design with respect to pressure losses and shunt currents flowing along distribution channels.

When compared to other rechargeable batteries, RFBs generally have lower energy density (both volumetric and gravimetric), which is caused by the limited solubility of electroactive species in electrolytes. This predetermines their application as stationary energy storage for large-scale applications, such as power quality and energy management (balancing the power supplies to the distribution grid). The energy density can be significantly increased, for example, by implementing metal plating electrode, although typically at the expense of reduced efficiency and durability. These so-called hybrid flow batteries are described in more detail in Section 14.2.4.

Systematic research in the RFB area started in the 1970s as a part of NASA's space program, with Fe–Ti and Fe–Cr redox couples [20]. In 1980s, the research activity in RFB moved mainly to Japan (Fe–Cr) and Australia (V–V). At present, RFBs of several chemistries (all-vanadium, Zn–Br, H–Br) are being successfully commercialized on industrial level. At the same time, intense R&D aims at the improvement of technical and economical parameters of the battery. The selected RFB types will be described in the following text.

## 14.2.1 Vanadium redox flow battery

Vanadium redox flow battery (VRFB) represents the mostly studied and matured RFB system nowadays [19, 21–23]. Presence of vanadium ion in four different oxidation states enables the usage of single electroactive element in both electrolytes ($V^{2+}/V^{3+}$ for anolyte and $VO^{2+}/VO_2^+$ for catholyte), which prevents their mutual contamination. The capacity decay due to asymmetric penetration of vanadium ions through the membrane, thus, does not cause irreversible degradation of the battery capacity, which can be easily recovered by their mixing and/or simple on-site rebalancing. This results in extreme durability of the technology (up to ten thousand cycles) when compared to other electrochemical energy storages [24].

VRFB electrode reactions in the discharge direction are as follows:

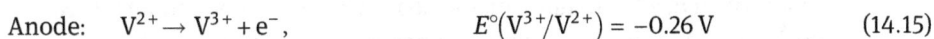

$$\text{Anode:} \quad V^{2+} \rightarrow V^{3+} + e^-, \qquad E°(V^{3+}/V^{2+}) = -0.26 \text{ V} \tag{14.15}$$

$$\text{Cathode:} \quad VO_2^+ + 2\,H^+ + e^- \rightarrow VO^{2+} + H_2O, \quad E°(VO_2^+, H^+/VO^{2+}, H_2O) = 1.00 \text{ V} \tag{14.16}$$

Graphitized polymeric fibrous materials (felt, fabric or paper) with large specific surface area, suitable texture properties and sufficient chemical and electrochemical stability are commonly used for construction of both electrodes. The surface of carbon fibres can be modified by various methods (typically, by thermal or chemical oxidation) to increase the wettability and to enhance the kinetics of the electrode reactions [25, 26]. Absence of platinum-group metals electrocatalysts in electrode construction significantly reduces the costs of stack VRFB, when compared to PEM hydrogen fuel cell/water electrolyser. However, higher area-specific resistance of the cell limits the practically used current densities to 100–250 mA cm$^{-2}$; however, high-power systems with minimized internal resistance have also been recently reported to provide maximal current and power densities in units of A cm$^{-2}$ and W cm$^{-2}$, respectively [27–29].

Cation-exchange or anion-exchange membranes are mostly used to separate individual battery half-cells [30, 31]. The proper choice of specific membrane properties depends on the required properties of the battery. Generally, higher ionic conductivity of cation-exchange membranes allows operation at higher current densities with low impact on voltage efficiency. On the other hand, anion-exchange membranes are less permeable for vanadium cations, which increase the coulombic efficiency of the charge/discharge cycle. The harsh acidic and oxidative environment of VRFB electrolytes limits the choice of usable membrane materials. Sufficient chemical stability was proven for homogeneous perfluorinated sulphonated acid (PFSA) membranes; however, their replacement by cheaper alternatives based on sulphonated aromatic polymers or their modifications would be appreciated [30]. Development of composite membranes (multi-layer, with inorganic fillers etc.) with optimized trade-off between conductivity and selectivity is, thus, vitally needed [32]. Relatively low current densities of VRFB operation allow the use of significantly cheaper but more resistive heterogeneous ionic-selective membranes and non-ionic porous separators [33]; however the chemical stability of these materials under the battery operating conditions needs to be proven. Besides the development of cheaper non-fluorinated hydrocarbon-based ion-exchange membranes with sufficient performance stability under harm operating conditions, the membraneless systems are also being investigated either based on laminar flow of electrolytes in microfluidic battery cell or two immiscible electrolytes.

In the battery stack, individual electrochemical cells are mutually separated by so-called bipolar plates, which are typically fabricated from carbon–polymer composites (similarly to PEM fuel cells). The application of metallic plates is disabled due to corrosive nature of electrolytes, unless a compact conductive protection layer is applied. Flat bipolar plates and porous electrodes with flow-through electrolyte feeding are most often used to obtain homogeneous electrolyte distribution at reasonable pressure drop. For larger stacks with thinner electrodes, however, the optimized flow field geometry can significantly improve the overall battery efficiency [34]. In the battery stack, the optimal geometry of electrolyte distribution channels needs to also take into account the phenomena of shunt currents, that is, the parasitic ionic currents bypassing the inner cells via distribution channels and manifolds, which can significantly decrease the efficiency and durability of battery operation [35–37].

The limited solubility of $VO_2^+$ (approx. 2 mol dm$^{-3}$) restricts the specific energy of VRFB to 25–35 Wh kg$^{-1}$, which is insufficiently low for mobile applications. The electrolyte stability at higher vanadium concentrations can be achieved, for example, by utilizing mixed HCl and $H_2SO_4$ supporting electrolytes [38] or adding solubilizing agents [38]. Various combinations of negative or positive vanadium half-cell with counter half-cells based on other chemistries (Br, $H_2$, $O_2$ and others) are also being developed [39].

Over the last few decades, many demonstration units have been installed with power/capacity in orders of MW/MWh pioneered by Japanese company Sumitomo Electric Industries. At present, other VRFB are being installed worldwide, primarily in combination with renewable energy sources (Prudent Energy, Cellennium Company Ltd., V-Fuel Pty Ltd., Ashlawn Energy and Enerox [19]). The biggest electrochemical energy storage in the world based on 200 MW/800 MWh VRFB is being constructed in Dalian Province by Chinese system manufacturer Rongke Power and UniEnergy Technologies.

## 14.2.2 Polysulphide-bromide battery (PSB)

In this type of RFB, aqueous solutions of sodium bromide (catholyte) and sodium polysulphide (anolyte) are used as electrolytes [20]. For battery discharging, the following electrochemical reactions occur on inert electrodes:

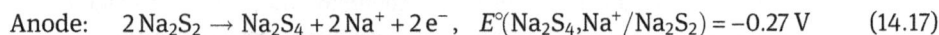

Anode:   $2\,Na_2S_2 \rightarrow Na_2S_4 + 2\,Na^+ + 2\,e^-$,   $E°(Na_2S_4,Na^+/Na_2S_2) = -0.27$ V     (14.17)

Cathode:   $NaBr_3 + 2\,Na^+ + 2\,e^- \rightarrow 3\,NaBr$,   $E°(NaBr_3,Na^+/NaBr) = 1.09$ V     (14.18)

The value of the equilibrium voltage of this system is approximately 1.5 V. The electrode half-cells are separated by a cation-exchange polymer membrane, which allows transporting sodium ions and prevents a direct reaction of sulphides with bromine. Although electrolytes are not directly toxic, there is substantial risk of releasing gaseous bromine and sulphane. The PSB system was developed in the 1990s by the company, Innogy Technologies, Regensis, Ltd. Due to low investment costs and relatively high energy density, many industrial-sized PSB were installed; for example, a 12-MW prototype built in 2002 at the power plant Little Barford in Great Britain.

## 14.2.3 Organic redox flow battery

The environmental and economic issues related to the mining, processing and recycling of metals, which are typically contained in RFB electrolytes, motivates the development of flow batteries based on organic electroactive elements. These could be cheaply synthesized, for example, from oil residues or bio-extracted. The R&D activities in the field of organic redox couples-based RFBs initiated with compounds of the quinone family (benzoquinone, anthraquinone and their derivatives), provide electrochemically reversible two-electron redox reaction on cheap carbon-based electrodes, according to the electrochemical reaction (14.19).

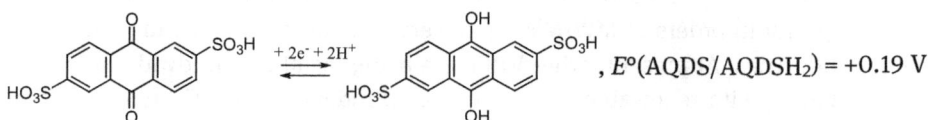

$$ \text{(structure)} \quad \underset{+ \, 2e^- + 2H^+}{\rightleftarrows} \quad \text{(structure)} \quad , \; E^\circ(\text{AQDS/AQDSH}_2) = +0.19 \text{ V} $$

This is a proposition of efficient battery operation at wide range of current densities. Negative half-cell based on sulphonated or hydroxylated 9,10-anthraquinone derivatives has been successfully combined with various organic (4,5-dihydroxy-1,3-disulphonic acid) [40] and inorganic ($Br_2/Br^-$, ferrocyanide/ferricyanide) [41, 42] positive half-cells.

At the same time, various other groups of organic redox materials has been reported for the application, including quinoxalines, radicals (such as 2,2,6,6-tetramethylpiperidine-1-oxyl (TEMPO)), alkoxybenzene-based compounds, viologens, ferrocens or redox polymer [39, 43]. Application of non-aqueous, organic supporting electrolytes with significantly wider potential stability window represents another promising method for increase of energy density. On the other hand, generally lower ionic conductivity of organic electrolytes, when compared to aqueous ones, limits current densities of battery operation.

## 14.2.4 Zinc-based flow batteries

Volumetric energy density of RFB can be significantly enhanced by its hybridization with metal-plating negative electrode using metal deposition during battery charge and its electrochemical dissolution during battery discharge. Apart from metals for aqueous electrolytes (Cu, Fe, Pb), zinc is mostly employed, either in acid or alkaline environment as it has low reduction potential, large volumetric capacity ($5.85 \text{ A h cm}^{-3}$) and low cost. Moreover, zinc-based systems are typically safe, non-flammable and non-explosive and environmentally friendly [35]. On the other hand, zinc-based FB exhibit a rather short lifetime (typically no more than hundreds of charge–discharge cycles), mainly due to uneven zinc deposition (dendrite formation) eventually leading to battery failure due to inner short-circuit of the cell. The morphology of the deposit can be partially controlled by optimizing charging conditions (electrolyte flow rate, current density) and/or use of additives [44]. Zinc-based anode can be coupled with various cathodic materials including metals (Ni, Ce, Pb, V) and non-metals (O, Br, I) and even organic redox couples (TEMPO) [45]. In the following section, we will briefly introduce the most promising zinc-based flow batteries based.

### 14.2.4.1 Zinc-bromine flow battery

Zinc-halogen batteries (especially chlorine and bromine) have been developed since the 1970s [20]. Of these systems, zinc-bromine flow battery represents the most matured and commercially available system (Ensync Energy systems, Redflow, Primus), currently. It uses acidic electrolytes and the following electrochemical reactions occur during battery discharge:

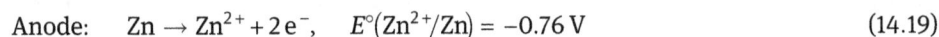

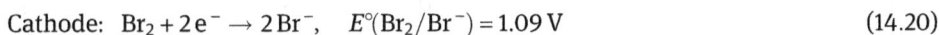

$$\text{Anode:} \quad Zn \rightarrow Zn^{2+} + 2e^-, \quad E°(Zn^{2+}/Zn) = -0.76 \text{ V} \tag{14.19}$$

$$\text{Cathode:} \quad Br_2 + 2e^- \rightarrow 2Br^-, \quad E°(Br_2/Br^-) = 1.09 \text{ V} \tag{14.20}$$

A porous separator that supresses the unwanted self-discharge due to zinc oxidation by bromine usually separates the half-cells. To prevent evaporation, bromine is extracted into the organic phase using quaternary amines (so-called sequestration). The resulting emulsion is easily separable from the aqueous electrolyte due to difference in density. The advantages of these systems include high standard cell voltage (1.85 V) and high theoretical energy density (440 Wh kg$^{-1}$). However, limitations associated with the dendritic growth of zinc limits the practical values to below 100 Wh kg$^{-1}$. The presence of gaseous halogens represents the general safety risks of these technologies.

### 14.2.4.2 Zinc-cerium flow battery

This type of flow battery, developed in 2004 by the company Plurion, Inc., is characterized by a high equilibrium voltage (around 2.4) due to high standard reduction potential of cerium [20]. The following electrochemical reactions occur in discharge direction:

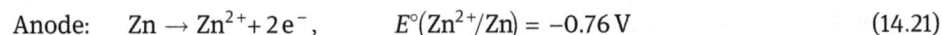

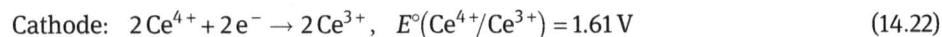

$$\text{Anode:} \quad Zn \rightarrow Zn^{2+} + 2e^-, \quad E°(Zn^{2+}/Zn) = -0.76 \text{ V} \tag{14.21}$$

$$\text{Cathode:} \quad 2Ce^{4+} + 2e^- \rightarrow 2Ce^{3+}, \quad E°(Ce^{4+}/Ce^{3+}) = 1.61 \text{ V} \tag{14.22}$$

The electrolyte consists of aqueous solution of methanesulphonic acid, which is less corrosive than sulphuric acid and allows achievement of higher concentrations of the cerium ion solutions. At the same time, the growth of zinc dendrites and hydrogen evolution are supressed during the battery charging. The high equilibrium voltage of the cell represents an increased risk of parasitic reactions of water splitting and component degradation, which must be considered, especially in case of the positive electrode, where platinized 3D electrodes are typically used. The electrode half-cells are usually separated by cation-selective membranes (PFSA type); however, undivided cells have been reported, too.

### 14.2.4.3 Zinc-air flow battery

The zinc-air flow battery combines zinc anode with gas diffusion cathode, utilizing reduction of atmospheric oxygen during discharge and oxygen evolution during charge, according to

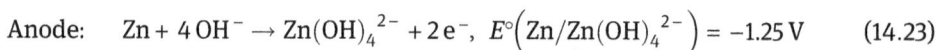

$$\text{Anode:} \quad Zn + 4\,OH^- \rightarrow Zn(OH)_4^{2-} + 2\,e^-, \quad E^\circ\!\left(Zn/Zn(OH)_4^{2-}\right) = -1.25\,V \quad (14.23)$$

$$\text{Cathode:} \quad O_2 + 2\,H_2O + 4\,e^- \rightarrow 4\,OH^-, \quad E^\circ(O_2/OH^-) = +0.40\,V \quad (14.24)$$

The outstanding theoretical energy density of 1,218 Wh kg$^{-1}$ makes zinc-air flow battery a promising candidate for both stationary and mobile energy storage. Alkaline aqueous electrolytes (mostly KOH) are typically used, which enables utilization of cheaper construction materials, including non-noble electro-catalysts. The half-cells are either undivided or separated by porous separator. The main drawbacks of the system coming from both electrodes are: limited cyclability due to nonhomogeneous zinc deposition and low energy efficiency due to high activation overvoltages of oxygen reactions. The latter can be overcome by use of transition metal-based electrocatalysts (similar to alkaline hydrogen fuel cells and water electrolysers). The efficient catalyst utilization requires the optimized content of ionic and electron conductive phases in the structure of gas diffusion electrode, together with good accessibility of active sites by the reactant. The oxygen reduction and evolution processes can be realized either on a single electrode, using bi-functional electrocatalyst, or they can be split to separate electrodes, using three-electrode configuration. In standard configuration, zinc is deposited from zincate solution directly in the cell on a suitable planar or porous support. Alternatively, zinc particles can be circulated through the battery stack as a suspension (slurry). The latter approach enables increasing the energy density and decoupling power and capacity of the storage, but problems of distribution channel clogging and lower zinc utilization need to be addressed.

## 14.3 Reverse electrodialysis

Reverse electrodialysis (RED) is a promising electromembrane technology designed to convert energy. It works on a principle that is a reverse of the electrodialysis process (see Chapter 7). The Gibbs free energy ($\Delta G_r < 0$) is obtained by mixing two solutions with different concentrations of dissolved salts (ions), separated by an ion-selective membrane and subsequently transformed into electrical energy (Fig. 14.11). The application possibilities of RED are, thus, in environments where two consider-able volumes of water with different concentrations are mixed. Suitable areas can be expected primarily in river deltas by the sea, with the natural occurrence of high con-centration gradient of salts. Another significant occurrence of the salt concentration gradient can be found in industrial concentrates (e.g. retentates of reverse osmosis, produced while processing drinking water from sea water), that can be, due to their high salinity, mixed not only with river water but also with sea water [27]. Technical literature [28, 29] states that the Gibbs free energy released by mixing two electrolytes with different concentrations and transforming to electrical energy theoretically rep-resents a renewable source in the order of TW worldwide.

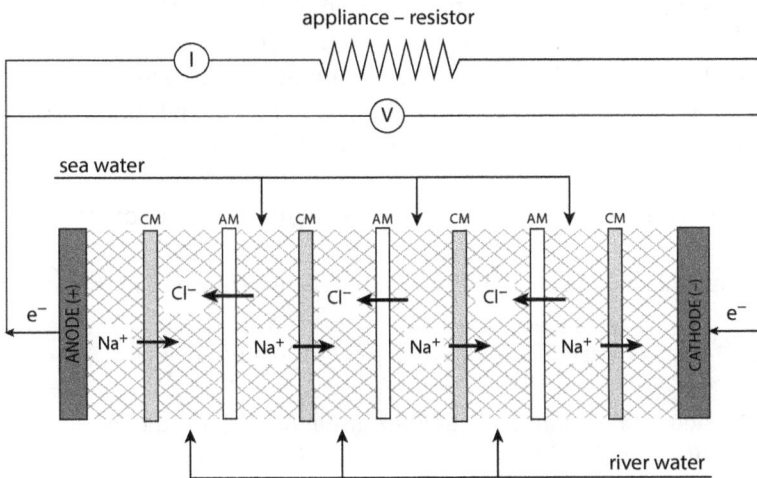

**Fig. 14.11:** Scheme of the principle of reverse electrodialysis (RED).

At this point, it must be noted that a process alternative to RED for the use of the concentration gradient energy is **direct osmosis**, which is based on the solvent transport through membranes by means of the difference of osmotic pressures in the system. This can be used, for instance, in pressure optimization of the standard reverse osmosis process.

In a RED module, the concentrate (sea water) solution and diluate (river water) solu-tion flow through regularly alternating cation-selective and anion-selective membranes that separate both solutions. The ion-selective membranes that are present selectively control the diffusion transport of individual ions according to their charge. The driving force of the entire process is the difference in the electrochemical potential of components of both solutions, defined by eq. (3.36). Therefore, anions migrate from the concentrate through the anion-selective membrane to the diluate towards the anode, and cations in the opposite direction from the concentrate through the cation-selective membrane towards the cathode. The difference of the electrochemical potentials of the solution components separated by the ion-selective membrane generates the membrane potential, $\Delta\varphi_M$ for each membrane. The total potential difference of the RED system, $\Delta\varphi_{RED}$ is determined by the sum of the membrane potential differences through all membranes present.

The metal electrodes included in the RED system serve to convert the electrochemical potential of components to electrical energy, based on redox reactions on the interface of both phases. The generated electrons pass through the outer circuit and, due to the potential difference between the two electrodes, perform electrical work.

Devices for technical implementation of RED are currently based on the experience gathered in designing equipment for electrodialysis, which mainly applies in the use of the plate arrangement of industrial electrodialysers for the RED module design. This particularly includes their internal geometry, total number of membrane chambers and arrangement of hydraulic distribution to minimize the total electrical resistance of the system. To describe the contributions and influences of individual components to the RED process, a simplified theoretical model can be used, based on the following assumptions:
- neglecting the concentration polarization near the surface of the ion-selective membrane,
- neglecting the resistance of electrodes compared to the resistance of ion-selective membranes,
- neglecting the change of ion concentration along the membrane.

This is met in the system under conditions of low current densities, large number of pairs and special configuration of flow hydrodynamics in membrane chambers of the RED stack. There is also another condition of maximum flow homogeneity inside individual chambers of the RED module and its outer and inner tightness.

Using these simplifications, the relations can be defined for the maximum energy output of the RED module, $W_{max,RED}$, which is a function of the average permselectivity of ion-selective membranes and total electrical resistance of the RED module [27]:

$$W_{max,RED} = \frac{N}{2R_{module}} \frac{\left[\psi_{av}\Delta\varphi_{RED}\right]^2}{4} \qquad (14.25)$$

where $N$ is the total number of membrane pairs in the RED stack, $\psi_{av}$ is the average permselectivity of ion-selective membranes, $\Delta\varphi_{RED}$ is the total potential gradient of the RED module and $R_{module}$ is the total electrical resistance of the RED module. This can be expressed as the sum of the resistances of its individual partial components:

$$R_{module} = R_M + R_{dil} + R_{conc} + R_{el} \tag{14.26}$$

where $R_M$ is the resistance of cation-selective and anion-selective membranes, $R_{dil}$ is the resistance of the diluate solution (river water), $R_{conc}$ is the resistance of the concentrate solution (sea) and $R_{el}$ is the resistance of the electrodes.

It is clear that ion-selective membranes represent the key component of the RED module. It follows from eqs. (14.25) and (14.26) that, if the maximum output of the module is to be achieved, the greatest emphasis must be laid on the proper characteristic properties of membranes and maximum decrease of the total electrical resistance $R_{module}$. The complex spectrum of membrane characteristics includes requirements for their high permselectivity and conductivity. Therefore, homogeneous types are preferred, and they currently exhibit better properties than heterogeneous membranes. The total resistance can be also decreased by the design reduction of the membrane chamber thickness, especially the diluate chambers (up to 0.2 mm) and the use of ion-conductive tubulization nets as spacers in the diluate chambers.

Although the theoretical possibilities of implementing the RED process for efficient energy conversion are significant, before they are broadly introduced into practice, the existing process limitations must be overcome. They mainly include the phenomenon of concentration polarization, which influences the transport of ions through membranes, high resistance of commercially available ion-selective membranes and, last but not the least, conductivity barriers associated with the design of the RED stack, in connection with the geometry of membrane chambers and hydraulic spacers used. These phenomena and problems currently attract most research potential in leading world institutions. We cannot forget the technological limitations of RED, which are to do with the necessity of pre-treatment of river and seawater with focus on microbial contamination, removal of solid particles and, last but not the least, removal of all components which can act as the so-called membrane poisons in the RED technology.

## 14.4 Supercapacitors

People have known the possibility of accumulating electric charge on capacitor plates since the mid-eighteenth century. The first capacitor, the so-called Leyden jar, was allegedly invented by Dutch scientist, Pieter van Musschenbroek. Conventional **capacitors** are currently used mainly in electrical engineering (signal smoothing and filtering, etc.). Their capacitance is directly proportional to the area of parallel plates $A$ and inversely proportional to the distance between plates $d$ according to equation

$$C = A\varepsilon_r\varepsilon_0/d \tag{14.27}$$

where $\varepsilon_r$ is the relative permittivity of dielectric material between plates, and $\varepsilon_0$ is the vacuum permittivity ($8.854 \times 10^{-12}$ F m$^{-1}$). With plate capacitors, small capacitances in the order of pF are achieved.

**Electric double-layer capacitors** (EDLC) are made of a pair of 3D electrodes, which are ionically connected by electrolyte. Compared to standard plate capacitors, significantly higher capacitances are achieved (up to 50 μF cm$^{-2}$ [46]). This is possible due to
- significantly larger electrode/electrolyte interface (> 1,500 m$^2$ g$^{-1}$ for activated carbons perfectly wetted with the electrolyte [47]),
- separation of the charge at the very small distance of the electrical double layer formed at electrode–electrolyte interface (in the case of compact double layers, reduced up to 0.3 nm).

The specific capacitance of real EDLC thus reaches the order of hundreds of farads per gram of the electrode material. Therefore, these devices are referred to as "supercapacitors" in literature. The electrodes are mutually separated to prevent their short-circuiting by macroporous separators based on cellulose, glass fibres or synthetic polymers (PVDF and its copolymers, FEP, PP and others [48]). The use of ion-selective membranes is required only for some types of asymmetrical pseudocapacitors with different electrolyte composition on the electrodes (see below).

The energy $E$ stored in a capacitor with the capacitance $C$ is expressed by the following equation:

$$E = 1/2CU^2 \tag{14.28}$$

where $U$ is the working voltage of the capacitor. If aqueous electrolytes are used (most often, KOH, $H_2SO_4$ solutions), the working voltage on a single cell is limited by the potential window of water stability (usually no more than 1.3 V). This fact motivates the development of supercapacitors with non-aqueous electrolytes (e.g. $N(C_2H_5)_4{}^+BF_4{}^-$ in acetonitrile solution or propylene carbonate [47]).

The specific capacitance of supercapacitors can be further increased by the contribution of faradaic processes occurring on the electrode/electrolyte interface or in the surface layer of the electrode. The processes can be divided into three groups:
- electrosorption (e.g. underpotential deposition of H atoms on a surface of platinum electrode),
- redox reaction (e.g. reversible oxidation of Ru to $RuO_2$),
- intercalation (e.g. intercalation of Li$^+$ into the hosting grid of a $MoS_2$ cathode) [46].

Such devices are referred to as "pseudocapacitors", and their specific capacitance reaches up to thousands of farads per gram. At the same time, these systems maintain some advantages of capacitors, that is, high power density, possibility of rapid charging/discharging and high cycling rate. Due to their properties, supercapacitors stand between conventional accumulators and capacitors, and they are frequently used, for instance, in electric vehicles, to regenerate the energy released on braking and to supply peak performance for start and acceleration. Other industrial applications of supercapacitors include back-up supply units of low-power devices, compensation of battery power fluctuations, grid voltage stabilization and others.

# References

[1]  M. Conte, et al. Hydrogen economy for a sustainable development: State-of-the-art and technological perspectives. J. Power Sources 100(1–2): 171–187, 2001.

[2]  Energy density. Wikipedia, the free encyclopaedia 2014 February 2014 [cited 25.2. 2014]; available from: http://en.wikipedia.org/wiki/Energy_density.

[3]  M. Inaba, Z. Ogumi, and Z.-I. Takehara, Application of the solid polymer electrolyte method to organic electrochemistry XIV. Effects of solvents on the electroreduction of nitrobenzene on Cu, Pt–Nafion. J. Electrochem. Soc 140(1): 19–22, 1993.

[4]  S. S. Kocha, J. D. Yang, and J. S. Yi, Characterization of gas crossover and its implications in PEM fuel cells. AIChE J 52(5): 1916–1925, 2006.

[5]  A. Hermann, T. Chaudhuri, and P. Spagnol, Bipolar plates for PEM fuel cells: A review. Int. J. Hydrogen Energy 30(12): 1297–1302, 2005.

[6]  G. Cabouro, et al. Opportunity of metallic interconnects for ITSOFC: Reactivity and electrical property. J. Power Sources 156: 39–44, 2006.

[7]  A. Heinzel, et al. Fuel Cells. In Ullmann's Encyclopaedia of Industrial Chemistry. 7th Edition. Wiley-VCH Verlag, 2000. ISBN 978-3-527-30673-2.

[8]  W. R. Grove, XXIV. On voltaic series and the combination of gases by platinum. Philos. Mag. Series 3 14(86): 127–130, 1839.

[9]  Z. Ogumi, et al. Application of the SPE method to organic electrochemistry; XIII. Oxidation of geraniol on Mn, Pt-Nafion. Electrochim. Acta 37(7): 1295–1299, 1992.

[10]  K. Kordesh and G. Simander. Fuel Cells and Their Applications. VCH Verlag, New York, 1996. ISBN 3527297774.

[11]  Q. Li, et al. Approaches and recent development of polymer electrolyte membranes for fuel cells operating above 100 °C. Chem. Mater. 15(26): 4896–4915, 2003.

[12]  K. C. Wincewicz and J. S. Cooper, Taxonomies of SOFC material and manufacturing alternatives. J. Power Sources 140(2): 280–296, 2005.

[13]  K. D. Kreuer, Proton-Conducting Oxides. Annu. Rev. Mater. Research 33: 333–3511, 2003.

[14]  J. Hnát, et al. Design of a zero-gap laboratory-scale polymer electrolyte membrane alkaline water electrolysis stack. Chem. Ing. Tech. 91(6): 821–832, 2019.

[15]  B. Bauer, H. Strathmann, and F. Effenberger, Anion-exchange membranes with improved alkaline stability. Desalination 79(2–3): 125–144, 1990.

[16]  D. E. Hall, Alkaline water electrolysis anode materials. J. Electrochem. Soc. 132(2): 41c–48c, 1985.

[17]  J. Divisek, J. Mergel, and H. Schmitz, Improvements of water electrolysis in alkaline media at intermediate temperatures. Int. J. Hydrogen Energy 7(9): 695–701, 1982.
[18]  S. H. Jensen, P. H. Larsen, and M. Mogensen, Hydrogen and synthetic fuel production from renewable energy sources. Int. J. Hydrogen Energy 32(15 spec. iss.): 3253–3257, 2007.
[19]  P. Alotto, M. Guarnieri, and F. Moro, Redox flow batteries for the storage of renewable energy: A review. Renewable and Sustainable Energy Rev. 29(0): 325–335, 2014.
[20]  C. Ponce De León, et al. Redox flow cells for energy conversion. J. Power Sources 160(1): 716–732, 2006.
[21]  Á. Cunha, et al. Vanadium redox flow batteries: a technology review. Int. J. Energy Res. 39(7): 889–918, 2015.
[22]  G. Kear, A. A. Shah, and F. C. Walsh, Development of the all-vanadium redox flow battery for energy storage: a review of technological, financial and policy aspects. Int. J. Energy Res. 36(11): 1105–1120, 2012.
[23]  K. Lourenssen, et al. Vanadium redox flow batteries: A comprehensive review. J. Energy Storage 23: 100844, 2019.
[24]  X.-Z. Yuan, et al. A review of all-vanadium redox flow battery durability: Degradation mechanisms and mitigation strategies. Int. J. Energy Res. 43(13): 6599–6638, 2019.
[25]  P. Mazúr, et al. Performance evaluation of thermally treated graphite felt electrodes for vanadium redox flow battery and their four-point single cell characterization. J. Power Sources 380: 105–114, 2018.
[26]  K. J. Kim, et al. The effects of surface modification on carbon felt electrodes for use in vanadium redox flow batteries. Mater. Chem. Phys. 131(1): 547–553, 2011.
[27]  D. S. Aaron, et al. Dramatic performance gains in vanadium redox flow batteries through modified cell architecture. J. Power Sources 206(0): 450–453, 2012.
[28]  J. Charvát, et al. Performance enhancement of vanadium redox flow battery by optimized electrode compression and operational conditions. J. Energy Storage 30: 101468. 2020.
[29]  H. R. Jiang, et al. A high power density and long cycle life vanadium redox flow battery. Energy Storage Mater. 24: 529–540, 2020.
[30]  X. Li, et al. Ion exchange membranes for vanadium redox flow battery (VRB) applications. Energy Environ. Sci. 4(4): 1147–1160, 2011.
[31]  A. Parasuraman, et al. Review of material research and development for vanadium redox flow battery applications. Electrochim. Acta. 141(10): 27–40, 2013.
[32]  H. Prifti, et al. Membranes for Redox Flow Battery Applications. Membranes 2(2): 275, 2012.
[33]  X. Wei, B. Li, and W. Wang, Porous Polymeric Composite Separators for Redox Flow Batteries. Polym. Rev. 55(2): 247–272, 2015.
[34]  Q. Xu, T. S. Zhao, and C. Zhang, Performance of a vanadium redox flow battery with and without flow fields. Electrochim. Acta 142: 61–67, 2014.
[35]  H. Fink and M. Remy, Shunt currents in vanadium flow batteries: Measurement, modelling and implications for efficiency. J. Power Sources 284(Suppl. C): 547–553, 2015.
[36]  A. Tang, et al. Investigation of the effect of shunt current on battery efficiency and stack temperature in vanadium redox flow battery. J. Power Sources 242: 349–356, 2013.
[37]  F. T. Wandschneider, et al. A multi-stack simulation of shunt currents in vanadium redox flow batteries. J. Power Sources 261(Suppl. C): 64–74, 2014.
[38]  C. Choi, et al. A review of vanadium electrolytes for vanadium redox flow batteries. Renewable and Sustainable Energy Rev. 69(C): 263–274, 2017.
[39]  J. Winsberg, et al. Redox-Flow Batteries: From Metals to Organic Redox-Active Materials. Angew. Chem. Int. Ed. Engl. 56(3): 686–711, 2017.
[40]  B. Yang, et al. High-Performance Aqueous Organic Flow Battery with Quinone-Based Redox Couples at Both Electrodes. J. Electrochem. Soc. 163(7): A1442–A1449, 2016.

[41] B. Huskinson, et al. A metal-free organic-inorganic aqueous flow battery. Nature 505(7482): 195–198, 2014.

[42] K. Lin, et al. Alkaline quinone flow battery. Science 349(6255): 1529–1532, 2015.

[43] P. Leung, et al. Recent developments in organic redox flow batteries: A critical review. J. Power Sources 360: 243–283, 2017.

[44] D. P. Trudgeon, et al. Screening of effective electrolyte additives for zinc-based redox flow battery systems. J. Power Sources 412: 44–54, 2019.

[45] J. Winsberg, et al. Poly(TEMPO)/Zinc Hybrid-Flow Battery: A Novel, "Green", High Voltage, and Safe Energy Storage System. Adv. Mater 28(11): 2238–2243, 2016.

[46] B. E. Conway, V. Birss, and J. Wojtowicz, The role and utilization of pseudocapacitance for energy storage by supercapacitors. J. Power Sources 66(1–2): 1–14, 1997.

[47] E. Frackowiak, Q. Abbas, and F. Béguin, Carbon/carbon supercapacitors. J. Energy Chem. 22(2): 226–240, 2013.

[48] D. Karabelli, Poly(vinylidene fluoride)-based macroporous separators for supercapacitors. Electrochim. Acta 57(0): 98–103, 2011.

[49] L. Li, et al. A stable vanadium redox-flow battery with high energy density for large-scale energy storage. Adv. Energy Mater. 1(3): 394–400, 2011.

# Index

https://doi.org/10.1515/9783110739466-018

www.ingramcontent.com/pod-product-compliance
Lightning Source LLC
Chambersburg PA
CBHW080702220326
41598CB00033B/5281